Pheromone Communication in Social Insects

Pheromone Communication in Social Insects

Ants, Wasps, Bees, and Termites

EDITED BY

Robert K. Vander Meer,
Michael D. Breed, Karl E. Espelie,
and Mark L. Winston

Westview Studies in Insect Biology

Published in 1998 in the United States of America by Westview Press, 5500 Central Avenue, Boulder, Colorado 80301-2877, and in the United Kingdom by Westview Press, 12 Hid's Copse Road, Cumnor Hill, Oxford OX2 9JJ

Library of Congress Cataloging-in-Publication Data
Pheromone communication in social insects : ants, wasps, bees, and
termites / edited by Robert K. Vander Meer ... [et al.].
p. cm.—(Westview studies in insect biology)
Includes bibliographical references and index.
ISBN 0-8133-8976-3 (hc)
1. Animal communication. 2. Pheromones. 3. Insect societies.
I. Vander Meer, Robert K. II. Series.
QL776.P485 1998
595.7159—dc21 97-29921
CIP

The paper used in this publication meets the requirements of the American National Standard for Permanence of Paper for Printed Library Materials Z39.48-1984.

10 9 8 7 6 5 4 3 2 1

Contents

Foreword

All social interactions involve communication, whether mutual attraction, repulsion, identification of species and kin, courtship and parental care, establishment of dominance and division of labor, mutualistic symbiosis or any other form of coexistence.

But what is communication? It is not easy to draw a line between stimuli produced by animals that are truly communicative signals and others that are not, and the definitions given by different authorities in the study of animal communication vary considerably.

Wilson (1975) defines biological communication relatively broadly as "the action on the part of one organism (or cell) that alters the probability pattern of behavior in another organism (or cell) in a fashion adaptive to either one or both of the participants."

Many authors, on the other hand, use the term communication only when the signalling behavior is believed to confer an average statistical benefit to both sender and receiver. In these cases, which Marler (1968) calls "true communication", the transfer of information is mutually adaptive to both participants. These communicative relationships can be exploited by predators and parasites that "intrude" into the signalling system of a prey or host species, either by decoding signals or by imitating signals and thereby manipulating the prey or host.

In a provocative paper Dawkins and Krebs (1978) suggested that all animal communication is a form of manipulation whereby communication or "signalling" is "characterized as a means by which one animal makes use of another animal's muscle power." They argue that "natural selection favors individuals who successfully manipulate the behavior of other individuals, whether or not this is to the advantage of the manipulated individual." These thoughts have led to a lively debate of whether or not animal signals should be "honest" or "dishonest" (see Markl 1985; Zahavi 1986; 1987; Grafen 1991; Maynard-Smith 1991). I do not disagree with the basic thesis of Dawkins and Krebs, which is primarily based on aggressive display behavior. I believe, however, that the "manipulation hypothesis" does not easily explain the evolution of complex communication systems which do promote true inter-individual cooperation, as often documented in the eusocial insects (Hölldobler 1984).

The tremendous evolutionary success and the ecological dominance of the social insects (mainly the ants, the termites, social bees and social wasps), are in large part due to their efficient social organizations and the underlying communication systems. Whereas a solitary animal can at any moment be in only one place and can be doing only one thing, a colony of social insects can be in many places by deploying its workers and can be doing many different things because of the size of the worker cohorts and their division of labor. The functional division into reproductive and sterile castes, the cooperation in rearing the young, gathering food, defending the nest, exploring new foraging grounds, establishing territorial borders, and discriminating and excluding foreigners from the society are regulated by the precise transmission of social signals in time and space.

Though several sensory modalities are involved in the communication of social insects, probably the most diverse and also best studied communication behavior is chemical communication. Chemical releasers (so called pheromones) are produced in a large variety of exocrine glands, and considerable progress has been made in chemically identifying many of these glandular secretions and in analyzing their behavioral functions. This book presents our current knowledge about pheromone communication in social insects.

Bert Hölldobler

Literature Cited

Dawkins, R. and Krebs, J. 1978. Animal signals: Information or manipulation? In: *Behavioral Ecology: An Evolutionary Approach.* (J.R. Krebs and N.D. Davies, eds.), pp. 282-389. Sinauer Assoc., Sunderland, Mass.

Grafen, A. 1991. Modelling in behavioral ecology. In: *Behavioural Ecology: An Evolutionary Approach* (J.R. Krebs and N.B. Davies, eds.). 3rd edn. Blackwell Scientific, Boston, Mass.

Hölldobler, B. 1984. Evolution of insect communication. In: *Insect Communication* (T. Lewis, ed.), pp. 349-377. Academic Press, London.

Markl, H. 1985. Manipulation, modulation, information, cognition: some of the riddles of communication. In: *Experimental Behavioral Ecology and Sociobiology* (B. Hölldobler and M. Lindauer, eds.), pp. 163-194. Sinauer Assoc., Sunderland, Mass.

Marler, P. 1968. Visual systems. In: *Animal Communication* (T.A. Sebeok, ed.), pp. 103-126. Indiana University Press, Bloomington, Indiana.

Maynard Smith, J. 1991. Honest signalling: The Philip Sidney game. *Animal Behavior* 42:1034-1035.

Wilson, E.O. 1975. *Sociobiology*. Harvard University Press, Cambridge, Mass.

Zahavi, A. 1986. Reliability in signalling motivation. *Behav. Brain* 9:741-772.

Zahavi, A. 1987. The theory of signal selection and some of its implications. In: *Internatl. Symp. of Biolog. Evol.* (V.P. Delftno, ed.). Adriatica Editrice, Bari.

Preface

This book is the result of two symposia held at the national meeting of the Entomological Society of America in Dallas, Texas in December, 1994. One of the symposia, organized by Michael Breed and Karl Espelie dealt with the chemistry of kin and nestmate recognition in social insects and the other, organized by Robert Vander Meer and Mark Winston, focused on pheromonal communication in social insects. After the symposia were presented the organizers hatched the idea of combining the two topics in book form with the goal of developing a comprehensive view of pheromones and chemical communication in social insects.

We solicited chapters from the symposium participants and added chapters in a few critical areas that had not been covered in the symposia. While the chapters reflect the content of the symposia, the authors all agreed to provide much more substantial reviews than they had in the symposia, and to include material that has appeared since the presentation of the symposia.

For us, as editors, the most striking feature of the work presented here is the high level of sophistication of behavioral and chemical understanding that the authors have achieved for a diverse set of communicatory systems. A thorough reading of the book yields several examples for which the biosynthetic pathways for a pheromone are understood, its glandular source elucidated, its behavioral role demonstrated, and the evolution of the signalling system traced.

Of course, not all the problems are solved. For example, understanding the evolution of nestmate recognition signals awaits further studies of olfactory nest recognition in solitary bees and wasps and comparative studies of the eusocial bees and wasps that extend beyond the familiar trio of *Lasioglossum*, *Apis*, and *Polistes*. Primer pheromones remain a more difficult area than releaser pheromones as pointed out in Part Four. The elegant work on ants and honeybees reviewed in chapters 12 and 14 is just a beginning in understanding the mechanisms of queen control and regulation in these social insects.

The future of studies of chemical communication in social insects rests in a better integration of the biochemical, physiological, anatomical and behavioral studies that are highlighted in this book with the ecology and evolution of signalling systems. What ecological and evolutionary factors drive the diversity of trail pheromones found in ants? Is nestmate recognition rooted in nest recognition mechanisms that evolved in solitary species, or does nestmate recognition evolve as a result of selection for kin identification in species that are already eusocial? And, is apparent queen control of worker reproduction the result of queen dominance or worker cooperation?

As with any collection of scientific results this book is both an end and a beginning. It captures the endpoint of a tremendous amount of effort by biologist and chemists to discover how chemicals are used in communication by social insects, but we also hope that the primary value of this book comes from the ability of its readers to see the next steps, and to pursue those steps.

Robert K. Vander Meer
Michael D. Breed
Karl E. Espelie
Mark L. Winston

Acknowledgments

The editors wish to thank all of the contributors to this book, whose patience, outstanding knowledge and professional skills have made our task much easier. Special thanks go to Johan Billen, David Morgan, and Abraham Hefetz, who enthusiastically added chapters that helped provide better coverage of the topic. We thank Bert Hölldobler for taking time out from his busy schedule to write an insightful Foreword to the book. We thank the U.S. Department of Agriculture, Agricultural Research Service for permitting Robert K. Vander Meer to work on this project.

Thanks go to the Westview staff, especially Janie McKenzie, Gabriela Zöller, and Elizabeth Lawrence, for their encouragement, patience, and guidance.

R.K.V.M.
M.D.B.
K.E.E.
M.L.W.

PART ONE

Introduction: Sources and Secretions

1

Pheromone Communication in Social Insects: Sources and Secretions

Johan Billen and E. David Morgan

Introduction

Although the general descriptions of insect anatomy and structure as found in entomological textbooks equally apply to social insects, the development of the exocrine apparatus in the latter clearly distinguishes them from solitary insects (Figure 1.1). An extremely diverse array of exocrine glands is found in all social insects, with 63 different glands described so far (39 if only considering the Formicidae, 21 for the Apidae, 14 for the Vespidae and 11 for the Isoptera) (Billen, 1994). Several of these glands serve 'individual' functions as the source of digestive enzymes or lubricant compounds, although the majority has a clear function related to the social organization of the colony (Hölldobler and Wilson, 1990). Some have a role in producing building material like the wax glands in bees, others secrete antibiotics like the metapleural glands of the ants, or elaborate sticky defensive substances like the frontal glands of some termite species. A major social function of exocrine glands, however, is the production of pheromones, for which many glands have become specialized.

The study of exocrine glands in general, and of pheromone producing glands in particular, has long been faced with a number of practical difficulties. Because of their ectodermal origin, all exocrine glands are associated with cuticle, which has put considerable constraints on the study of gland structure. The development of plastic embedding techniques has allowed much better sectioning conditions, which have resulted in a clearer picture of the structural organization of the exocrine system compared with the information obtained from paraffin

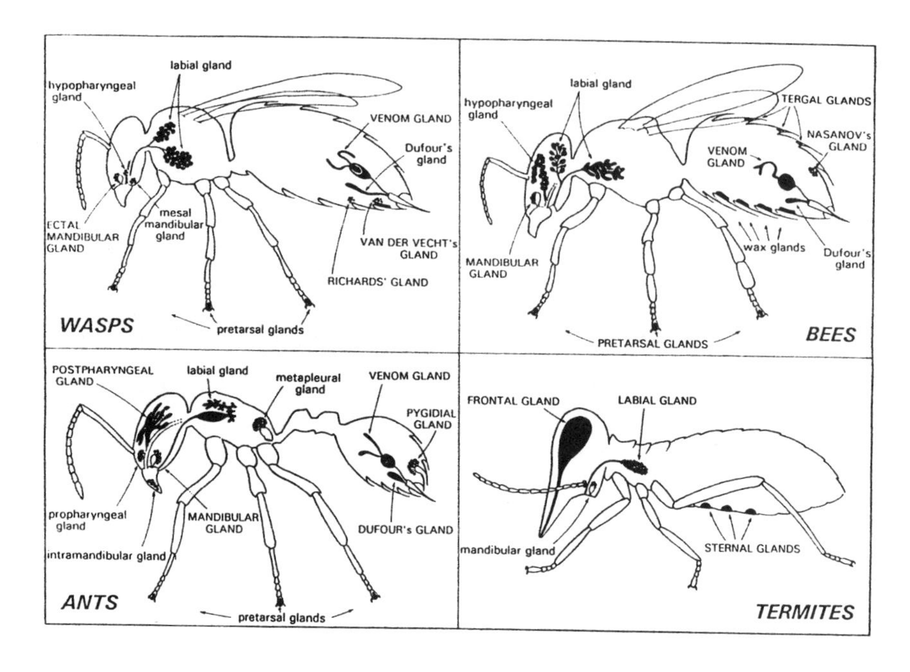

FIGURE 1.1 Schematical profile drawings showing the commonly found exocrine glands in wasps, bees, ants and termites. Glands with a pheromonal function are indicated with capital lettering.

sections. The small size of insects, on the other hand, for long represented a considerable drawback in our chemical understanding of the glandular secretions. The availability of more sophisticated equipment and techniques in the past decades has made analysis at the nanogram level possible, thus resulting in the identification of many glandular products. In this chapter, we focus on the structural and chemical complexity of the pheromone producing exocrine glands of the social insects.

Structural Organization of Pheromone-Producing Glands

Exocrine glands can be classified into two major types according to the structural organization of their secretory cells (Billen, 1991), corresponding with types I and III in the pioneer paper on insect glands by Noirot and Quennedey (1974). Glands with secretory cells of the first type (type I) are directly derived from the tegumental epidermis as is reflected in their epithelial organization. The secretory cells form a monolayered epithelium either as part of the external body tegument through which they directly discharge their secretory products (e.g.

the trail producing sternal glands of termites, Figure 1.2), or as the lining of an internalized reservoir where secretion can be temporarily stored (e.g. the postpharyngeal gland in ants, that is involved in recognition mechanisms, Figure 1.3).

More complicated are the glands with secretory cells of the second type (type III according to Noirot and Quennedey, 1974), where the gland is formed by a variable number of bicellular units, each comprising a secretory cell and a duct cell. Each unit of this gland type originates via tetrad formation by an epithelial stem cell through two mitoses, and subsequent differentiation of one daughter cell into a slender duct cell and one into the secretory cell, while the remaining two daughter cells degenerate (Sreng and Quennedey, 1976). The contact area between the remaining duct cell and secretory cell is known as the 'end apparatus', that represents a specialized region to allow secretory products to find their way to the outside. Glands of this type can equally open, by means of their duct cells, either directly through the tegument (e.g. the Van der Vecht gland in wasps, which is the source of repellent substances, Figure 1.4), or into a reservoir (e.g. the mandibular gland, which is the source of alarm substances in many social insects, Figures 1.5,1.6). The reservoir in these glands is formed by flattened, generally non-secretory epithelial cells.

Secretory cells of type II were described as basally located epithelial cells without contact with the apical cuticle, as is sometimes found in sternal glands of termites (Noirot and Quennedey, 1974). They probably do not represent another secretory cell type, but are to be considered as oenocytes (Noirot and Quennedey, 1991).

Many glands are common to all social insects and occur in the queens, workers and males, as is for example the case for the mandibular and salivary glands. Apart from these standard exocrine glands that are also found in solitary insects, some glands represent neoformations that are characteristic for the family, subfamily, genus or even for the species. In this way, wax glands are characteristic for the Apidae, and are not found elsewhere. Likewise, Van der Vecht's gland and Richards' glands are specifically found in wasps, while ants are characterized by the presence of postpharyngeal, metapleural and pygidial glands. The exocrine glands of the heterometabolous termites cannot be homologized with the glands of the Hymenoptera, and therefore can be considered as rather specific.

The number of known exocrine glands in social insects becomes more and more impressive, and reflects the evolution of sectioning techniques. Several hitherto unknown glands have recently been described (Figure 1.7), although their function often still remains unknown so far. Because of this steadily increasing variety of exocrine

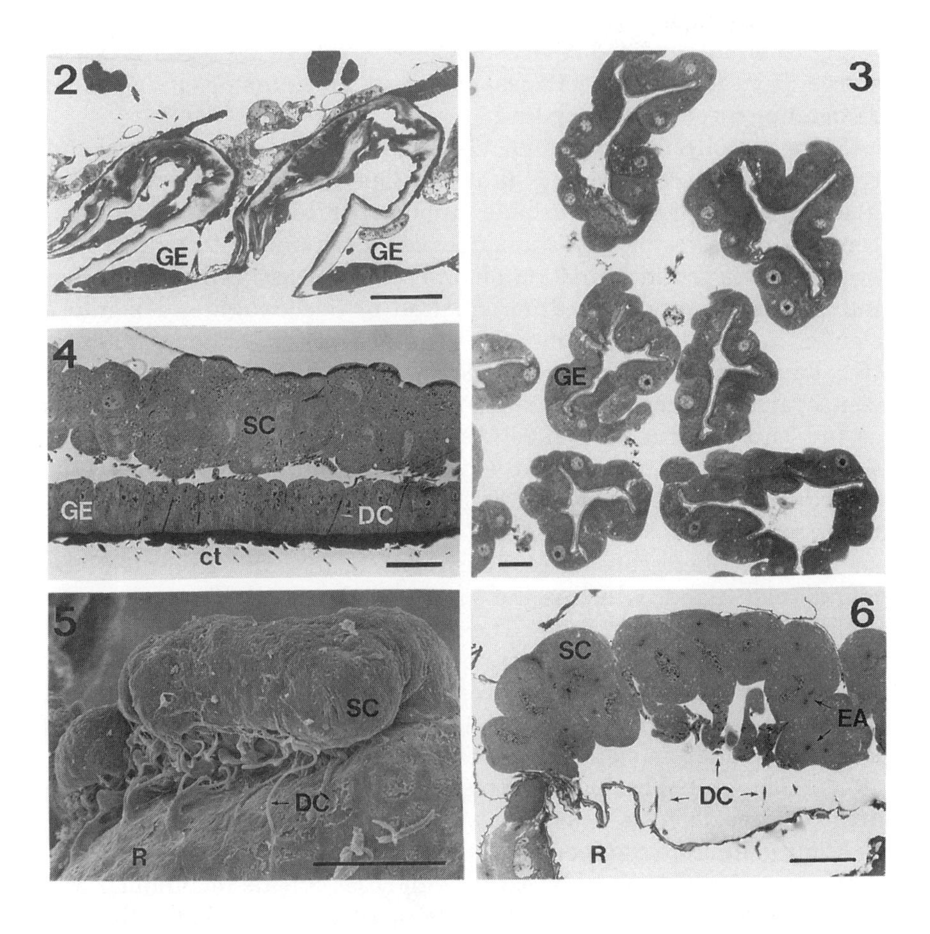

FIGURES 1.2-1.6 Semi-thin sections through various types of pheromone producing glands. 1.2 - epithelial glands directly opening through the tegument (sternal glands of the termite *Schedorhinotermes lamanianus,* scale bar 50 µm); 1.3 - epithelial gland with reservoir (postpharyngeal gland of the ant *Cataglyphis niger,* scale bar 20 µm); 1.4 - bicellular units opening directly through the tegument (Van der Vecht's gland of a worker of the wasp *Polistes annularis*. Also note associated epithelial gland, scale bar 20 µm); 1.5 - scanning micrograph showing the mandibular gland of a worker of the ant *Formica sanguinea* with bicellular units opening into common reservoir (scale bar 50 µm); 1.6 - bicellular units opening into common reservoir (mandibular gland of a worker of the bumblebee *Bombus pratorum*, scale bar 50 µm). ct: cuticle, DC: duct cell, EA: end apparatus, GE:glandular epithelium, R reservoir, SC: secretory cell.

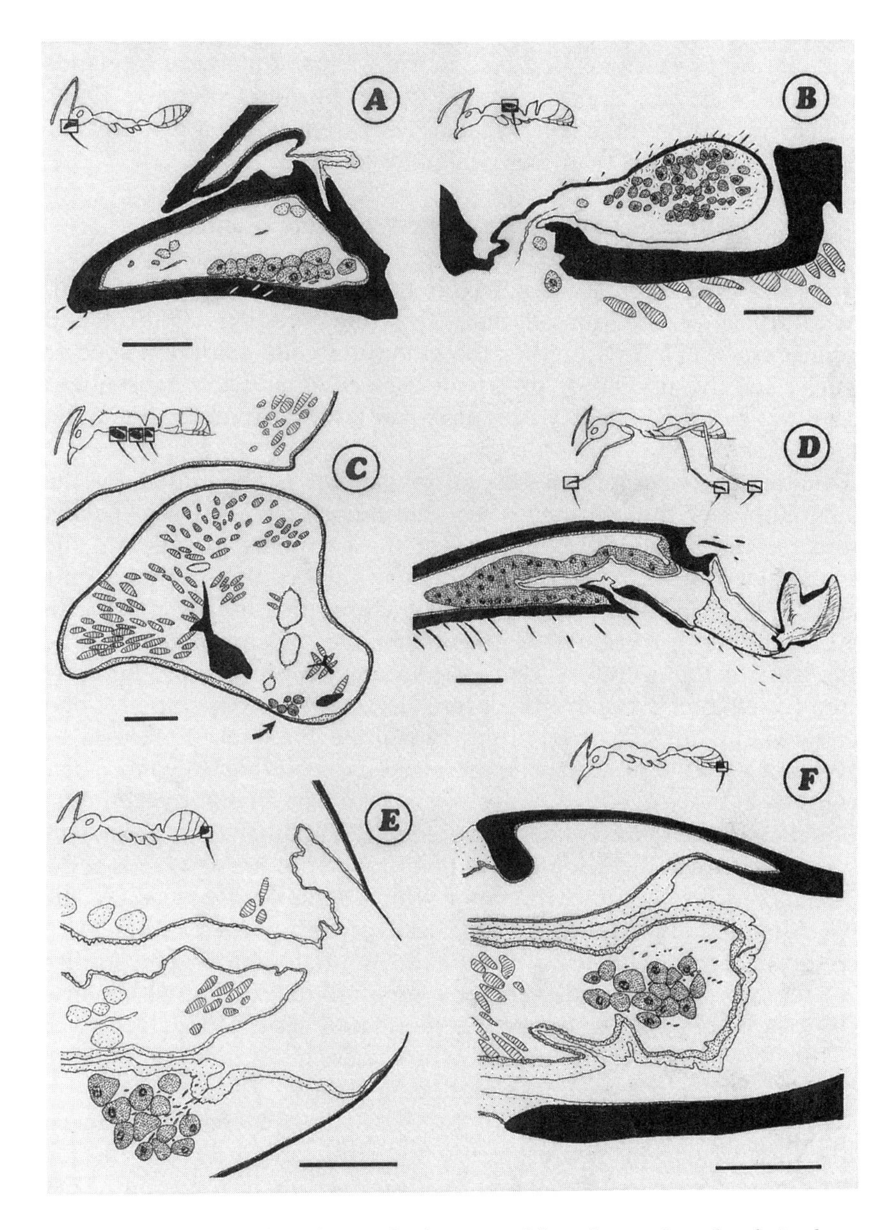

FIGURE 1.7 Examples of recently discovered "new" exocrine glands in the Formicidae (A. intramandibular gland; B. gemma gland *Diacamma*; C. coxal glands *Pachycondyla*; D. pretarsal glands; E. cloacal gland *Cataglyphis*; F. sting bulb gland *Myrmecia* and *Nothomyrmecia*). Scale bar 100 μm.

glands, a clear designation of the various glands is necessary. The terminology 'sternal glands', for instance, includes at least 6 different glands in the Formicidae, and also in the wasps, a plethora of glands associated with the abdominal sternites is found (Downing, 1991). Similarly, recent research has revealed the existence of a much broader variety of 'leg glands' than was originally thought.

Ultrastructure of Pheromone-Producing Glands

Glands are structures specialized for the storage and the emission, spreading or evaporation of their secretory products (Noirot and Quennedey, 1991). This implies the glandular cells display a specific capacity for the uptake of precursor molecules and the subsequent biosynthesis of the secretory products, and can effectively regulate the release of secretion.

Uptake of precursor substances in general is facilitated by the invaginations of the cell membrane that increase the surface. This is found for the basal cell membrane of the epithelial glands and the peripheral cell membrane of the secretory cells of the bicellular unit glands, which in both cases represent the area that is in contact with the haemolymph, from where precursor molecules are obtained. The cytoplasm of the secretory cells of pheromone producing glands is generally characterized by the presence of a well developed Golgi apparatus and smooth endoplasmic reticulum (Figs. 1.8,1.9). This is in agreement with the production of non-proteinaceous and low molecular weight substances, which are characteristic of many pheromone-producing glands (Noirot and Quennedey, 1974; Billen, 1991). Another common feature in the cytoplasm of pheromone producing glands is the occurrence of various inclusions, of which lamellated bodies are the most conspicuous (Figure 1.13). These probably correspond with secretory material, as may be concluded from autoradiography studies in vertebrates, where lipid secretions were found to occur as lamellar inclusions within the glandular cells (Boudreau et al., 1983; Hidalgo et al., 1985). Mitochondria generally are also abundant among the cellular organelles in pheromone producing glands.

Once the secretory products are formed via the metabolic machinery of the glandular cell, they are ready for transport to the outside for immediate release or for temporary storage in a reservoir. This pathway, however, produces for these molecules a double barrier as they have to cross both the cell membrane of the secretory cell and the cuticle. In epithelial glands, the apical cell membrane is generally differentiated into microvilli that create a considerable increase of surface (Figs. 1.8,1.9). Tubular extensions of smooth endoplasmic reticulum may continue into the microvilli and thus bring the secretory products straight to the site of release (Billen, 1991; Figure 1.8). The cuticle overlying the epithelium may either be permeable for the secretory molecules to diffuse through (Figure 1.8), or may display conspicuous cuticular pores (Figure 1.9; see Blomquist et al., this book). Glands formed by bicellular secretory units have the end apparatus in

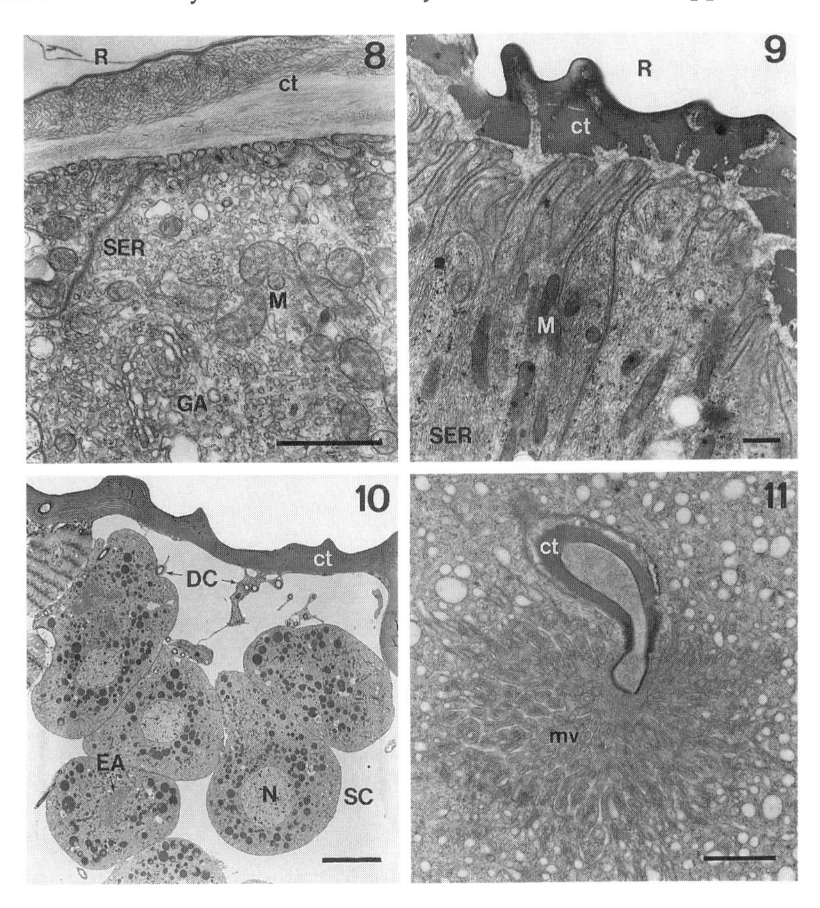

FIGURES 1.8-1.11 Electron micrographs illustrating the ultrastructural organization of pheromone-producing gland cells. 1.8 - apical region in the postpharyngeal gland of a callow worker of *Formica sanguinea* (scale bar 1 µm); 1.9 - apical region in the tibial gland of *Crematogaster scutellaris* showing cuticular pores (scale bar 1 µm); 1.10 - secretory cells of the pygidial gland of the amazon ant *Polyergus rufescens* (scale bar 10 µm); 1.11 - contact area between duct cell and secretory cell in Nasanov's gland of a *Apis mellifera carnica* worker, showing end apparatus with regular microvillar organization (scale bar 1 µm).

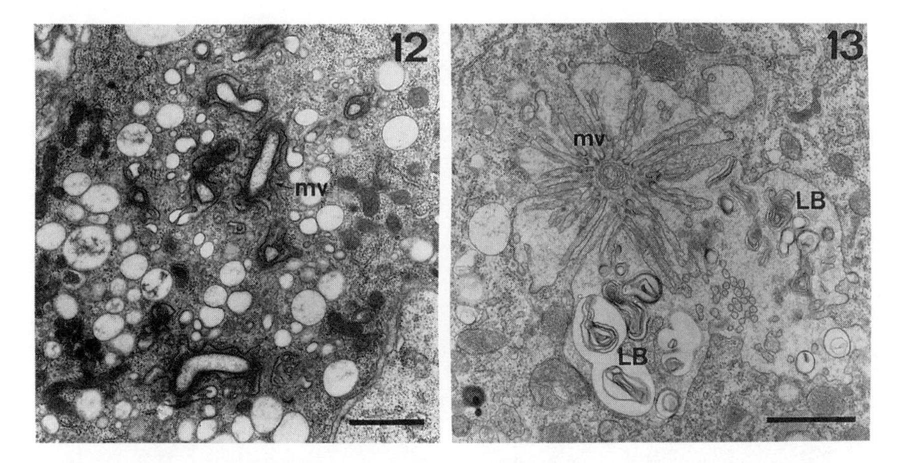

FIGURES 1.12-1-13. 1.12 - Branched end apparatus with blunt microvilli in the venom gland of a *Vespula vulgaris* worker (scale bar 1 μm); 1.13 - end apparatus with distorted microvillar pattern and multilamellar bodies, venom gland *Nothomyrmecia macrops* (scale bar 1 μm) ct: cuticle, DC: duct cell, EA: end apparatus, GA: Golgi apparatus, LB: lamellar bodies, M: mitochondria, mv: microvilli, N: nucleus, R: reservoir, SC: secretory cell, SER: smooth endoplasmic reticulum.

each unit to assure efficient release of the secretory products from the glandular cells, from where associated duct cells carry these products to the reservoir or to the outside (Figure 1.10). This specialized region is formed by an invaginated microvillar differentiation of the cell membrane of the secretory cell, surrounding a central cuticular canal that continues into the duct cell (Figure 1.11). The microvillar sheath again offers a considerable surface increase, while the cuticular lining of the end apparatus always shows pore canals. The lining cuticle considerably thickens where the end apparatus joins the duct cell, thus forming an efficient porous structure in the secretory cell and a discontinuous sheath in the duct cell to prevent leaking (Figure 1.11). The end apparatus often has a sinuous course in the secretory cells, while it may occur with many branches only in exocrine glands of the Vespidae (Delfino et al., 1979; Figure 1.12). The microvillar pattern may vary from an orderly arrangement (Figure 1.11) to very distorted (Figure 1.13). This appearance of extracellular space may create an additional storage capacity for the secretory cells, as was first documented by Bazire-Bénazet and Zylberberg (1979). Lamellar bodies often accumulate in these intermicrovillar spaces of the end apparatus (Figure 1.13), which supports their secretory nature. Also in epithelial glands, lamellar bodies can be found in association with the apical microvilli and cuticle, from where they reach the reservoir or the outside (Billen, 1991).

The mechanisms that regulate the discharge of secretion are not very well known. Glands without reservoir release their secretion directly to the outside, and do not appear to have a direct control mechanism over their secretory activity. Glands with a reservoir often have muscle fibers surrounding the latter, contraction of which will result in discharge of secretion. In addition to the reservoir musculature, also the duct region may be equipped with a muscular supply that may open or close the duct, and so affecting the secretory activity of the gland, as is the case for the venom gland and Dufour's gland (Billen, 1982). The mandibular gland of bees (Nedel, 1960) and wasps (Hermann et al., 1971) appears to possess a muscular sphincter around its duct, although in ants we could not find any muscular equipment in association with the mandibular gland duct (Billen and Schoeters, 1994). Nervous control of the discharge of secretion by social insect glands is poorly understood. Innervation is fairly commonly found, although the physiological mechanisms remain unknown. A very complex 'multiple innervation' system has been described for the salivary glands of *Kalotermes*, with mixed nerves containing up to 6 different categories of axons (Alibert, 1983).

Occurrence of Pheromone-Producing Glands

Pheromone producing glands are distributed over all regions of the social insect's body, with a location that is directly linked with their function. Alarm pheromones are primarily produced by the exocrine glands associated with the mandibles and sting, that are the defensive weapons par excellence. The mandibular gland has been reported to be the source for alarm substances in a wide variety of ants and bees (Maschwitz, 1964). In termites, the mandibles themselves also appear to be used as mechanical weapons (Deligne et al., 1981), although no pheromonal function could be attributed to their mandibular glands so far. Many termite species, however, possess an unpaired frontal gland, which opens on the anterior part of the head (Noirot, 1969). The gland is especially developed in the soldiers, and is known as an efficient ant repellent that also offers protection to the termites in their galleries (Kaib, 1985). The evolution of the hymenopteran ovipositor towards a sting equally resulted in a change of the venom and Dufour's gland, being the female accessory reproductive glands, from an initially reproductive function to a defensive role. The venom gland, which always releases its secretion through the sting, especially plays an important role in the alarm-defence system. The complex compartmentalized organization of the venom gland, that comprises an internalized convoluted gland, may prevent self-toxication (Schoeters and Billen, 1995a,b). The Dufour's gland of wasps and bees opens ventrally of the sting, so that secretion is released into the oviduct (Billen, 1987). This anatomical organization may explain why a pheromonal function for Dufour's gland in these social insects is generally not known. In ants, on the other hand, both the venom and Dufour's gland open through the sting, which represents an excellent anatomical position for the production of trail pheromones.

The use of trail pheromones represents a commonly known communication system for terrestrial social insects like ants and termites. In termites, these are elaborated in the sternal glands. The variety of exocrine glands producing trail substances in the Formicidae is absolutely astonishing. At least 9 different abdominal glands may be involved, where their anatomical position allows efficient deposition of the active substances onto the substrate (Figure 1.14). Several of these glands open through the sting or the anal opening, or occur in association with the sternites. Also tergal glands may produce trail substances, however, although in this case their opening site is very close to the abdominal tip, as is found for the pygidial gland in some

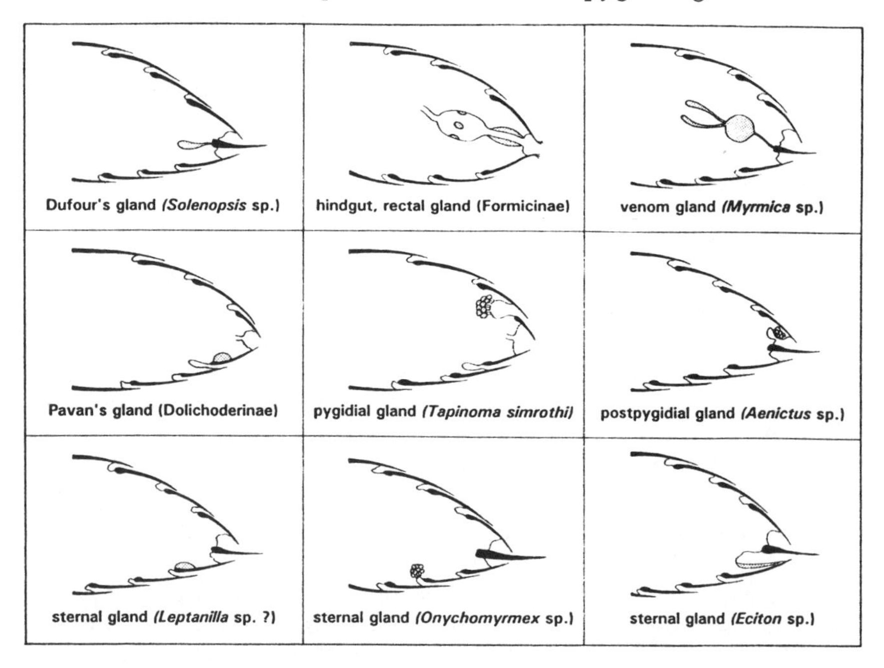

FIGURE 1.14 Schematical illustration of the various abdominal glands in ants that are known to produce trail pheromones.

dolichoderines (Simon and Hefetz, 1991) or the postpygidial gland in *Aenictus* army ants (Oldham et al., 1994a). Besides abdominal glands, the legs represent an obvious position for the eventual production and release of trail substances. Until recently, the tibial gland in *Crematogaster* (Figure 1.9; Leuthold, 1968; Billen, 1984) was reported as the only non-abdominal source for trail pheromones in ants. A combination of behavioural and morphological examination, however, has also revealed a trail marking function in some ponerine ants for the pretarsal glands (Hölldobler and Palmer, 1989) and for the newly discovered basitarsal glands (Hölldobler et al., 1992).

A gland that has very much come into focus recently is the postpharyngeal gland, that represents a unique exocrine structure for the Formicidae only. Although earlier papers generally attributed digestive functions to the postpharyngeal gland (Delage-Darchen, 1976), the finding that the chemical profile of the gland matches to that of the epicuticular hydrocarbons (Bagnères and Morgan, 1991) has raised the question on the role of this gland in nestmate recognition (Soroker et al., 1994; Vienne et al., 1995; Meskali et al., 1995). For this function, the postpharyngeal glands occupy an obviously suited position as their secretion can conveniently be exchanged via oral contact during grooming with nestmates.

Chemical Aspects of Pheromone-Producing Glands

In an attempt to reduce the various kinds of pheromones encountered in insects to some kind of order, they have been classified in various ways. That of Dethier et al. (1960) is heavily based on movement, and was intended for other kinds of semiochemicals in addition to pheromones. It classifies them as locomotory stimulant, arrestant, attractant, repellent, a stimulant (for feeding, mating, oviposition, etc.) and a deterrent (for similar behaviours). Another more widely used system divides pheromones first into primer pheromones and releaser pheromones (Wilson and Bossert, 1963). So few primer pheromones have as yet been identified, that they have not been further divided into categories. Most of the pheromones we know are releaser pheromones, and at present, these are usually divided into nine groups: sex; aggregation; dispersal; alarm; recruitment or trail following; territorial or home range; surface; funeral and invitation pheromones. For a detailed discussion, see Ali and Morgan (1990).

These categories were largely devised for simple types of behaviour, exemplified by the excitement and attraction of males by the female lepidopteran sex pheromones. Compared to the relative simplicity of the sexual attraction of many lepidopterans and dipterans, the behaviour induced by pheromones of the social insects is intricate and relatively little understood. The potential complexity of social insect signalling is indicated by the great variety of glands described here that are capable of emitting pheromones even after making allowance for many exocrine glands which serve individual functions as indicated earlier.

Pheromone signals may be multicomponent in all insect orders and that is equally true for social insects, but the latter may have a blend of substances from more than one gland (Silverstein and Young, 1976; Hölldobler and Carlin, 1987; Hölldobler, 1995). For example, the trail pheromone of the ant *Messor bouvieri* is derived from both the poison gland (3-ethyl-2,5-dimethylpyrazine) and the Dufour gland (a mixture of C_{13} to C_{19} alkenes with one, two or three double bonds) (Jackson et al., 1991). Similarly, queen dominance in *Apis* appears to involve the tergal and pretarsal glands as well as the mandibular glands (Vierling and Renner, 1977; Lensky and Slabezski, 1981). Some pheromone blends may comprise a primer and a releaser component, as we have encountered in the trail pheromone of an *Aenictus* species (Oldham et al., 1994a) where methyl nicotinate acts as a primer substance which must be detected before ants can follow the releaser component methyl anthranilate. Neither substance is effective without the other but they do not have to be released at the same time.

The simple picture of a lepidopteran imago coming to maturity emitting or detecting pheromone, leading to fertilization, bears little resemblance to the situation in social insects where worker castes are normally separated from reproduction and may perform different tasks, dependant upon their size or age. Tengö et al. (1991) have recorded that the Dufour gland secretion of bumblebees varies with time of year. We have observed changes in the mandibular gland and Dufour gland secretions of the ant *Formica sanguinea* (Moens et al., 1990; Ali et al., 1988) as these workers grow older.

We have also found in *Atta sexdens rubropilosa*, a sharp variation in mandibular gland pheromone with size of workers (do Nascimento et al., 1993). There are numerous examples where there are distinct differences between worker and soldier caste, or between sexuals and workers.

On the other hand, we have examples where different glands contain the same or very similar blends of substances. Bagnères and Morgan (1991) showed that the postpharyngeal gland of several ant species selected randomly contained the same hydrocarbon substances as found on the cuticle, which receives its hydrocarbons from the epidermal glands (Blomquist et al., Chapter 2, this book). Oldham et al. (1994b) have found the mixture of hydrocarbons on the cuticle of some bumblebee species is the same as that found in the Dufour glands. The pheromonal function is understood in neither case.

Trail or Recruitment Pheromones

Among social insects, the trail following or recruitment pheromones are probably the most clearly defined because they still form a fairly small group. They have been reviewed from a chemical viewpoint by Morgan (1990). Ant trail pheromones have been reviewed by Attygalle and Morgan (1985) and Jackson and Morgan (1993).

The first of these to be studied were termite trail pheromones, in the 1960's when three substances were identified. The first to be isolated was Neocembrene A **1** (now (E)-6-cembrene A) from *Nasutitermes exitiosus* (Moore, 1966). Evidence at present suggests that termite trail pheromones are by no means species specific. This same cembrene is apparently used by *N. walkeri* and *N. graveolus* (Moore, 1966) and by

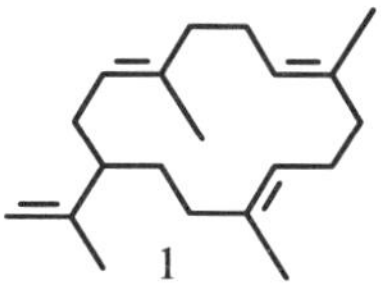

1

OH

2

Trinervitermes bettonianus (McDowell and Oloo, 1984). (3Z,6Z,8E)-3,6,8-dodecatrienol **2** was first identified as the trail pheromone of *Reticulitermes virginiacus* (Matsumura et al., 1968; Tai et al., 1969).

There was then a long pause in the study of termites, but recently the subject has regained attention. Toroko et al. (1989) identified the same dodecatrienol as the pheromone of *Coptotermes formosanus*, and of *Reticulitermes speratus* (Toroko et al., 1991), while Laduguie et al. (1994) identified it in *Reticulitermes santonensis*, and Grace et al. (1995) have identified it in *R. hesperus*. These two compounds remain the only firmly identified trail pheromones of Isoptera.

The number of trail pheromones of ants that has been identified is relatively large. They are gathered together in Table 1.1 and Figure 1.15 (**3** to **27**). Some of them appear to be species-specific, others are widely distributed through several genera. The claim of identification of the trail pheromone of *Lasius fuliginosus* by Huwyler et al. (1975) has been dropped from the list since Quinet and Pasteels (1995) have shown the mixture of fatty acids is inactive for this species. Species specificity is probably imparted by other glandular secretions, such as the Dufour gland contents laid down at the same time. It is noteworthy how frequently the alkylpyrazines **4,13,14** occur among the Myrmicinae and that the polar (water soluble) compounds come from the venom gland while the non-polar (lipophilic) compounds tend to come from the Dufour gland. Almost all the substances are volatile, with small molecules, and only geranylgeranyl acetate **22** and possibly hexadecenal **25** (relatively large molecules) and N,N-dimethyluracil **23** (relatively involatile) are exceptions to these rules so far. These substances are, in every case recorded, present only in nanogram quantities. Those in the venom glands are little more than impurities in the venom. The small quantities would make them very difficult to identify were it not for the relatively simple bioassays available. See also Vander Meer and Alonso, Chapter 7 this book.

There are numerous reports of spot-marking by male bumblebees close to good food sources, but these are essentially to attract females. Similar marking behaviour has been observed in stingless bees, using the mandibular gland. The mandibular pheromone of *Trigona*

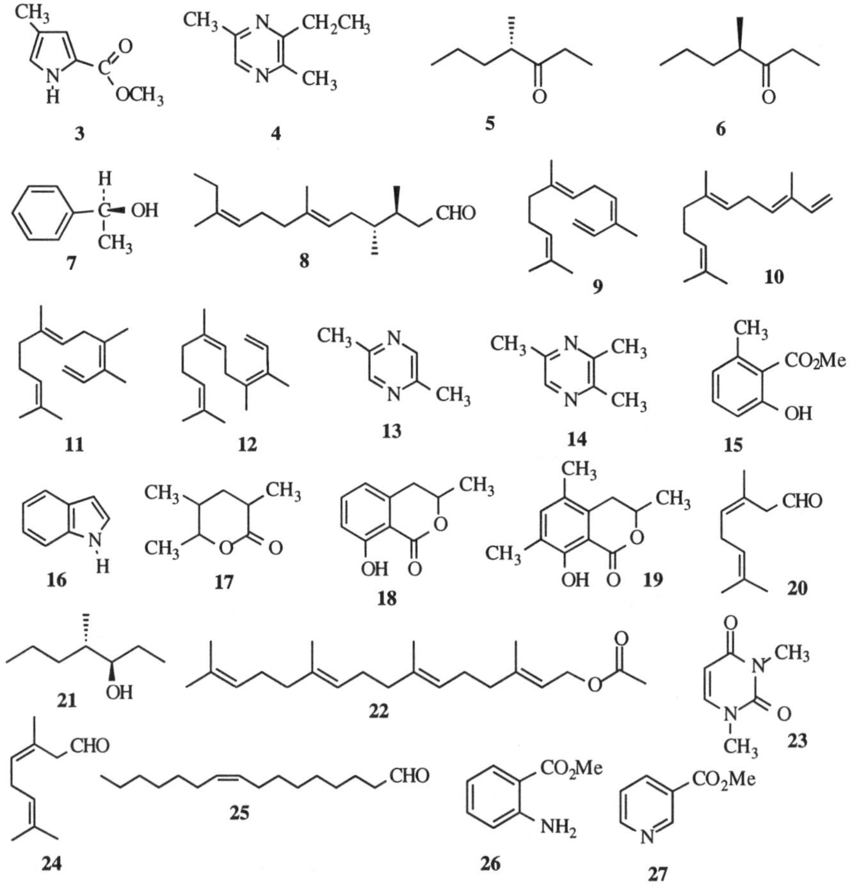

FIGURE 1.15 Trail pheromone substances of ants.

TABLE 1.1 Trail pheromone substances identified in ants.

SPECIES	COMPOUND[1]	SOURCE[2]	NOTE/REFERENCE
Myrmicinae			
Acromyrmex octospinosus	M4MPC **3**	VG	Robinson *et al.* 1974,
A. subterraneus subterraneus	M4MPC **3**	VG	a/do Nascimento *et al.* 1994
Aphaenogaster albisetosus	4-Methyl-3-heptanone **5,6**	VG	b/Hölldobler *et al.* 1995A
A. cockerelli	(R)-1-Phenylethanol **7**	VG	Hölldobler *et al.* 1995a
Atta cephalotes	M4MPC **3**	VG	Riley *et al.* 1974
A. sexdens rubropilosa	EDMP **4**	VG	Cross et al. 1979
A. sexdens sexdens	EDMP **4**	VG	Evershed and Morgan 1983
A. texana	M4MPC **3**	VG	Tumlinson *et al.* 1972
Manica rubida	EDMP **4**	VG	Attygalle *et al.* 1986
Messor bouvieri	EDMP **4**	VG	c/Jackson *et al.* 1989, 1991
Monomorium pharaonis	Faranal **8**	DG	Ritter *et al.* 1977
Myrmica rubra	EDMP **4**	VG	d/Cammaerts-Tricot *et al.* 1977
M. lobicornis, M. lonae, M. ruginodis, M. rugulosa, M. sabuleti, M. scabrinodis, M. schencki, M. sulcinodis	EDMP **4**	VG	Evershed *et al.* 1982
M. gallieni, M. specioides	EDMP **4**	VG	Jackson *et al.* 1989
M. sp. nova from Idaho	EDMP **4**	VG	Jackson *et al.* 1991
Pheidole pallidula	EDMP **4**	VG	e/Ali *et al.* 1988
Solenopsis invicta	(Z,E)-α-farnesene **9**	DG	Vander Meer *et al.* 1981, 1988
	(E,E)-α-farnesene **10**		
	(Z,E)-homofarnesene **11**,		Alvarez *et al.* 1987
	(E,E)-homofarnesene **12**		
	homoselenine		f/Vander Meer, pers. comm.
Tetramorium caespitum	EDMP **4**, DMP **13**	VG	Attygalle and Morgan 1984
T. impurum	Methyl-6-methylsalicylate **15**	VG	Morgan and Ollett 1987
T. meridionale	Indole **16**	VG	g/Jackson et al. 1990
Formicinae			
Camponotus herculeanus	2,4-dimethyl		
Formica rufa	Mellein **18**	Hindgut	Bestmann *et al.* 1992
Lasius niger	*(R)*-8-hydroxy-3,5,7-trimethyl dihydro-isocoumarin **19**	Hindgut	Bestmann *et al.* 1992
Ponerinae			
Leptogenys diminuta	*(3R,4S)*-4-methyl-3-heptanol **21**	Pyg	g/Attygalle *et al.* 1988, Steghaus-Kovac *et al.* 1992
Megaponera foetens	N,N-dimethyluracil **23**	Pyg	Janssen *et al.* 1995
Pachycondyla marginata	Citronellal **24**	Pyg	h/Hölldobler *et al.* 1995b
Dolichoderinae			
Linepithema humile	*(Z)-9-Hexadecenal* **25**	PavG	Cavill *et al.* 1979, 1980
Aenictinae			
Aenictus sp. nova	methyl anthranilate **26**, methyl nicotinate **27**	Postpyg	i/Oldham *et al.* 1994a

[1]For structures, see Fig. 1.15. [2]VG is venom gland, DG is Dufour gland, Pyg is pygidial gland, PavG is Pavan's gland and Postpyg is postpygidial gland. a=1.2 ng/ant; b=8:2 mixture of *S* and *R* enantiomers; c=Dufour gland is more active; d=5 ng/ant; e=other unidentified substances are active; f=structure not yet complete; g=small quantities of DMP **13**, TMP **14** and EDMP **4** also present and active; h=*C. ligniperda, C. vagus, C. pennsylvanicus* and *C. socius* followed same substance; g=*cis*-isogeraniol **20** acts as a recruitment pheromone, Attygalle *et al.* 1991; h=isopulegol may act as a synergist; i=methyl nicotinate is a primer,required either before or with methyl anthranilate.

subterranea is dominated by geranial and neral (Blum et al., 1970) but the necessary proof that these are the pheromone is lacking. A number of volatile substances are found in the mandibular glands of *Trigona spinipes*, of which only 2-heptanol **28** successfully marked trails that were followed by *T. spinipes* workers.

The Nasanov gland in *Apis mellifera* is a well recognized source of a marking pheromone, consisting of geraniol **29** and geranial **30** and other minor related components (Pickett et al., 1980). No other trail of marking pheromones of *Apis* species, nor of any wasps have yet been identified.

Alarm Pheromones

Alarm pheromones are a rather ill-defined class because it is not easy to distinguish or define alarm in a bioassay. The effect might as correctly be called alerting, dispersing or attracting in some of the bioassays recorded and many substances in a concentrated form, placed near a group of insects can agitate or alarm them (but see Vander Meer and Alonso, Chapter 7 this book).

From early work on nasute termites, it was learned that large amounts of monoterpenes (α-pinene **31**, β-pinene **32**, limonene **33**, terpinolene **34**, α-phellandrene **35** and myrcene **36**) are present in the frontal gland secretion of soldiers, in addition to the rapidly oxidizing sesquiterpenes which immobilize predators. It was suggested by Moore (1968) that the monoterpenes may function as alarm pheromones. It would appear that this idea has not been followed up by any recent experimental work.

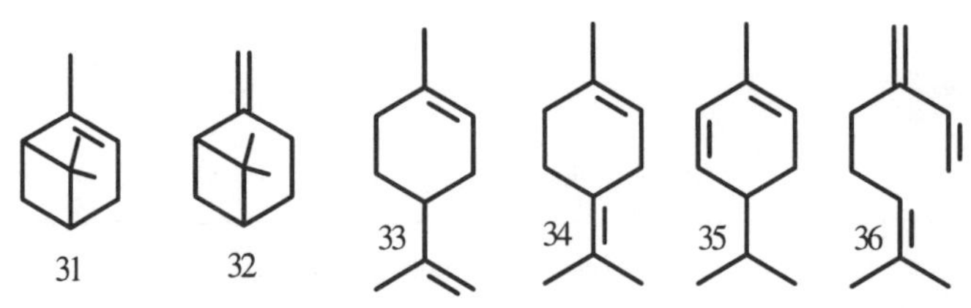

Hölldobler and Wilson (1990, p. 263,264) have assembled a table of those substances identified as alarm pheromones of ants, together with

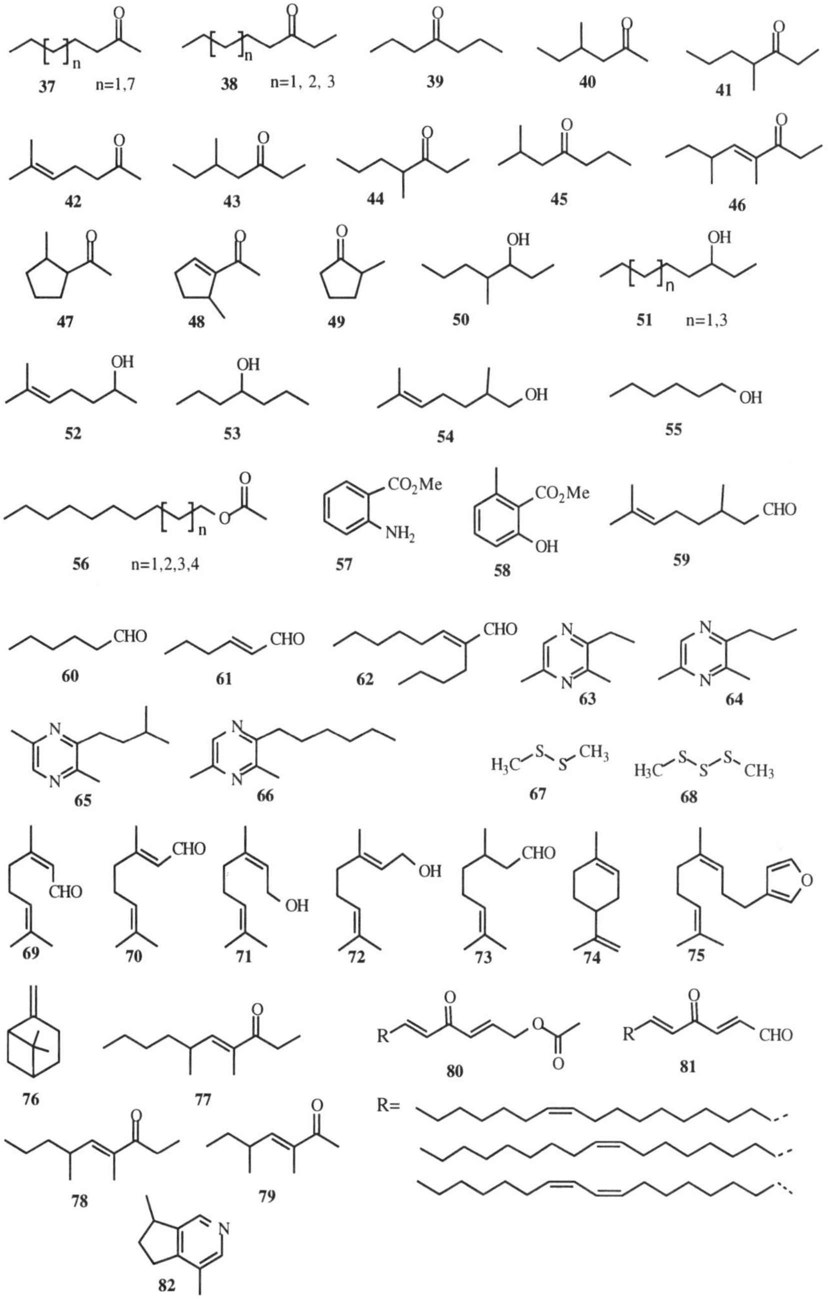

FIGURE 1.16 Alarm pheromone substances in ants.

their glandular sources. The data is not reproduced here since it is comparatively up-to-date and readily available. The list contains 16 ketones **37-49**, 7 alcohols **50-55**, 6 esters **56-58**, 4 aldehydes **59-62** and 4 alkylpyrazines **63-66**, together with dimethyl disulphide **67** and trisulphide **68**, 8 terpenes **69-76**, formic acid, decane and undecane (Figure 1.16). The mandibular gland is the commonest source, followed by the pygidial gland. Although mandibular gland secretions are generally thought to be alarm substances, Cammaerts and Mori (1987) found that the 3-octanol (**38**, n=1) of *Myrmica* mandibular glands was an attractant or arrestant, rather than causing alarm. There is no correlation between type of compound and glandular source, except that the cyclopentyl compounds **47-49**, which are related to iridoids, come from the pygidial gland of a dolichoderine. Three additions are necessary to the list of Hölldobler and Wilson (1990), first, the unusual group of compounds **77-79**, represented by manicone **46**, in the mandibular glands of *Manica rubida* (Bestmann et al., 1988), and secondly, the unusual contact poison and alarm pheromone **80, 81** from the Dufour glands of *Crematogaster scutellaris* (Pasteels et al., 1989). Thirdly, Janssen et al. (1995) recently found that actinidine **82** from the pygidial gland of *Megaponera foetens* stimulated workers to leave the nest and possibly functions as an alarm pheromone.

Alarm behaviour was long recognized in bees and the pheromone elicits an easily recognized response. The composition of the secretion, from the Koschevnikov's gland in the sting apparatus presents more of a problem. The principal component was soon identified as isopentyl acetate **83** (Boch et al., 1962) but another 22 components have been subsequently identified, including 7 acetate esters **84-86**, 11 alcohols **87-**

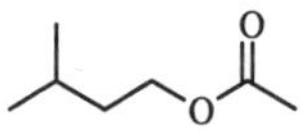

83 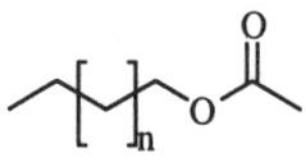 84 n=1,3,5,7,17 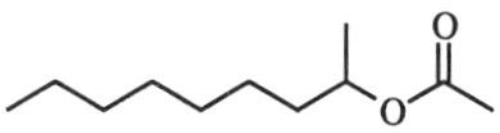 85 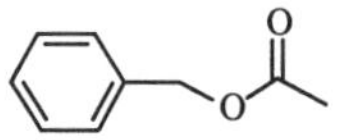 86

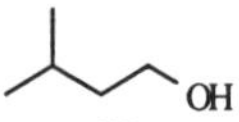

87 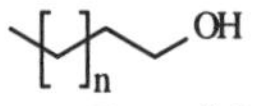 88 n=1,2,5,15 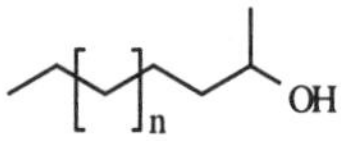 89 n=1,3 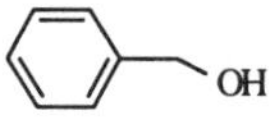 90 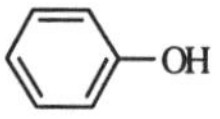 91

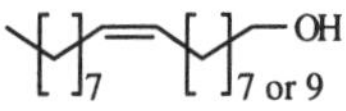

92 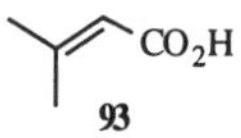 93 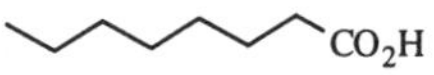 94 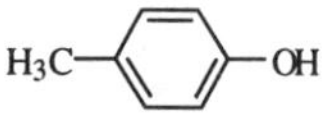 95

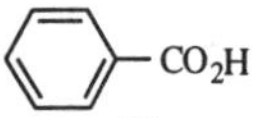

96 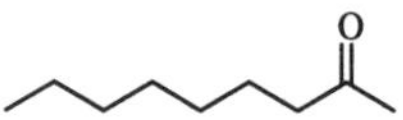 97 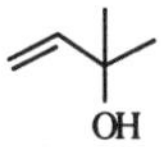 98 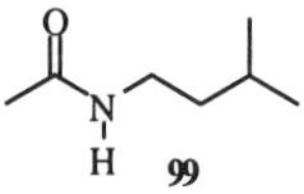 99

90, 3 acids **93, 94, 96** and phenol **91** and p-cresol **95**, identified either by Blum (1982) or by Pickett et al. (1982). 2-Heptanone from *A. mellifera* worker mandibular glands is widely quoted as an alarm pheromone, but Vallet et al. (1991) have shown that it does not induce alarm and claim that it is a forage marking pheromone.

Only one alarm pheromone from a stingless bee has been identified certainly. The mandibular secretion of *Trigona pectoralis* contains 2-heptanol, 2-heptanone, 2-nonanol and 2-nonanone. The 2-nonanone **97** alone released attacking response (quoted in Free, 1987; without source). Veith et al. (1984) have identified the major component of the alarm pheromone of the hornet *Vespa crabro* as 2-methyl-3-buten-2-ol **98**. N-3-Methylbutylacetamide **99** has been recognized as an alarm pheromone from the venom sac in two species of wasps, *Vespula squamosa* (Heath and Landolt, 1988) and *Vespula maculifrons* (Landolt et al., 1995).

Sexual Pheromones

The trail pheromone substance 3,6,8-dodecatrien-1-ol (**2**) already described as a trail pheromone of termites from the sternal gland, has been suggested to be also the sexual attractant pheromone of the termites *Pseudacanthotermes spiniger* (Bordereau et al., 1991) and *Reticulitermes santonensis* (Laduguie et al., 1994). *P. spiniger* has ten times more of the substance in the glands of alate females than of males, and more than in the glands of workers.

Although a number of substances have been identified in the mandibular glands of sexual ants, where these are different from the substances found similarly in workers, we are not aware of any examples where these have definitely been shown to act as sexual attractant pheromones. On the other hand there is plenty of evidence that they exist and for their origins. The difficulties of bioassays have held back their isolation.

By far the best understood of the sexual pheromones of social insects is the queen substance of honeybees. By a fortunate chance the queen substance used for "queen control" inside the nest is also the substance used by virgin queens to attract drones for mating. (E)-9-Oxodec-2-enoic acid (9-ODA) **100** was identified in queen mandibular glands by Callow and Johnston (1960) and Barbier and Lederer (1960). The pure 9-ODA is not as attractive to the drones as whole mandibular glands. Later Callow et al. (1964) identified 13 further components from crushed heads of queens, to which Barbier added methyl p-hydroxybenzoate **101**. Bioassay of these showed activity only in 9-ODA and (E)-9-hydroxydec-2-enoic acid **102, 103**, a secondary component in the gland.

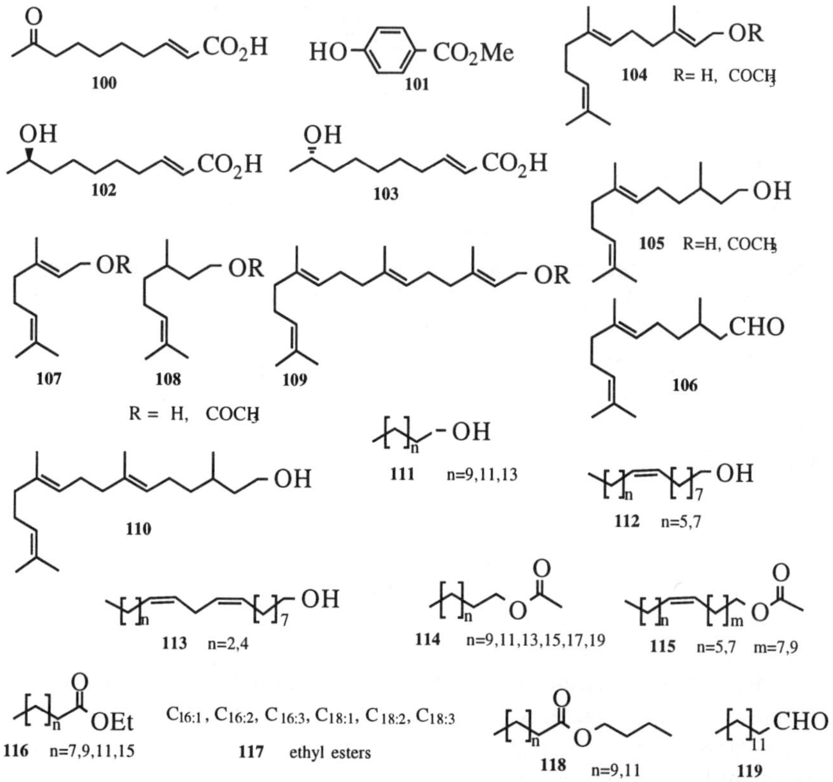

100   101   104 R= H, $COCH_3$

102   103   105 R=H, $COCH_3$

107   108   109   R = H, $COCH_3$ 106

110   111 n=9,11,13   112 n=5,7

113 n=2,4   114 n=9,11,13,15,17,19   115 n=5,7 m=7,9

116 n=7,9,11,15 $C_{16:1}$, $C_{16:2}$, $C_{16:3}$, $C_{18:1}$, $C_{18:2}$, $C_{18:3}$ 117 ethyl esters   118 n=9,11   119

The quantity of queen pheromone is enormous, with 150-200 μg in mated queens of *A. mellifera* and 200-300 μg in *A. cerana* and *A. dorsata*.

Drones home in on the 9-ODA and other substances of the mandibular gland, but they are also attracted to the queen's abdomen and the tergal glands probably emit an unidentified copulatory pheromone also.

Sexual attractants are as yet unknown in stingless bees. In social wasps, females are very attractive to conspecific males, although the chemical substances involved nor their anatomical origin could yet be identified (Downing, 1991). The spot marking by male bumblebees with their labial gland secretion has been mentioned earlier. The scent markings attract females to the sites and lead to copulation. So they are essentially sexual attractants. Farnesol **104** was first identified as the attractant of *Bombus terrestris* by Stein (1963). Labial gland pheromones have been the subject of an extended study by Bergström and Kullenberg (Kullenberg et al., 1970). In 18 species of bumblebees, they identified some 12 terpenes **105-110**, 7 alcohols **111-113**, 8 acetates **114, 115**, 11 ethyl esters **116, 117**, two butyrates **118** and tetradecanal **119** (Bergström, 1981).

Queen Recognition Pheromones

Whether queens control workers or workers control queens, it is evident that workers are attracted to queens and the absence of queens causes change of organization in at least some social insects.

We have no examples among termites, but two examples from ants. Rocca et al. (1983) and Glancey (1986) have identified the queen recognition pheromone from the poison glands of *Solenopsis invicta* as a mixture of two lactones **120, 121**. A third component, dihydro-actinidolide **122**, is inactive. The virgin alate females of *Monomorium pharaonis* attract and stimulate males with an unidentified sexual pheromone from their Dufour glands (Hölldobler and Wust, 1973), but the dealate queens produce cembrene A **123**, an isomer of the trail substance of some termites, and this acts as a queen recognition pheromone for workers (Edwards and Chambers, 1984). Cembrene A is not present in the Dufour glands of virgin females or workers.

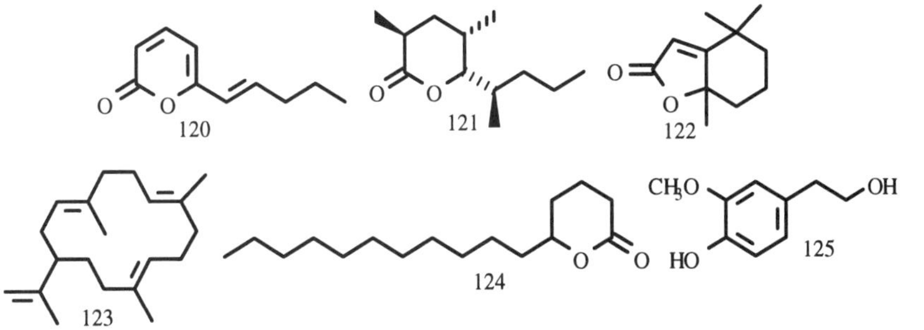

δ-n-Hexadecalactone **124**, isolated from head extracts as the queen pheromone of the oriental hornet *Vespa orientalis* (Ikan et al., 1969), has influence on worker behaviour, although suppression of ovarian development in workers has not been shown experimentally (Spradbery, 1991). Queen control in polistine wasps has been reported with a possible role for the Dufour's and fifth sternal glands, but also here there is no conclusive experimental evidence to support these data so far (Spradbery, 1991).

The queen substance of the honeybee, first investigated by Butler and Callow (1968), has been thoroughly analysed by Slessor's group so that we now have the full blend of substances, which added to 9-ODA **100**, reproduce the effect of the queen's mandibular gland. The response is due to five substances, methyl p-hydroxybenzoate **101** (13 μg) (E)-9-oxodec-2-enoic acid **100** (150 μg), 9-hydroxydec-2-enoic acid **102, 103** (71% R-(-) enantiomer **102**, 55 μg) and 4-hydroxy-3-methoxyphenyl-ethanol **125** (1-5 μg) per queen. Each of these substances was weakly active alone but the blend was active at as little as 10^{-7} queen equivalents (Slessor et al., 1988). Both isomers of the hydroxydecenoic acid are necessary to give highest activity. On the other hand, Velthuis (1985) has shown that the queen pheromone is a product of the tergal glands and the tarsal or Arnhard glands, in addition to the mandibular glands. Worker ovary development is normally inhibited by the queen. It is noteworthy that in queenless colonies the worker ovaries are inhibited from development by the presence of larvae. Arnold et al. (1994) have shown this inhibition is caused by methyl and ethyl esters of the common fatty acids on the larval cuticle. They describe it as a primer brood pheromone. In ants, several categories of 'brood pheromones' have been defined that are involved in brood discrimination, but the chemical nature of these could not yet be identified (Vander Meer and Morel, 1988).

Home Range Marking

There are territorial social insects that defend a territory marked with their glandular secretions. To demonstrate that secretion is deposited is not to conclude it is the necessary pheromone. There are also home range pheromones that mark a territory without implying defence. Cammaerts has shown that first foragers of *Myrmica rubra, M. rugulosa* and *M. schencki* mark territories with their Dufour glands so that subsequent foragers move more quickly over areas marked with their own secretion (Attygalle et al., 1983a). The secretion of *M. rubra* consists of C_{13} to C_{19} hydrocarbons with a small amount of (Z,E)-α-farnesene (**9**)and two higher homologues. That of *M. rugulosa* consists of almost equal proportions of the terpenes and the same hydrocarbons and the *M. schencki* Dufour gland contains almost only terpenes (Attygalle et al., 1983b). Recent work on the desert ant *Cataglyphis niger* revealed that secretion from the cloacal gland, which is a common exocrine structure in the genus *Cataglyphis,* acts as a home marking pheromone (Wenseleers et al., 1996).

This brief review has, of necessity, confined itself to identified pheromone compounds and excluded venoms, defensive compounds and secretions which may serve to maintain the exoskeleton in a healthy and waterproof condition. The variety of compounds used as pheromones and their various biosynthetic origins are striking. But it is also noteworthy that the same odorous compounds turn up frequently in insects, and sometimes in plants, serving different functions.

Concluding Remarks

The exocrine system in social insects is extremely well developed. Thanks to the modern sectioning techniques, new glands will probably continue to be discovered, and thanks to micro-chemical techniques, the composition of more secretions will be identified. Unraveling their respective functions will remain a major challenge for the entomologist. Pheromonal communication, as illustrated in this book, will need further and more careful study, both of the contribution of the active substances of individual glands, as well as of the synergistic effects of multicomponent pheromones, that may occur as sister components within a particular gland or that find their origin in different glands. Attention must also be directed to investigate the effect on pheromonal messages of acoustic and tactile communication cues. In a rapidly advancing subject we can expect many more exciting discoveries ahead.

Acknowledgements

We are very grateful to D. Corstjens and J. Cillis for their assistance in specimen preparation for morphological examination.

References

Ali, M.F. and E.D. Morgan, 1990. Chemical communication in insect societies: a guide to insect pheromones with special emphasis on social insects. Biol. Rev. 65: 227-247.

Ali, M.F., A.B. Attygalle, J.P.J. Billen, B.D. Jackson and E.D. Morgan, 1988. Change of Dufour gland contents with age of workers of *Formica sanguinea* (Hymenoptera: Formicidae). Physiol. Entomol. 13: 249-255.

Alibert, J., 1983. Innervation de l'appareil salivaire du termite *Kalotermes flavicollis* Fabr. Histologie et ultrastructure: relation des axones avec les cellules de la glande et du réservoir. Arch. Anat. microsc. 72: 133-162.

Alvarez, F. M., R.K. Vander Meer and C.S. Lofgren, 1987. Synthesis of homofarnesenes: Trail pheromone components of the fire ant, *Solenopsis invicta*. *Tetrahedron,* 43: 2897-2900.

Arnold, G., Y. Leconte, J. Trouiller, H. Hervet, B. Chappe and C. Masson, 1994. Inhibition of worker honeybee ovary development by a mixture of fatty acid esters from larvae. C.R. Acad. Sci. Series III 317: 511-515.

Attygalle, A.B. and E.D. Morgan, 1984. Chemicals from the glands of ants. Chem. Soc. Reviews 13: 245-278.

Attygalle, A.B. and E.D. Morgan, 1985. Ant trail pheromones. Adv. Insect Physiol. 18: 1-30.

Attygalle, A.B., M.C. Cammaerts and E.D. Morgan, 1983a. Dufour gland secretion of *Myrmica rugulosa* and *Myrmica schencki* workers. J. Insect Physiol. 29: 27-32.

Attygalle, A.B., R.P. Evershed, E.D. Morgan and M.C. Cammaerts, 1983b. Dufour gland secretions of workers of the ants *Myrmica sulcinodis* and *Myrmica lobicornis*, and comparison with six other species of *Myrmica*. Insect Biochem. 13: 507-512.

Attygalle, A.B., M.C. Cammaerts, R. Cammaerts, E.D. Morgan and D.G. Ollett, 1986. Chemical and ethological studies of the trail pheromone of the ant *Manica rubida* (Hymenoptera: Formicidae). Physiol. Entomol. 11: 125-132.

Attygalle, A.B., O. Vostrowsky, H.J. Bestmann, S. Steghaus-Kovac and U. Maschwitz, 1988. (3R,4S)-4-methyl-3-heptanol, the trail pheromone of the ant *Leptogenys diminuta*. Naturwissenschaften 75: 315-317.

Attygalle, A.B., S. Steghaus-Kovac, V.U. Ahmad, U. Maschwitz, O. Vostrowsky and H.J. Bestmann, 1991. cis-Isogeraniol, a recruitment pheromone of the ant *Leptogenys diminuta*. Naturwissenschaften 78: 90-92.

Bagnères, A.G. and E.D. Morgan, 1991. The postpharyngeal glands and the cuticle of Formicidae contain the same characteristic hydrocarbons. Experientia 47: 106-111.

Barbier, J. and E. Lederer, 1960. Structure chimique de la substance royale de la reine d'abeille (*Apis mellifica* L.). C.R. Acad. Sci. Paris 251: 1131-1135.

Bazire-Bénazet, M. and L. Zylberberg, 1979. An integumentary gland secreting a territorial pheromone in *Atta* sp.: detailed structure and histochemistry. J. Insect Physiol. 25: 751-765.

Bergström, G., 1981. Chemical aspects of insect exocrine signals as a means for systematic and phylogenetic discussions in aculeate Hymenoptera. Entomol. Scand. Supp. 15: 213-224.

Bestmann, H.J., A.B. Attygalle, J. Glasbrenner, R. Riemer, O. Vostrowsky, M.G. Constantino, G. Melikian and E.D. Morgan, 1988. Identification of the volatile components of the mandibular gland secretion of the ant *Manica rubida*: structure elucidation, synthesis, and absolute configuration of manicone. Liebigs Ann. Chem.: 55-60.

Bestmann, H.J., F. Kern, D. Schäfer and M.C. Witschel, 1992. 3,4-Dihydroisocoumarins, a new class of ant trail pheromones. Angew. Chem. Internat. Ed. English 31: 757-758.

Bestmann, H.J., U. Haak, F. Kern and B. Hölldobler, 1995a. 2,4-Dimethyl-5-hexanolide, a trail pheromone component of the carpenter ant *Camponotus herculeanus*. Naturwissenschaften 82: 142-144.

Bestmann, H.J., E. Janssen, F. Kern, B. Liepold and B. Hölldobler, 1995b. All-trans geranylgeranyl acetate and geranylgeraniol. Recruitment pheromone components in the Dufour gland of the ponerine ant *Ectatomma ruidum*. Naturwissenschaften 82: 334-336.

Billen, J., 1982. The Dufour gland closing apparatus in *Formica sanguinea* Latreille (Hymenoptera, Formicidae). Zoomorphology 99: 235-244.

Billen, J., 1984. Morphology of the tibial gland in the ant *Crematogaster scutellaris*. Naturwissenschaften 71: 324-325.

Billen, J., 1987. New structural aspects of the Dufour's and venom glands in social insects. Naturwissenschaften 74: 340-341.

Billen, J., 1991. Ultrastructural organization of the exocrine glands in ants. Ethol. Ecol. Evol., special issue 1: 67-73.

Billen, J., 1994. Morphology of exocrine glands in social insects: an up-date 100 years after Ch. Janet. In: "Les Insectes Sociaux" (A. Lenoir, G. Arnold and M. Lepage, Eds.), Publications Université Paris Nord, p. 214.

Billen J. and E. Schoeters, 1994. Morphology and ultrastructure of the mandibular gland in *Formica* L. ants (Hymenoptera, Formicidae). Memorabilia Zool. 48: 9-16.

Blum, M.S., 1982. Pheromonal bases of insect sociality: communication, conundrums and caveats. In: Les Médiateurs Chimiques. Versailles, 1981, INRA Publications.

Blum, M.S., R.M. Crewe, W.E. Kerr, L.H. Keith, A.W. Garrison and M.M. Walker, 1970. Citral in stingless bees: isolation and function in trail-laying and robbing. J. Insect Physiol. 16: 1637-1648.

Boch, R., D.A. Shearer and B.C. Stone, 1962. Identification of isoamylacetate as an active component in the sting pheromone of the honey bee. Nature, London 195: 1018-1020.

Bordereau, C., A. Robert, O. Bonnard and J.L. Lequere, 1991. (3Z,6Z,8E)-3,6,8-dodecatrien-1-ol - sex pheromone in a higher fungus-growing termite *Pseudacanthotermes spiniger* (Isoptera: Macrotermitinae). J. Chem. Ecol. 17: 2177-2191.

Boudreau, J., A.R. Beaudoin and D. Nadeau, 1983. Sequential isolation of lamellar bodies and surfactant fractions from rat lungs. Can. J. Biochem. Cell Biol. 61: 231-239.

Butler, C.G. and R.K. Callow, 1968. Pheromones of the honeybee (*Apis mellifera* L.): the "inhibitory scent" of the queen. Proc. Roy. Entomol. Soc. London (A) 43: 62-65.

Callow, R.K. and N.C. Johnston, 1960. The chemical constitution and synthesis of queen substance of honeybees (*Apis mellifera* L.). Bee World 41: 152-153.

Callow, R.K., J.R. Chapman and P.N. Paton, 1964. Pheromones of the honeybee: chemical studies of the mandibular gland secretion of the queen. J. Apicult. Res. 3: 77-89.

Cammaerts, M.C. and K. Mori, 1987. Behavioural activity of pure chiral 3-octanol for the ants *Myrmica scabrinodis* Nyl. and *Myrmica rubra* L. Physiol. Entomol. 12: 381-385.

Cammaerts-Tricot, M.C., E.D. Morgan and R.C. Tyler, 1977. Isolation of the trail pheromone of the ant *Myrmica rubra*. J. Insect Physiol. 23: 511-515.

Cavill, G.W.K., P.L. Robertson and N.W. Davies, 1979. An Argentine ant aggregation factor. Experientia 35: 989-990.

Cavill, G.W.K., N.W. Davies and F.J. McDonald, 1980. Characterization of aggregation factors and associated compounds from the Argentine ant *Iridomyrmex humilis*. J. Chem. Ecol. 6: 371-384.

Cross, J.H., R.C. Byler, U. Ravid, R.M. Silverstein, S.W. Robinson, P.M. Baker, J. Sabino de Oliveira, A.R. Jutsum and J.M. Cherrett, 1979. The major component of the trail pheromone of the leaf-cutting ant, *Atta sexdens rubropilosa* Forel. J. Chem. Ecol. 5: 187-203.

Cross, J.H., J.R. West, R.M. Silverstein, A.R. Jutsum and J.M. Cherrett, 1982. Trail pheromone of the leaf-cutting ant *Acromyrmex octospinosus* (Reich) (Formicidae: Myrmicinae). J. Chem. Ecol. 8: 1119-1124.

Delage-Darchen, B., 1976. Les glandes post-pharyngiennes des fourmis. Connaissances actuelles sur leur structure, leur fonctionnement, leur rôle. Ann. Biol. 15: 63-76.

Delfino, G., M.T. Marino Piccioli and C. Calloni, 1979. Fine structure of the glands of Van der Vecht's organ in *Polistes gallicus* (L.) (Hymenoptera Vespidae). Monitore zool. ital. (N.S.) 13: 221-247.

Deligne, J., A. Quennedey and M.S. Blum, 1981. The enemies and defense mechanisms of termites. In: Social Insects, volume 2 (H.R. Hermann, Ed.), Academic Press, New York, pp. 1-76.

Dethier, V.G., L.B. Browne and C.N. Smith, 1960. The designation of chemicals in terms of the responses they elicit from insects. J. Econ. Entomol. 53: 134.

do Nascimento, R.R., E.D. Morgan, J. Billen, E. Schoeters, T.M.C. Della Lucia and J.M.S. Bento, 1993. Variation with caste of the mandibular gland secretion in the leaf-cutting ant *Atta sexdens rubropilosa*. J. Chem. Ecol. 19: 907-918.

do Nascimento, R.R., E.D. Morgan, D.D.O. Moreira and T.M.C. Della Lucia, 1994. Trail pheromone of the leaf-cutting ant *Acromyrmex subterraneus subterraneus* (Forel). J. Chem. Ecol. 20: 1719-1724.

Downing, H.A., 1991. The function and evolution of exocrine glands. In: The Social Biology of Wasps (K.G. Ross and R.W. Matthews, Eds.), Comstock Publ. Ass., Ithaca and London, pp. 540-569.

Edwards, J.P. and J. Chambers, 1984. Identification and source of a queen-specific chemical in the Pharaoh's ant, *Monomorium pharaonis* (L.). J. Chem. Ecol. 10: 1731-1747.

Evershed, R.P. and E.D. Morgan, 1983. The amounts of trail pheromone substances in the venom of workers of four species of Attine ants. Insect Biochem. 13: 469-474.

Evershed, R.P., E.D. Morgan and M.C. Cammaerts, 1982. 3-Ethyl-2,5-dimethylpyrazine, the trail pheromone from the venom gland of eight species of *Myrmica* ants. Insect Biochem. 12: 383-391.

Free, J.B., 1987. Pheromones of Social Bees. Chapman and Hall, London, p.154.

Glancey, B.M., 1986. The queen recognition pheromone of *Solenopsis invicta*. In: Fire Ants and Leaf-cutting Ants: Biology and Management (C.S. Lofgren and R.K. Vander Meer, Eds.). Westview Press, Boulder, pp. 223-230.

Grace, J.K., D.L. Wood, I. Kubo and M. Kim, 1995. Behavioural and chemical investigation of trail pheromone from the termite *Reticulitermes hesperus* Banks (Isoptera: Rhinotermitidae). J. Appl. Entomol. 119: 501-505.

Heath, R.R. and P.J. Landolt, 1988. The isolation, identification and synthesis of the alarm pheromone of *Vespula squamosa* (Drury) (Hymenoptera: Vespidae) and associated behaviour. Experientia 44: 82-83.

Hermann H.R., A.N. Hunt and W.F. Buren, 1971. Mandibular gland and mandibular groove in *Polistes annularis* (L.) and *Vespula maculata* (L.) (Hymenoptera: Vespidae). Int. J. Insect Morphol. & Embryol. 1: 43-49.

Hidalgo, J., A. Velasco, A. Gomez Pascual and J.L. Lopez-Campos, 1985. Glycoconjugates in the alveolar epithelium of *Rana ridibunda*. Arch. Biol. (Bruxelles) 96: 179-185.

Hölldobler, B., 1995. The chemistry of social regulation: multicomponent signals in ant societies. Proc. Natl. Acad. Sci. USA 92: 19-22.

Hölldobler, B. and Carlin, N., 1987. Anonymity and specificity in the chemical communication signals of social insects. J. Comp. Physiol 161A: 567-581.

Hölldobler, B. and J.M. Palmer, 1989. Footprint glands in *Amblyopone australis* (Formicidae, Ponerinae). Psyche 96: 111-121.

Hölldobler, B. and E.O. Wilson, 1990. The Ants. Cambridge, Mass.: Harvard University Press, 732 pp.

Hölldobler, B. and M. Wust, 1973. Ein Sexualpheromon bei der Pharaoameise *Monomorium pharaonis* (L.). Z. Tierphysiol. 32: 1-9.

Hölldobler, B., M. Obermayer and E.O. Wilson, 1992. Communication in the primitive cryptobiotic ant *Prionopelta amabilis* (Hymenoptera: Formicidae). J. Comp. Physiol. 170A: 9-16.

Hölldobler, B., N.J. Oldham, E.D. Morgan and W.A. König, 1995a. Recruitment pheromones in the ants *Aphaenogaster albisetosus* and *A. cockerelli* (Hymenoptera: Formicidae). J. Insect Physiol. 41: 739-744.

Hölldobler, B., E. Janssen, H.J. Bestmann, I.R. Leal, P.S. Oliveira, F. Kern and W.A. König, 1995b. Communication in the migratory termite-hunting ant *Pachycondyla (= Termitopone) marginata* (Hymenoptera, Formicidae). J. Comp. Physiol. 178A: 47-53.

Huwyler, S., K. Grob and M. Viscontini, 1975. The trail pheromone of the ant *Lasius fuliginosus*: identification of six components. J. Insect Physiol. 21: 299-304.

Ikan, R., R. Gottlieb, E.D. Bergmann and J. Ishay, 1969. The pheromone of the queen of the oriental hornet, *Vespa orientalis*. J. Insect Physiol. 15: 1709-1712.

Jackson, B.D. and E.D. Morgan, 1993. Insect chemical communication: pheromones and exocrine glands of ants. Chemoecology 4: 125-144.

Jackson, B.D., E.D. Morgan and P.J. Wright, 1989. 3-Ethyl-2,5-dimethylpyrazine, a component of the trail pheromone of the ant *Messor bouvieri*. Experientia 45: 487-489.

Jackson, B.D., P.J. Wright and E.D Morgan, 1991. Chemistry and trail-following of a harvester ant. In: Insect Chemical Ecology (I. Hrdy, Ed.). Academia, Prague and SPB Academic Publishing, The Hague, pp. 109-112.

Jackson, B.D., S.J. Keegans, E.D Morgan, M.C. Cammaerts and R. Cammaerts, 1990. Trail pheromone of the ant *Tetramorium meridionale*. Naturwissenschaften 77: 294-296.

Janssen, E., H.J. Bestmann, B. Hölldobler and F. Kern, 1995. N,N-dimethyluracil and actinidine, two pheromones of the ponerine ant *Megaponera foetens* (Fab.) (Hymenoptera: Formicidae). J. Chem. Ecol. 21: 1947-1955.

Kaib, M., 1985. Defense strategies of termites: a review exemplified by *Schedorhinotermes lamanianus*. Mitt. dtsch. Ges. allg. angew. Ent. 4: 302-306.

Kullenberg, B., G. Bergström and S. Ställberg-Stenhagen, 1970. Volatile components of the cephalic marking secretion of male bumblebees. Acta Chem. Scand. 24: 1481-1483.

Laduguie, N., A. Robert, O. Bonnard, F. Vieau, J.L. Lequere, E. Semon and C. Bordereau, 1994. Isolation and identification of (3Z,6Z,8E)-3,6,8-dodecatrien-1-ol in *Reticulitermes santonensis* Feytaud (Isoptera: Rhinotermitidae) - roles of worker trail-following and in alate sex-attraction behaviour. J. Insect Physiol. 40: 781-787.

Landolt, P.J., R.R. Heath, H.C. Reed and K. Manning, 1995. Pheromonal mediation of alarm in the eastern yellowjacket (Hymenoptera: Vespidae). Florida Entomol. 78: 101-108.

Lensky, Y. and Y. Slabezski, 1981. The inhibiting effect of the queen bee (*Apis mellifera* L.) foot-print pheromone on the construction of swarming queen cups. J. Insect Physiol. 27: 313-323.

Leuthold, R.H., 1968. A tibial gland scent-trail and trail-laying behavior in the ant *Crematogaster ashmeadi* Mayr. Psyche 75: 233-248.

Maschwitz, U., 1964. Gefahrenalarmstoffe und Gefahrenalarmierung bei sozialen Hymenopteren. Z. Vergl. Physiol. 47: 596-655.

Matsumura, F., H.C. Coppel and A. Tai, 1968. Isolation and identification of termite trail-following pheromone. Nature, London 219: 963-964.

McDowell, P.G. and G.W. Oloo, 1984. Isolation, identification and biological activity of trail-following pheromone of termite *Trinervitermes bettonianus* (Sjostedt) (Termitidae: Nasutitermitinae). J. Chem. Ecol. 10: 838-851.

Meskali, M., E. Provost, A. Bonavita-Cougourdan and J.L. Clément, 1995. Behavioural effects of an experimental change in the chemical signature of the ant *Camponotus vagus* (Scop.). Ins. Soc. 42: 347-258.

Moens, N., J. Billen, B.D. Jackson and E.D. Morgan, 1990. Evolution ontogénétique chimique de la glande mandibulaire de *Formica sanguinea* (Latr.) (Hymenoptera, Formicidae). Actes Coll. Insectes Soc. 6: 173-177.

Moore, B.P., 1966. Isolation of the scent-trail pheromone of an Australian termite. Nature, London 211: 746-747.

Moore, B.P., 1968. Studies of the chemical composition and function of the cephalic gland secretion in Australian termites. J. Insect Physiol. 14: 33-39.

Morgan, E.D., 1990. Insect trail pheromones: a perspective of progress. In: Chromatography and Isolation of Insect Hormones and Pheromones (A.R. McCaffery and I.D. Wilson, Eds.), Plenum Press, New York and London, pp. 259-270.

Morgan, E.D. and D.G. Ollett, 1987. Methyl 6-methylsalicylate, trail pheromone of the ant *Tetramorium impurum*. Naturwissenschaften 74: 596-597.

Nedel, J.O., 1960. Morphologie und Physiologie der Mandibeldrüse einiger Bienen-Arten (Apidae). Z. Morph. Ökol. Tiere 49: 139-183.

Noirot, C., 1969. Glands and secretions. In: Biology of Termites (K. Krishna and F.M. Weesner, Eds.), Academic Press, New York, pp. 89-123,

Noirot, C. and A. Quennedey, 1974. Fine structure of insect epidermal glands. Ann. Rev. Entomol. 19: 61-80.

Noirot, C. and A. Quennedey, 1991. Glands, gland cells, glandular units: some comments on terminology and classification. Annls. Soc. ent. Fr. (N.S.) 27: 123-128.

Oldham, N.J., E.D. Morgan, B. Gobin and J. Billen, 1994a. First identification of a trail pheromone of an army ant (*Aenictus* species). Experientia 50: 763-765.

Oldham, N.J., J. Billen and E.D. Morgan, 1994b. On the similarity of the Dufour gland secretion and the cuticular hydrocarbons of some bumblebees. Physiol. Entomol. 19: 115-123.

Pasteels, J.M., D. Daloze and J.L. Boeve, 1989. Aldehydic contact poisons and alarm pheromone of the ant *Crematogaster scutellaris* (Hymenoptera: Myrmicinae). Enzyme-mediated production from acetate precursors. J. Chem. Ecol. 15: 1501-1511.

Pickett, J.A., I.H. Williams, A.P. Martin and M.C. Smith, 1980. Nasanov pheromone of the honeybee, *Apis mellifera* L. (Hymenoptera: Apidae). Part I. Chemical characterization. J. Chem. Ecol. 6: 425-434.

Pickett, J.A., I.H. Williams and A.P. Martin, 1982. (Z)-11-Eicosen-1-ol, an important new pheromonal component from the sting of the honeybee *Apis mellifera* L. (Hymenoptera: Apidae). J. Chem. Ecol. 8: 163-175.

Quinet, Y. and J.M. Pasteels, 1995. Trail following and stowaway behaviour of the myrmecophilous staphylinid beetle *Homoeusa acuminata* during foraging trips of its host *Lasius fuliginosus* (Hymenoptera: Formicidae). Insectes Soc. 42: 31-44.

Riley, R.G., R.M. Silverstein, B. Carroll and R. Carroll, 1974. Methyl 4-methylpyrrole-2-carboxylate: a volatile trail pheromone from the leaf-cutting ant, *Atta cephalotes*. J. Insect Physiol. 20: 651-654.

Ritter, F.J., I.E.M. Bruggemann-Rotgans, P.E.J. Verwiel, C.J. Persoons and E. Talman, 1977. Trail pheromone of the pharaoh's ant, *Monomorium pharaonis*: isolation and identification of faranal, a terpenoid related to juvenile hormone II. Tetrahedron Lett. 30: 2617-2618.

Robinson, S.W., J.C. Moser, M.S. Blum and E. Amante, 1974. Trail following responses of four leaf-cutting ants with notes on the specificity of the trail pheromone of *Atta texana*. Insectes Soc. 21: 87-94.

Rocca, J.R., J.H. Tumlinson, B.M. Glancey and C.S. Lofgren,1983. The queen recognition pheromone of *Solenopsis invicta*, preparation of (E)-6-(1-pentenyl)-2H-pyran-2-one. Tetrahedron Lett. 24: 1889-1892.

Schoeters, E. and J. Billen, 1995a. Morphology and ultrastructure of the convoluted gland in the ant *Dinoponera australis* (Hymenoptera: Formicidae). Int. J. Insect Morphol. & Embryol. 24: 323-332.

Schoeters, E. and J., Billen, 1995b. Morphology and ultrastructure of a secretory region enclosed by the venom reservoir in social wasps (Hymenoptera, Vespidae). Zoomorphology 115: 63-71.

Silverstein, R.M. and Young, J.C., 1976. Insects generally use multicomponent pheromones. In: Pest Management with Insect Sex Attractants and other Behavior-Controlling Chemicals (M. Beroza, Ed.), ACS Symposium series 23, Washington, D.C., pp. 1-29.

Simon, T. and A. Hefetz, 1991. Trail-following responses of *Tapinoma simrothi* (Formicidae: Dolichoderinae) to pygidial gland extracts. Insectes Soc. 38: 17-25.

Slessor, K.N., L.A. Kaminski, G.G.S. King, J.H. Borden and M.L. Winston, 1988. Semiochemical basis of the retinue response to queen honey bees. Nature, London 332: 354-356.

Soroker, V., C. Vienne, E. Nowbahari and A. Hefetz, 1994. The postpharyngeal glands as a 'Gestalt' organ for nestmate recognition in the ant *Cataglyphis niger*. Naturwissenschaften 81: 510-513.

Spradbery, J.P., 1991. Evolution of queen number and queen control. In: The Social Biology of Wasps (K.G. Ross and R.W. Matthews, Eds.), Comstock Publ. Ass., Ithaca and London, pp. 336-388.

Sreng, L. and A. Quennedey, 1976. Role of a temporary ciliary structure in the morphogenesis of insect glands. J. Ultrastruct. Res. 56: 78-95.

Steghaus-Kovac, S., U. Maschwitz, A.B. Attygalle, R.T.S. Frighetto, N. Frighetto, O. Vostrowksy and H.J. Bestmann, 1992. Trail-following response of *Leptogenys diminuta* to stereo isomers of 4-methyl-3-heptanol. Experientia 48: 690-694.

Stein, G. 1963. Uber den Sexuallockstoff von Hummelmännchen. Naturwissenschaften 50: 305.

Tai, A., F. Matsumura and H.C. Coppel, 1969. Chemical identification of the trail-following pheromone of a southern subterranean termite. J. Org. Chem. 34: 2180-2182.

Tengö, J., A. Hefetz, A. Bertsch, U. Schmitt, G. Lübke and W. Francke, 1991. Species specificity and complexity of Dufour's gland secretion of bumble bees. Comp. Biochem. Physiol. 99B: 641-646.

Toroko, M., M. Takahashi, K. Tsunoda and R. Yamaoka, 1989. Isolation and primary structure of trail pheromone of the termite, *Coptotermes formosanus* Shiraki (Isoptera: Rhinotermitidae). Wood Res. 76: 29-38.

Toroko, M., M. Takahashi, K. Tsunoda, R. Yamaoka and K. Hayashiya, 1991. Isolation and identification of trail pheromone of the termite *Reticulitermes speratus* (Kolbe) (Isoptera: Rhinotermitidae). Wood Res. 78: 1-14.

Tumlinson, J.H., J.C. Moser, R.M. Silverstein, R.G. Brownlee and J.M. Ruth, 1972. A volatile trail pheromone of the leaf-cutting ant, *Atta texana*. J. Insect Physiol. 18: 809-814.

Vallet, A., P. Cassier and Y. Lensky, 1991. Ontogeny of the fine structure of the mandibular glands of the honeybee (*Apis mellifera* L.) workers and the pheromonal activity of 2-heptanone. J. Insect Physiol. 37: 789-804.

Vander Meer, R.K. and L. Morel, 1988. Brood pheromones in ants. In: Advances in Myrmecology (J.C. Trager, Ed.), E.J. Brill, Leiden, pp. 491-513.

Vander Meer, R.K., F.D. Williams and C.S. Lofgren, 1981. Hydrocarbon components of the trail pheromone of the red imported fire ant *Solenopsis invicta*. Tetrahedron Lett. 22: 1651-1654.

Vander Meer, R.K., F. Alvarez and C.S. Lofgren, 1988. Isolation of the trail recruitment pheromone of *Solenopsis invicta*. J. Chem. Ecol. 14: 825-838.

Veith, H.I., N. Koeniger and U. Maschwitz, 1984. 2-Methyl-3-buten-2-ol, a major component of the alarm pheromone of the hornet *Vespa crabro*. Naturwissenschaften 71: 328-329.

Velthuis, H.H.W., 1985. The honeybee queen and the social organization of her colony. Fortschnitte Zool. 31: 343-357.

Vienne, C., V. Soroker and A. Hefetz, 1995. Congruency of hydrocarbon patterns in heterospecific groups of ants: transfer and/or biosynthesis? Insectes Soc. 42: 261-277.

Vierling, G. and M. Renner, 1977. Die Bedeutung des Sekretes der Tergittaschendrüsen für die Attraktivität der Bienenkönigin gegenüber jungen Arbeiterinnen. Behav. Ecol. Sociobiol. 2: 185-200.

Wenseleers, T., E. Schoeters, J. Billen and A. Hefetz, 1996. Morphologie et ultrastructure de la glande cloacale chez *Cataglyphis niger*. Actes Coll. Insectes Soc. 10: 189-194.

Wilson, E.O. and W.H. Bossert, 1963. Chemical communication among animals. Rec. Progr. Hormone Res. 19: 673-716.

2

The Cuticle and Cuticular Hydrocarbons of Insects: Structure, Function, and Biochemistry

Gary J. Blomquist, Julie A. Tillman, Shuping Mpuru, and Steven J. Seybold

Introduction

Significance of the Cuticle to Social Insects

Cuticular hydrocarbons provide insects with the chemical equivalent of the visually variable colored plumage of birds. These waxes occur predominantly near the surface of the insect integument, which is a bipartite organ composed of a monolaminar cellular epidermis and the non-cellular cuticle (Wigglesworth, 1972; Hadley, 1985; Hepburn, 1985; Lockey, 1988). Of the two organ components, the structure, function, and biochemical origin of the wax covered cuticle is the subject of this chapter. For several reasons, the cuticle is a significant structure for all insects. First in importance to the theme of this book, the cuticular surface of insects represents a rich reservoir of chemicals. Some of these chemicals appear to be superfluous to the organism, prompting a reference to the cuticle as a "metabolic dustbin" (Hepburn, 1985), while others have tremendous informational value. These informational molecules are often hydrocarbons and they function as intra- and interspecific signals for insects (Howard, 1993), particularly among the social insects. Most recently, the informational content of the insect cuticle has also been exploited by scientists, as insect chemotaxonomists search for new methods to classify the world's exceedingly diverse collection of hexapodous arthropods (Howard, 1993).

The lipid-rich surface of the cuticle also plays an essential role as an anti-desiccatory barrier for a class of animals that in some cases occupies some of the driest terrestrial ecological niches. Conservation of water is the primary challenge faced by terrestrial animals with a high surface area to volume ratio such as occurs with insects (Wigglesworth, 1964). The anti-desiccatory function of the cuticular waxes is crucial in meeting this need.

The process of sclerotization or tanning stabilizes the insect cuticle so that an organic exoskeleton is formed that is structurally tough, yet flexible and light (Hopkins and Kramer, 1992). Sclerotization has been defined as the enzyme-catalyzed incorporation of low molecular weight phenolic material into the cuticular structure, and the resulting increase in stiffness and resistance to digestion and degradation (Andersen, 1991). Thus, sclerotized cuticle protects an insect's internal organs and provides muscle attachment sites, while affording a lightweight body plan that has permitted the evolution of flight in the Insecta. In some cases the stiffness of sclerotized cuticle has been counterbalanced by the presence of the rubber-like cuticular protein resilin, which provides potential energy storage and elasticity for ambulation and flight (Wigglesworth, 1972). In addition, sclerotized cuticle is highly resistant to enzymatic degradation, thereby presenting insects with a barrier to pathogenic invasion (St. Leger, 1991; Hopkins and Kramer, 1992).

Structure and Origin of Insect Cuticle

Structure of Insect Cuticle

The insect cuticle is a heterogenous membranous outer skin composed of several morphologically distinguishable layers (Hadley, 1985; Hepburn, 1985; Lockey, 1988). Initially, the cuticle can be subdivided into a proximal procuticle and a distal epicuticle (Figure 2.1). The procuticle contains chitin (a polysaccharide polymer of N-acetyl-glucosamine units) and proteins, while the epicuticle contains protein, but not chitin (Lockey, 1988). The procuticle can be further subdivided into a proximal endocuticle and a distal exocuticle. The proteins of the exocuticle are sclerotized (Lockey, 1988), while those in the endocuticle are not. Thus, the exocuticle is rigid and amber or black colored, while the endocuticle is elastic and colorless (Wigglesworth, 1972). Although it is relatively thin (0.03-4.0 μm) and does not contain polysaccharides, the insect epicuticle is structurally complex (Wigglesworth, 1972; Hadley, 1985). It can be subdivided into an inner epicuticle (0.5-1.0 μm), or dense layer, and an outer epicuticle (10-18 nm). In addition, several

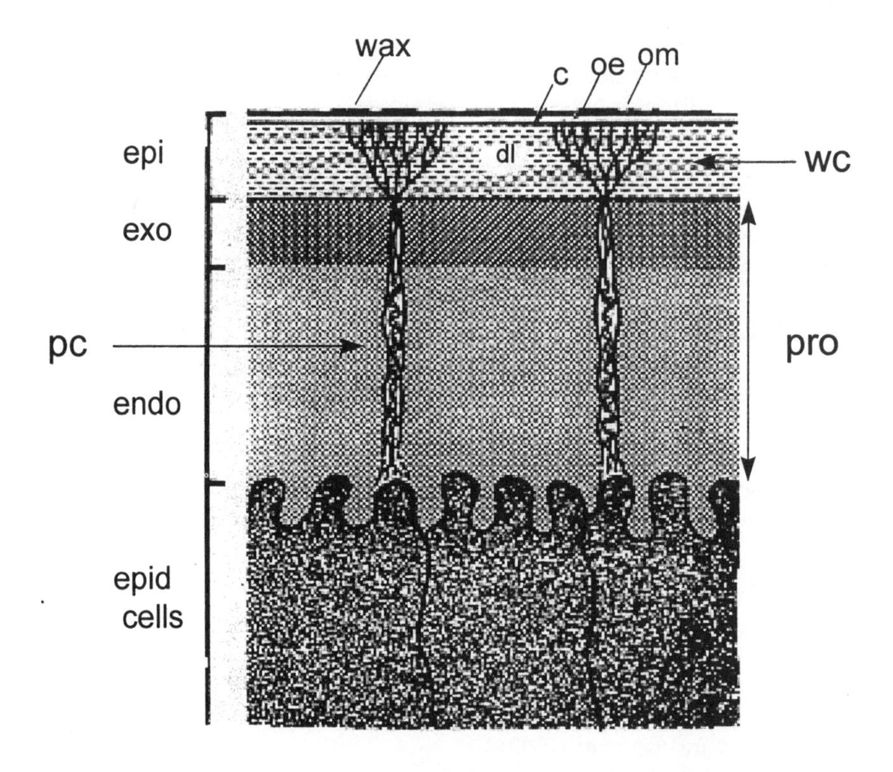

FIGURE 2.1. Diagrammatic representation of the insect integument, including epidermal cell layer (epid cells) and cuticular membrane composed of the procuticle (pro) and epicuticle (epi). The procuticle is subdivided into endocuticle (endo) and exocuticle (exo), while the epicuticle is subdivided into a dense layer (dl), a cuticulin layer (c), an outer epicuticle (oe), and an outer membrane (om) covered by a wax layer (wax). Pore canals (pc) traverse the cuticle from the epidermal cells and divide into smaller wax canals (wc) in the epicuticle. Figure adapted from Hadley (1985) with the permission of the author and publisher.

extremely thin extracuticular layers (lipid or wax layer; cement layer or outer membrane) rest on the surface of the epicuticle (Figure 2.1).

Origin of Insect Cuticle

The ultimate source of the monomeric units required to construct the polysaccharide, protein, and lipid components of insect cuticle is the hemolymph. However, the insect epidermis, which synthesizes and secretes the cuticle, is the central tissue, both functionally and spatially, for the formation of cuticle (Locke, 1991). In short, through the hormonally regulated process of ecdysis, the epidermal cells progressively secrete a new cuticle and digest the old cuticle under the aegis of the newly synthesized epicuticle (Wigglesworth, 1972). The molecular events underlying the process are dynamic and complex, revealing an insect tissue that has evolved a high degree of developmental orchestration (Willis, 1991; Riddiford, 1991).

Lipid is thought to occur throughout the cuticle, and to be intimately associated with sclerotin or tanned cuticle protein (Wigglesworth, 1991). The lipid of the cuticle has been functionally defined as "free lipid" (extractable from the cuticle by organic solvents) and "structural lipid" (insoluble in organic solvents directly, but soluble following destructive oxidation in warm concentrated nitric acid and potassium chlorate) (Lockey, 1988; Wigglesworth, 1991). Lockey (1988) states that structural (or bound) lipid is a major component of the inner epicuticle, but can also be found in the exocuticle and even on the surface of the epicuticle as a "waterproofing layer of cuticulin" (Wigglesworth, 1972, 1991). However, removal of the cuticular lipids with organic solvents greatly increases the transpiration rate from the insect (Nelson and Blomquist, 1995), calling into question the role of the structural lipid. Free lipid is thought to occur entirely in the epicuticle and may exist as a "loose, but thick covering of free wax" (Wigglesworth, 1991). Nearly all analyses of insect cuticular lipids have dealt with "free lipids" and the complete absence of chemical information on "structural lipids" of the insect cuticle highlights a much-needed area of research. Determination of the molecular identity and taxonomic distribution of "structural lipids" in the Insecta are mandatory steps to proving their existence.

Informational Content of Insect Cuticle

Aliphatic surface fingerprints of insects serve as inter- and intraspecific distinguishing characteristics that guide a variety of insect behaviors. The role of hydrocarbons as behavioral chemicals such as pheromones and allomones will be explored in greater depth in subsequent chapters, thus we will only introduce this topic here.

In social insects, chemical communication results from glandularly or cuticularly released semiochemical signals, and the line of demarcation between these two modes is not always clear. Insect exocrine glands consist of modified epidermal cells that are generally involved in the *de novo* biosynthesis and secretion of behavioral chemicals (Blum, 1985). These glands may be found in all body regions. For example, Blum (1985) lists ten glands of the Argentine ant,

Iridomyrmex humilis. Of these, three glands are located in the head, two are in the thorax, and five are in the abdomen. The mandibular and postpharyngeal glands are two highly developed structures in the head. Many glandular secretions of social insects are defensive allomones (e.g. α-pinene from the frontal gland of termite soldiers), although some, like the queen substance, 9-oxo-(*E*)-2-decenoic acid, from the mandibular glands of the honey bee, *Apis mellifera*, are true sex pheromones. Semiochemical secretions from glandular epidermal cells can simply move to the surface through the cuticle as in female lepidopteran sex pheromone glands, or they may pass to the surface via elaborate ducts or canals, as in the dermal gland secretions of the yellow mealworm, *Tenebrio molitor* (Coleoptera: Tenebrionidae) (Blum, 1985). In all cases, the epidermal cells secrete the glandular cuticle as well as the behavioral chemical exudate.

Chemical communication based on cuticular hydrocarbons without apparent glandular origin is a frequent phenomenon among the eusocial termites, bees, wasps, and ants (Howard, 1993). In termites, cuticular hydrocarbons appear to regulate species-specific and, depending on the genus, colony-specific agonistic behavior (Haverty and Thorne, 1989; Clément and Bagneres, Chapter 6). Similarly, in the European carpenter ant, *Camponotus vagus*, both cuticular and postpharyngeal gland hydrocarbons elicit agonistic behavior from non-nestmates. Of the hydrocarbon components, the dimethylalkanes provide the most variation between nests (Bonavita-Cougourdan *et al.*, 1987). Brood and caste recognition are additional biological attributes of *C. vagus* that appear to be modulated by cuticular hydrocarbons (reviewed in Howard, 1993: see Vander Meer and Morel, Chapter 4, for a discussion of nestmate recognition in ants). The role of hydrocarbons in nestmate recognition for wasps is discussed in Chapter 5 (Singer *et al.*), for honey bees in Chapter 3 (Breed), and for termites in Chapter 6 (Clément and Bagneres).

Unique to insect societies is the occurrence of invasion by heterospecific organisms, culminating in the phenomenon of inquilinism, where the invader chemically mimics the cuticular hydrocarbon profile of the occupants of the host colony. The role of hydrocarbons in this ironic ecological drama has been widely studied for inquilines of ants and termites, and to a lesser extent, for bees (Howard, 1993; Dettner and Liepert, 1994; Chapters 3 and 5).

One of the difficulties in the study of hydrocarbon semiochemistry of social insects has been the development of adequate behavioral assays for the proposed roles of hydrocarbons. Thus, the recent study by Takahashi and Gassa (1995) is particularly important, as it outlines a direct assay to monitor recognition behavior. The cuticular hydrocarbons of workers of the subterranean termite *Reticulitermes speratus* were applied in a Triton-X 100 solution to workers of the Formosan subterranean termite, *Coptotermes formosanus*. Soldiers of *C. formosanus* consistently exhibited aggressive behavior toward conspecific workers treated with heterospecific worker hydrocarbons. Similar results were obtained with the inverse experiment using *R. speratus* soldiers and *R. speratus* workers treated with *C. formosanus* hydrocarbons. Development of analogous assays for other social insect taxa will accelerate our understanding of social insect behavioral chemistry.

The hydrocarbons utilized by insects to recognize con- and heterospecifics are also employed by humans using chemical instrumentation to differentiate insect species. Methyl-branching patterns, number and placement of double bonds, and chain length combine to form a plethora of chemical characters for the insect chemotaxonomist. For example, in the non-social pine engraver beetles (*grandicollis* subgeneric group of *Ips*; Coleoptera: Scolytidae), the seven members of the *grandicollis* group and three related outgroup species contained a total of 248 identifiable cuticular hydrocarbon components (Page *et al.*, 1997). Of this total, 12 were *n*-alkanes, 16 were monounsaturated alkenes, 6 were alkadienes, and 214 were methyl-branched alkanes. Generally abundant among the methyl-branched alkanes were di- and trimethylalkanes, and in rare cases, tetramethylalkanes. Due to the great variability afforded by combining various chain lengths and branch positions, these three groups of methyl-branched hydrocarbons tend to have the most taxonomic utility. In most instances, the methyl branches occur on odd-numbered carbons in the parent chain, with either odd- or even-length parent carbon chains. In the rare occurrences of methyl branches on even-numbered carbons in this survey, the parent chains also had an even number of carbons (Page *et al.*, 1997). This appears to be a general phenomenon among all insects (Nelson, 1993). The positions of the methyl-branch groups are determined by capillary gas chromatography-mass spectrometry (CGC-MS) (Nelson and Blomquist, 1995) and the positions of the double bonds by one of several derivatization steps followed by CGC-MS (Howard, 1993).

Over the past several decades the biosynthetic pathways for cuticular hydrocarbons have been determined, and work is now in progress to determine how the process is regulated. The next section of this chapter will give insight into how methyl-branch positions, points of unsaturation and the length of the parent hydrocarbon chain are regulated during biosynthesis of cuticular hydrocarbons. The differences in hydrocarbon composition among different species have a genetic basis (Coyne *et al.* 1994), and these phenotypic variations are likely due to the variable occurrence and activities of enzymes that produce hydrocarbons with different methyl-branching positions, different patterns of unsaturation, and different parent chains. Three groups of enzymes appear to be involved. Based on work in a limited number of insects, it appears that the location of methyl branching is controlled by a unique microsomal fatty acid synthase; insertion of double bonds is regulated by specific desaturases; and the chain length is regulated by the specificity of fatty acyl-CoA elongases, and to a lesser extent by the reduction of the acyl-CoA to aldehyde. Much of the work in this area has been done with the house fly, *Musca domestica*, and the German (*Blattella germanica*) and the American (*Periplaneta americana*) cockroaches, but the results certainly apply to social insects as well.

Biosynthesis of Insect Cuticular Hydrocarbons: Enzyme Activities That Influence the Specific Blends of Hydrocarbons Involved in Chemical Communication

Site of Synthesis

Much of the work on the biosynthesis of cuticular lipids has been directed toward hydrocarbons. Results from *in vitro* experiments in various laboratories with widely diverse species indicate that surface lipids are synthesized by cells associated with the epidermis, with the oenocytes often implicated in hydrocarbon production (Nelson and Blomquist, 1995). Nelson (1969) first presented evidence that cuticle-associated tissue synthesized hydrocarbons in the American cockroach, *P. americana*, and this has been confirmed in a number of insect species, including honey bees, termites and ants (Nelson and Blomquist, 1995). Romer (1980) demonstrated that in *T. molitor*, isolated oenocytes efficiently and specifically incorporated labeled acetate into hydrocarbon. Thus, the available evidence strongly points to the epidermal cells as the site of hydrocarbon synthesis with the oenocytes specifically indicated in some species.

Methyl-Branched Hydrocarbons

Methyl-branched hydrocarbons predominate in and are a unique feature of insect cuticular hydrocarbons. Differences in methyl-branching patterns are critical in recognition phenomena among social and non-social insects. The formation of these rather insect-specific compounds involves a methylmalonyl-CoA being inserted in place of

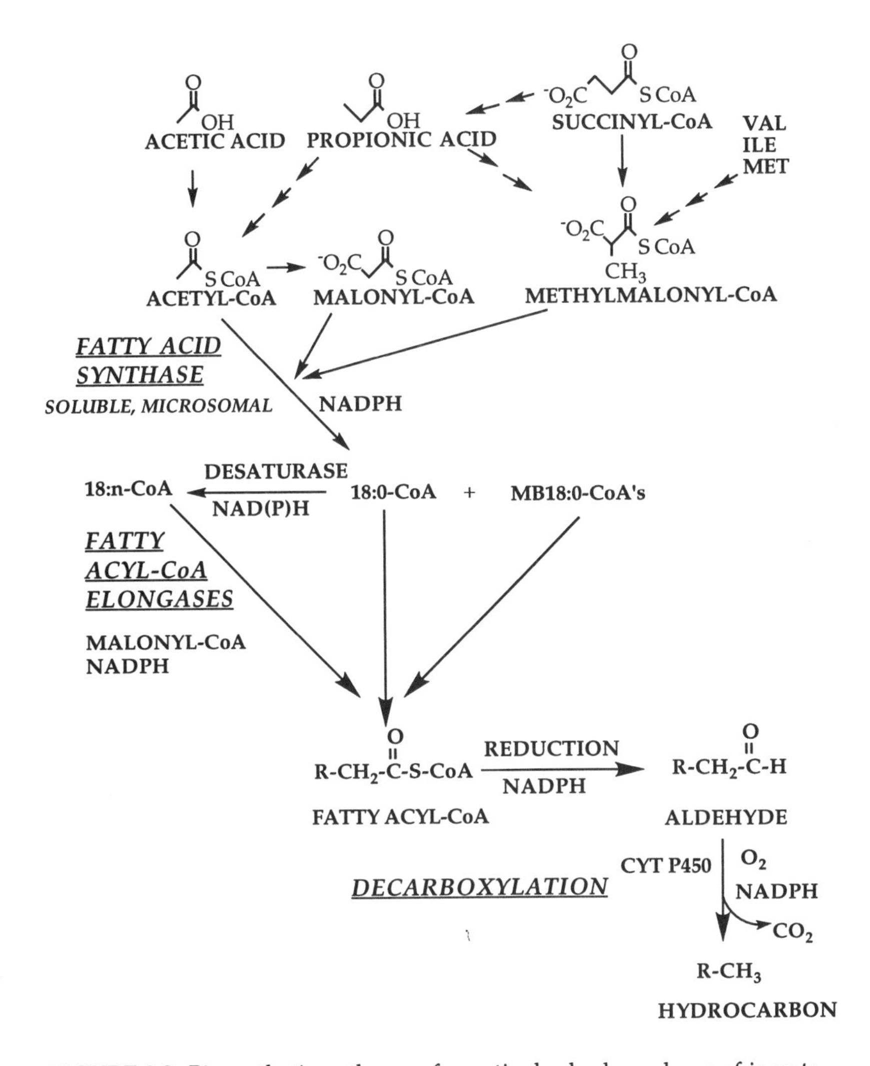

FIGURE 2.2 Biosynthetic pathways for cuticular hydrocarbons of insects. MB=methyl-branched; 18:n, n=number of double bonds; VAL, ILE, and MET = valine, isoleucine, and methionine, respectively; and CYT P450 = cytochrome P450.

malonyl-CoA at specific points early in the chain elongation process to result in formation of the methyl-branched alkanes (Figure 2.2). This was demonstrated for the methyl-branched sex pheromone components of *M. domestica*. ^{13}C-NMR, mass spectrometry, and radiochemical studies (Dwyer *et al.*, 1981; Dillwith *et al.*, 1982) demonstrated that propionate, as the methylmalonyl-CoA derivative, was inserted in place of malonyl-CoA during the initial stages of the elongation process in *P. americana* and *M. domestica*. The methylmalonyl-CoA is derived from the amino acids valine, isoleucine and methionine in *M. domestica* (Dillwith *et al.*, 1982) and *B. germanica* (Chase *et al.*, 1990), whereas it arises from succinate in the dampwood termite, *Zootermopsis angusticollis* (later identified as *Z. nevadensis*) (Chu and Blomquist, 1980a; Blomquist *et al.*, 1980) (Figure 2.2). The conversion of succinate to methylmalonyl-CoA in *Z. nevadensis* apparently involves gut tract microorganisms (Guo *et al.*, 1991). Recently, a novel microsomal fatty acid synthase (FAS) was shown to be involved in the synthesis of *B. germanica* (Juarez *et al.*, 1992; Gu *et al.*, 1993) and *M. domestica* methyl-branched fatty acids (Blomquist *et al.*, 1994). This microsomal FAS demonstrates higher *in vitro* velocity when assayed with methylmalonyl-CoA as an elongating substrate and, in contrast to the soluble FAS, can use methylmalonyl-CoA alone and obtain chain elongation. Also, this microsomal FAS is present in the epidermal tissue. In both *B. germanica* (Juarez *et al.*, 1992) and *M. domestica* (Blomquist *et al.*, 1994), methyl-branched fatty acids with methyl-branching patterns corresponding to the methyl-branched alkanes were characterized by GC-MS. These observations and other data show that methyl-branched alkanes are formed via elongation of methyl-branched fatty acyl-CoAs and that methylmalonyl-CoA insertions occur during the initial stages of the elongation process. Thus, it appears that the abundance and variety of methyl-branch points in insect hydrocarbons arise from the specificity of methylmalonyl-CoA insertion. The regulation of the insertion of methylmalonyl-CoA at specific stages during hydrocarbon biosynthesis to give rise to the species specific pattern of methyl-branched alkanes is not understood.

Acyl-CoA Desaturation in Alkene Formation

Double bonds in insect alkenes can be found in almost any position and account for some of the differences in the hydrocarbons of the social insects (Nelson and Blomquist, 1995). They are most commonly found in the (Z)-9- and (Z,Z)-6,9-positions, arising from the elongation of oleic and linoleic acids respectively. The Δ^9 desaturase is ubiquitous among all aerobic organisms and has been characterized in *M. domestica* (Wang *et al.*, 1982). The Δ^{12} desaturase inserts a second double bond in oleoyl-CoA, resulting in a $\Delta^{9,12}$ fatty acid, linoleic acid, which when elongated and converted to hydrocarbon, gives rise the (Z,Z)-6,9-alkadienes. Most animals are unable to synthesize linoleic acid, but a number of insect species, including termites, have been shown to have this ability (Blomquist *et al.*, 1982; de Renobales *et al.*, 1987). A Δ^{11}

desaturase has been characterized in the pheromone biosynthesis gland of lepidopterans (Bjostad *et al.*, 1987), and presumably other as yet uncharacterized desaturases give rise to the precursors of alkenes with double bonds in a variety of positions. Very is little is known of the biosynthetic details of the insertion of double bonds in alkene biosynthesis by insects. Clearly, more work is needed to determine how insects form products with one or more double bonds at almost any position in the carbon chain.

Acyl-CoA Elongation Steps Regulate Hydrocarbon Chain Length

Differences in the chain length of hydrocarbons is one of the major distinguishing features of hydrocarbon profiles of different species. Chain lengths range from C21 to C45+ (Nelson and Blomquist, 1995). Control of the chain length could be accomplished by the chain length specificity of either the fatty acyl-CoA elongases or the enzyme which reductively converts the very long chain fatty acyl-CoA to hydrocarbon (Figure 2.2). Tillman-Wall *et al.* (1992) investigated these two possibilities in *M. domestica*, where females switch from making alkenes of 27 carbons and longer to a 23 carbon alkene as they become vitellogenic under the influence of the hormone 20-hydroxyecdysone. The roles of the two enzyme activities in regulating the chain length of the hydrocarbon products were investigated by comparing fatty acyl-CoA elongation and reductive decarboxylation activities in microsomes from vitello-genic females (major product is (Z)-9-tricosene [Z9-23:Hy]) to those from immature (i.e. previtellogenic) females and mature males (major product is Z9-27:Hy). There was no significant difference between vitellogenic females, immature females, and males in reductive conversion of radiolabeled 24:1-CoA to 23:1-Hy at low, and presumably closer to physiological, 24:1-CoA concentra-tions. This indicates that the reductive decarboxylation of 24:1-CoA to 23:1-Hy is not the ecdysone-controlled step in the biosynthetic pathway for sex pheromone production in *M. domestica*. However, *in vitro* elongation studies comparing vitellogenic females, previtello-genic females, and males showed a difference in 24:1-CoA elongation capability for vitellogenic females relative to males and previtello-genic females. Both males and previtellogenic females were able to efficiently elongate both 18:1-CoA and 24:1-CoA substrates to fatty acyl-CoA chains 28 carbons and longer. In contrast, vitellogenic females could not elongate 18:1-CoA efficiently beyond 24:1-CoA and showed virtually no elongation of 24:1-CoA. The results from these studies indicate that endocrine regulation of sex pheromone production in *M. domestica* occurs at the elongation reactions rather than the reductive decarboxylation that converts of 24:1-CoA to 23:1-Hy. These data suggest that ecdysone represses elongation of 24:1-CoA to induce sex pheromone production in vitellogenic (ecdysone-producing) females. These data also allow one to hypothesize that there may exist two or more separate elongation systems, with one elongase functioning in elongation of 18:1-CoA to 24:1-CoA and another in 24:1-CoA elongation (Tillman-Wall *et al.,* 1992). Ecdysone may repress this second, separate 24:1-CoA elongase to induce pheromone production in female houseflies. Thus, based on these data from *M. domestica,* it can be speculated that the chain length specificities of the fatty acyl-CoA elongation systems are important in regulating the chain length distribution of the hydrocarbons of insects.

Further *in vitro* elongation studies investigated the specific site or step [condensation, reduction, dehydration, or reduction] of 24:1-CoA elongation that is targeted for repression by ecdysone to result in the induction of sex pheromone production (Blomquist *et al.,* 1995). Because the first step in many biochemical pathways is often the regulated, rate-determining step, studies began by investigating the condensation step of the 24:1-CoA elongation reaction as the potential site of ecdysone regulation. *In vitro* condensation assays were performed in the same manner as previous elongation assays (Tillman-Wall *et al.,* 1992), with the exception that NADPH was omitted to prevent catalysis of the second step (reduction) in the elongation reaction. Production of the resulting condensation product (ß-keto-acyl-CoA) was assayed and compared using 24:1-CoA in vitellogenic females (day 4), immature females (day 1), and males of both ages (day 1 and day 4). Immature females and males of both ages showed notably higher 24:1-CoA condensation activity in comparison to vitellogenic females (Blomquist *et al.,* 1995). This work suggests that the condensation step of the elongation reaction is the specific site of ecdysone-mediated repression of 24:1-CoA elongation during the induction of pheromone production in vitellogenic females (Blomquist *et al.,* 1995).

Besides work in *M. domestica,* elongation reactions have also been studied in *P. americana* (Vaz *et al.,* 1988). The major hydrocarbon components of *P. americana* are *n*-pentacosane, 3-methylpentacosane and (Z,Z)-6,9-heptacosadiene. In a comparison of the elongation of 18:0-CoA, 18:1-CoA, 18:2-CoA and 18:3-CoA, only 18:0-CoA and 18:2-CoA were efficiently elongated beyond 20 carbons, and 18:0-CoA was elongated up to 26:0 and 18:2-CoA up to 28:2. Thus, the fatty acids 18:0 and 18:2 were both elongated to the appropriate chain lengths to give rise to the major *n*-C25 and 27:2 hydrocarbons. These data indicate that the elongation reactions play key roles in regulating both the degree of unsaturation and the chain length of the hydrocarbons of insects.

Mechanism of Hydrocarbon Formation in Insects

Several hypotheses have been advanced during the past seven decades to explain the biological formation of hydrocarbons. The condensation of two fatty acids in a head-to-head manner to form a ketone that is subsequently reduced to the alkane was proposed by Channon and Chibnall (1929). This hypothesis was based on structural similarities between ketones and hydrocarbons in plants. The subject remained dormant until the 1960's when Kolattukudy and co-workers presented the first evidence that fatty acids with *n* carbons were precursors to hydrocarbons with *n*-1 carbons (Kolattukudy *et al.*, 1976). Based on Kolattukudy's work in plants, it was shown in insects that very long chain fatty acids were converted to hydrocarbons one carbon shorter (Major and Blomquist, 1978; Chu and Blomquist, 1980b). An uncharacterized decarboxylation mechanism was presumed to carry out the conversion of fatty acid to hydrocarbon.

In extending the *in vivo* work to cell-free preparations, Khan and Kolattukudy (1974) found an unexpected result when they showed that a ^3H-C_{32} fatty acid was converted to a C_{30} hydrocarbon in addition to the expected C_{31} hydrocarbon in the presence of O_2 and ascorbate by particulate and solubilized enzyme preparations from pea leaves. This suggested the occurrence of both direct decarboxylation and α-oxidation followed by decarboxylation. Further experiments showed that an aldehyde with *n*-1 carbons might have been the immediate precursor of the C_{30} hydrocarbon (Bognar *et al.*, 1984) during biosynthesis of alkanes.

The role of aldehyde intermediates during the conversion fatty acids to hydrocarbon was established in the 1980's by Kolattukudy and co-workers, who also proposed that the aldehyde was then reductively decarbonylated to a hydrocarbon in cell-free preparations from plants (Cheesbrough and Kolattukudy, 1984), a vertebrate (Cheesbrough and Kolattukudy, 1988), a microorganism (Dennis and Kolattukudy, 1991; 1992) and in the flesh fly *Sarcophaga crassipalpis* (Yoder *et al.*, 1992). The decarbonylation of aldehydes was based on the observation that fatty acids were reduced to aldehydes in the presence of CoASH, ATP and NADH whereas conversion of the aldehyde to alkane and CO did not require O_2 nor the addition of cofactors. The decarbonylation mechanism was a novel idea as no other enzymatic analog has been described. In contrast, Gorgen and co-workers presented evidence that in an asteraceaous plant (Gorgen *et al.*, 1989) and in the confused flour beetle, *Tribolium confusum* (Coleoptera: Tenebrionidae) (Gorgen *et al.*, 1990), 1-alkenes were formed via a decarboxylation mechanism.

Work in our laboratory (Reed *et al.*, 1994) confirmed that hydrocarbon formation in *M. domestica* occurs via the reduction of very long chain fatty acyl-CoAs to aldehydes. However, only very low activity in the conversion of the aldehyde to hydrocarbon occurred in microsomal preparations in the absence of O_2 and NADPH (Reed *et al.*, 1994). The addition of O_2 and NADPH (but not NADH) resulted in the conversion of long chain aldehydes to the corresponding hydrocarbon and CO_2.. The conclusion that the carbonyl carbon was removed as CO_2 and not CO was based on radio-GLC results from the microsomal metabolism of [1-^{14}C](Z)-15-tetracosenal in the house fly (Reed *et al.*, 1994) and [1-^{14}C]octacosanal in a number of insect species (Mpuru *et al.*, 1996). The requirement for NADPH and O_2, inhibition by CO (which could be partially overcome by white light) and the inhibition by antibody to cloned and expressed housefly cytochrome P450 reductase (Reed *et al.*, 1994) provided strong evidence that the mechanism in *M. domestica* involved a cytochrome P450. GC-MS analyses of the microsomal products of deuterated substrates showed that the protons on positions 2 and 3 of the fatty acyl-CoA are retained during the conversion to hydrocarbon and that the proton on position 1 of the aldehyde is transferred to the adjacent carbon and retained during hydrocarbon formation (Reed *et al.*, 1995). Hydrogen peroxide, cumene hydroperoxide and iodosobenzene were able to support hydrocarbon production from aldehyde in place of O_2 and NADPH for short incubation times (Reed *et al.*, 1995). From these data, a cytochrome P450 mechanism was proposed in which the perferryl iron-oxeme, resulting from heterolytic cleavage of the O-O bond of the iron-peroxy intermediate, abstracts an electron from the C-O double bond of the carbonyl group of the aldehyde. The reduced perferryl attacks the 1-carbon of the aldehyde to form a thiyl-iron-hemiacetal diradical. The latter intermediate can fragment to form an alkyl radical and a thiyl-iron-formyl radical. The alkyl radical then abstracts the formyl hydrogen to produce the hydrocarbon and CO_2. Investigation of hydrocarbon biosynthesis in microsomal preparations from a number of other insect species, including *S. crassipalpis*, showed that a C_{18} aldehyde was converted to the C_{17} alkane and CO_2 only in the presence of O_2 and NADPH, indicating that the cytochrome P450 mechanism for hydrocarbon formation may be general among Insecta (Mpuru *et al.*, 1996).

Physiology

Much of the temporal work on the synthesis and transport of hydrocarbons has been performed in lepidopterans and a dictyopteran, but the results from these insects undoubtedly apply to the social insects. All of the hydrocarbon present on the surface of the cabbage looper, *Trichoplusia ni* and the southern armyworm, *Spodoptera eridania*, (both: Lepidoptera: Noctuidae) and presumably other insects, is lost on the cast skin during intra-larval and larval-pupal molts (Dwyer *et al.*, 1986; de Renobales *et al.*, 1988; Guo and Blomquist, 1991; de Renobales *et al.*, 1991). Thus, insects must synthesize a new complement of cuticular hydrocarbon at each molt. Extensive studies correlating the timing of synthesis and transport of hydrocarbons have been performed with *T. ni* and *S. eridania*. Within a few hours after shedding the cast skin during an intra-larval molt, the *in vivo* incorporation of labeled acetate into hydrocarbon increases dramatically. During the first part of the next instar (feeding stages), some of the newly synthesized hydrocarbon is transported to the surface, and then during the later part of the instar, less hydrocarbon is synthesized, but almost all of the newly synthesized hydrocarbon remains internal. Of the hydrocarbon that is found within the insect prior to a molt, about 1/3 to 1/2 is present on the surface of the newly emerged larva, and the remainder remains internal. Thus, it appears that the hydrocarbon present on the surface of the newly molted insect was synthesized primarily during the previous instar, stored internally, and then became the hydrocarbon that provided a full complement of surface wax for the newly molted insect. The critical role of hydrocarbon in preventing a lethal rate of desiccation has apparently dictated the evolution of a system such that the newly emerged instar is fully protected because of biosynthetic activity during the previous stadium.

A number of studies have shown that much of the "internal" hydrocarbon is associated with a transport protein called lipophorin and that the "internal" hydrocarbons have the same composition as the cuticular hydrocarbons (Chino and Downer, 1982; Katase and Chino, 1982; 1984; Gu *et al.*, 1995). Furthermore, work done in *B. germanica* demonstrates that the newly synthesized hydrocarbon first appears in the epidermal associated cells, then in the hemolymph associated with lipophorin, and then later on the surface of the insect (Gu *et al.*, 1995). This implies a route of transport of hydrocarbons from epidermal cells via lipophorin in the hemolymph and then deposition on the surface of the insect by a yet unknown route. This pathway of hydrocarbon transport is quite different from the pathway accepted for many years in which the newly synthesized hydrocarbon was thought to be transported to the surface of the insect through pore canals directly from epidermal cells (Figure 2.1) (Hadley, 1985; Blomquist and Dillwith, 1985). The necessity for storing and transporting hydrocarbons internally may be explained in part by the need for the newly emerged insect, after an intra-larval, larval-pupal, or pupal-imaginal molt, to have a full complement of hydrocarbon on the surface. By storing relatively large amounts of hydrocarbon with lipophorin, there would a ready source of hydrocarbon to deposit on the newly forming cuticle, even at times when hydrocarbon synthesis is minimal. This model would partially explain the relatively large amount of hydrocarbon often found internally or specifically with lipophorin. More recent studies (Young and Schal, personal communication) in which the deposition of hydrocarbons on the newly forming cuticle was examined, show that a portion of the newly synthesized hydrocarbon is present in the hemolymph associated with lipophorin, and that increasing amounts are present on the newly forming cuticle as the insect approaches ecdysis.

The transport of hydrocarbons in the hemolymph by lipophorin may also serve as a route to deliver hydrocarbons to specific organs and glands, such as the ovary and postpharyngeal gland. In *B. germanica*, large amounts of hydrocarbon were associated with the ovary, and it appears that lipophorin carried the hydrocarbon from the site of synthesis in the epidermis to the ovary (Gu *et al*., 1995). The process of hydrocarbon accumulation and exchange is likely to be even more complex in some social insects. In many ants the postpharyngeal glands contain the same characteristic hydrocarbons as does the cuticle (Bagnères and Morgan, 1991). In a study of the formicine ant, *Cataglyphis niger*, Soroker *et al.* (1995) showed that following [1-^{14}C] acetate injection, labeled hydrocarbons appeared first internally and after 24 hr accumulated on the cuticular surface and in the postpharyngeal gland. The conclusion that the postpharyngeal gland sequesters rather than synthesizes hydrocarbons was supported by data which showed that decapitated ants, which therefore lacked their postpharyngeal glands, synthesized as much hydrocarbon as did intact ants (Soroker *et al*., 1994). In addition, Soroker *et al.* (1995) stated that "preliminary *in vitro* experiments revealed that the gland failed to biosynthesize *de novo* hydrocarbons from acetate." The route by which the postpharyngeal gland accumulated hydrocarbon was examined by sealing the mouthparts prior to injection of labeled acetate into the ant. This procedure decreased but did not abolish the accumulation of labeled hydrocarbon in the postpharyngeal gland. These results were interpreted as evidence for both a hemolymph source and a self-grooming and/or trophallaxis source for hydrocarbon accumulation in the postpharyngeal gland. The data strongly indicated that the postpharyngeal gland serves as a pool for mixing colonial hydrocarbons and may serve to attain a unified colony odor (Soroker *et al.*, 1995). In an artificially mixed colony of the ants, *Formica selysi* and *Manica rubida*, a mixed odor was obtained by hydrocarbon transfer by trophallaxis, grooming and body contact (Vienne *et al.*, 1995).

Thus, a physiological model for hydrocarbon transport is evolving in which the newly synthesized hydrocarbon first enters the hemolymph bound to lipophorin, and is then transported to the sites of deposition, with the added complexity in social insects that an exchange of hydrocarbons may occur among individuals.

Concluding Remarks

The voluminous growth of literature in recent years reflects a growing appreciation for the roles of cuticular hydrocarbons in insect chemical communication, especially among the social insects. The advances are due in part to what has become an almost routine analysis of cuticular hydrocarbons by CCG-MS coupled with increasingly sophisticated behavioral assays. Along with the increased recognition of the roles of insect hydrocarbons has been continued work on the biochemistry and identification of these informationally rich compounds. Current studies are emphasizing the regulation of key enzymes that result in the complex and species-specific hydrocarbon blends that guide behavior and distinguish sexes and species. Future work will likely emphasize the characterization, purification and cloning of the key enzymes in hydrocarbon production. This should lead to a deeper understanding of how the process is regulated and ultimately how the tremendous diversity of cuticular hydrocarbons is achieved in the world of insects.

Acknowledgments

The work from the authors' laboratory was supported in part by National Science Foundation (IBN-9220092 and IBN-9630916) and by U. S. Department of Agriculture Research Initiative Grants (9401822, 9502551, and 9601832). This is a contribution of the Nevada Agricultural Experiment Station.

Literature Cited

Andersen, S.O. 1991. Sclerotisation. In The Physiology of the Insect Epidermis. Ed. by Retnakaran, A. and Binnington, K. CSIRO, Melbourne. pp. 123-140.

Bagnères, A.-G. and E.D. Morgan. 1991. The postpharyngeal glands and the cuticle of Formicidae contain the same characteristic hydrocarbons. *Experientia* 47:106-11

Bjostad, L.B., W.A. Wolf, and W.L. Roelofs. 1987. Pheromone biosynthesis in lepidopterans: desaturation and chain shortening. In Pheromone Biochemistry, Ed. by Prestwich, G.D. and Blomquist, G.J. Academic Press, N. Y. pp. 77-120.

Blomquist, G.J., A.J. Chu, J.H. Nelson and J.G. Pomonis. 1980. Incorporation of [2,3-^{13}C]succinate into the methyl-branched alkanes in a termite. *Arch. Biochem. Biophys.* 204:648-650.

Blomquist, G.J. and J.W. Dillwith. 1985. Cuticular lipids. In Comprehensive Insect Physiology, Biochemistry and Pharmacology. Vol. 3. Integument, Respiration and Circulation. Ed. by Kerkut, G.A. and Gilbert, L.I. Pergamon Press, Oxford. pp. 117-154.

Blomquist, G.J., L.A. Dwyer, A.J. Chu, and R.O. Ryan. 1982. Biosynthesis of linoleic acid in a termite, cockroach and cricket. *Insect Biochem.* 12:349-353.

Blomquist, G.J., L. Guo, P. Gu, C. Blomquist, R.C. Reitz,and, J.R. Reed. 1994. Methyl-branched fatty acids and their biosynthesis in the housefly, *Musca domestica* L. (Diptera:Muscidae). *Insect Biochem. Molec. Biol.* 24:803-810.

Blomquist, G.J., J.A. Tillman, J.R. Reed, P. Gu, D. Vanderwel, S. Choi, and R.C. Reitz. 1995. Regulation of enzymatic activity involved in sex pheromone production in the housefly, *Musca domestica*. *Insect Biochem. Molec. Biol.* 25:751-757.

Blum, M.S. 1985. Exocrine Systems. In Fundamentals of Insect Physiology, Ed. by M.S. Blum, John Wiley & Sons, Inc., New York. pp. 535-579.

Bognar, A.L., G. Paliyath, L. Rogers and P.E. Kolattukudy. 1984. Biosynthesis of alkanes by particulate and solubilized enzyme preparations from pea leaves (*Pisum sativum*). *Arch. Biochem. Biophys.* 235:8-17.

Bonavita-Cougourdan, A., J.-L. Clément and C. Lange. 1987. Nestmate recognition: the role of cuticular hydrcarbons in the ant *Camponotus vagus* Scop. *J. Entomol. Sci.* 22:1-10.

Channon, H.J. and A.C. Chibnall. 1929. XXII. The ether-soluble substances of cabbage leaf cytoplasm. V. The isolaton of *n*-nonacosane and di-*n*-tetradecyl ketone. *Biochem. J.* 23:168-175.

Chase, J., R.A. Jurenka, C. Schal, P.P. Halarnkar, and G.J. Blomquist. 1990. Biosynthesis of methyl branched hydrocarbons in the German cockroach, *Blattella germanica* (L.)(Orthoptera, Blattellidae). *Insect Biochem.* 20:149-156.

Cheesbrough, T.M. and P.E. Kolattukudy. 1984. Alkane biosynthesis by decarbonylation of aldehydes catalyzed by a particulate preparation from *Pisum sativum*. *Proc. Natl. Acad. Sci. USA* 81:6613-6617.

Cheesbrough, T.M. and P.E. Kolattukudy. 1988. Microsomal preparations from animal tissue catalyzes release of carbon monoxide from a fatty aldehyde to generate an alkane. *J. Biol. Chem.* 263:2738-2743.

Chino, H. and R.G.H. Downer. 1982. Insect hemolymph lipophorin: a mechanism of lipid transport in insects. *Adv. Biophys.* 15:67-92.

Chu, A.J. and G.J. Blomquist. 1980a. Biosynthesis of hydrocarbons in insects: Succinate is a precursor of the methyl branched alkanes. *Arch. Biochem. Biophys.* 201:304-312.

Chu, A.J. and G.J. Blomquist. 1980b. Decarboxylation of tetracosanoic acid to *n*-tricosane in the termite *Zootermopsis angusticollis*. *Comp. Biochem. Physiol.* 66B:313-317.

Coyne, J.A., A.P. Crittenden, and K. Mah. 1994. Genetics of a pheromonal difference contributing to reproductive isolation in *Drosophila*. *Science* 265:1461-1464.

Dennis, M.W. and P.E. Kolattukudy. 1991. Alkane biosynthesis by decarbonylation of aldehyde catalyzed by a microsomal preparation from *Botrycocus brauni*. *Arch. Biochem. Biophys.* 287:268-275.

Dennis, M.W. and P.E. Kolattukudy. 1992. A cobalt-porphyrin enzyme converts a fatty aldehyde to a hydrocarbon and CO. *Proc. Natl. Acad. Sci. USA* 89:5306-5310.

de Renobales, M., C. Cripps, D.W. Stanley-Samuelson, R.A. Jurenka, and G.J. Blomquist. 1987. Biosynthesis of linoleic acid in insects. *Trends Biochem. Sci.* 12: 364-366.

de Renobales, M., D.R. Nelson, and G.J. Blomquist. 1991. Cuticular lipids. In The Physiology of the Insect Epidermis. Ed. by Retnakaran, A. and Binnington, K. CSIRO, Melbourne. pp. 240-251.

de Renobales, M., D.R. Nelson, M.E. Mackay, A.C. Zamboni, and G.J. Blomquist. 1988. Dynamics of hydrocarbon biosynthesis and transport to the cuticle during pupal and early adult development in the cabbage looper *Trichoplusia ni* (Lepidoptera: Noctuidae). *Insect Biochem.* 18:607-613.

Dettner, K. and C. Liepert. 1994. Chemical mimicry and camouflage. *Ann. Rev. Entomol.* 39:129-154.

Dillwith, J.W., J.H. Nelson, J.G. Pomonis, D.R. Nelson, and G.J. Blomquist. 1982. A ^{13}C-NMR study of methyl-branched hydrocarbon biosynthesis in the housefly. *J. Biol. Chem.* 257:11305-11314.

Dwyer, L.A., A.C. Zamboni, and G.J. Blomquist. 1986. Hydrocarbon accumulation and lipid biosynthesis during larval development in the cabbage looper, *Trichoplusia ni*. *Insect Biochem.* 16:463-469.

Dwyer, L.A., G.J. Blomquist, J.H. Nelson, and J.G. Pomonis. 1981. A ^{13}C-NMR study of the biosynthesis of 3-methylpentacosane in the American cockroach. *Biochim. Biophys. Acta* 663:536-544.

Gorgen, G., C. Frobl, W. Boland, and K. Dettner. 1989. Biosynthesis of 1-alkenes in higher plants: stereochemical implications. A model study with *Carthamus tinctorius* (Asteraceae). *Eur. J. Biochem.* 185:237-242.

Gorgen, G., C. Frobl, W. Boland, and K. Dettner. 1990. Biosynthesis of 1-alkenes in the defensive secretions of *Tribolium confusum* (Tenebrionidae); stereochemical implications. *Experientia* 46:700-704.

Gu, X., D.R. Quilici, P. Juarez, G.J. Blomquist, and C. Schal. 1995. Biosynthesis of hydrocarbons and contact sex pheromones and their transport by lipophorin in females of the German cockroach (*Blatella germanica*). *J. Insect Physiol.* 41: 257-267.

Gu, X., W.H. Welch and G.J. Blomquist. 1993. Methyl-branched fatty acid biosynthesis in the German cockroach, *Blattella germanica*: kinetic studies comparing a microsomal and soluble fatty acid synthetase. Insect Biochem. Mol. Biol. 23: 263-271.

Guo, L. and G.J. Blomquist. 1991. Identification, accumulation and biosynthesis of the cuticular hydrocarbons of the southern armyworm *Spodoptera eridania* (Lepidoptera: Noctuidae). *Arch. Insect Biochem. Physiol.* 16:19-30.

Guo, L., D.R. Quilici, J. Chase, and G.J. Blomquist. 1991. Gut tract microorganisms supply the precursors for methyl-branched hydrocarbon biosynthesis in the termite, *Zootermopsis nevadensis. Insect Biochem.* 21:327-333.

Hadley, N.F. (1985) The Adaptive Role of Lipids in Biological Systems. John Wiley & Sons, Inc., New York, pp. 117-128 and pp. 247-289.

Haverty, M.I. and B.L. Thorne. 1989. Agonistic behavior correlated with hydrocarbon phenotypes in dampwood termites, *Zootermopsis* (Isoptera: Termopsidae). *J. Insect Behav.* 2:523-543.

Hepburn, H.R. (1985) The Integument. In Fundamentals of Insect Physiology. Ed. by M.S. Blum, John Wiley & Sons, Inc., New York. pp. 139-183.

Hopkins, T.L. and K.J. Kramer. 1992. Insect cuticle sclerotization. *Ann. Rev. Entomol.* 37:273-302.

Howard, R.W. (1993) Cuticular hydrocarbons and chemical communication. In Insect Lipids. Ed. by D.W. Stanley-Samuelson and D.R. Nelson, University of Nebraska Press, Lincoln. pp. 177-226.

Juarez, P., J. Chase, and G.J. Blomquist. 1992. A microsomal fatty acid synthetase from the integument of *Blattella germanica* synthesizes methyl-branched fatty acids, precursors to hydrocarbon and contact sex pheromone. *Arch. Biochem. Biophys.* 293:333-341.

Katase, H. and H. Chino. 1982. Transport of hydrocarbons by the lipophorin of insect hemolymph. *Biochim. Biophys. Acta* 710:341-348.

Katase, H. and H. Chino. 1984. Transport of hydrocarbons by haemolymph lipophorin in *Locusta migratoria. Insect Biochem.* 14:1-6.

Khan, A.A. and P.E. Kolattukudy. 1974. Decarboxylation of long chain fatty acids to alkanes by cell free preparations of pea leaves (*Pisum sativum*). *Biochem. Biophys. Res. Comm.* 61:1379-1386.

Kolattukudy, P.E., R. Croteau, and J.S. Buckner. (1976) Biochemistry of plant waxes. In Chemistry and Biochemistry of Natural Waxes. Ed. by Kolattukudy, P.E. Elsevier, Amsterdam. pp. 289-347.

Locke, M. (1991) Insect epidermal cells. In The Physiology of the Insect Epidermis. Ed. by Retnakaran, A. and Binnington, K. CSIRO, Melbourne. pp. 1-22.

Lockey, K.H. 1988. Lipids of the insect cuticle: Origin, composition, and function. *Comp. Biochem. Physiol.* 89B: 595-645.

Major, M.A. and G.J. Blomquist. 1978. Biosynthesis of hydrocarbons in insects: decarboxylation of long chain acids to *n*-alkanes in *Periplaneta. Lipids* 13:323-328.

Mpuru S., J.R. Reed, R.C. Reitz, and G.J. Blomquist. 1996. Mechanism of hydrocarbon biosynthesis from aldehyde in selected insect species: Requirement for O_2 and NADPH and carbonyl group released as CO_2. *Insect Biochem. Molec. Biol.* 26: 203-208.

Nelson, D.R. 1969. Hydrocarbon synthesis in the American cockroach. *Nature Lond.* 221:854-855.

Nelson, D.R. 1993. Methyl-branched lipids in insects. In Insect Lipids. Ed. by D.W. Stanley-Samuelson and D.R. Nelson, University of Nebraska Press, Lincoln. pp. 271-315.

Nelson, D.R. and G.J. Blomquist. 1995. Insect Waxes. In Waxes: Chemistry, Molecular Biology and Functions. Ed. by R.J. Hamilton and W.W. Christie, The Oily Press, Ltd. England. pp. 1-90.

Page, M., L.J. Nelson, G.J. Blomquist, and S.J. Seybold. 1997. Cuticular hydrocarbons as chemotaxonomic characters of pine engraver beetles (*Ips* spp.) (Coleoptera: Scolytidae) in the *grandicollis* subgeneric group. *J. Chem. Ecol.* (in press).

Reed, J.R., D. Vanderwel, S. Choi, J.G. Pomonis, R.C. Reitz, and G.J. Blomquist. 1994. Unusual mechanism of hydrocarbon formation in the housefly: cytochrome P450 converts aldehyde to the sex pheromone component (Z)-9-tricosene and CO_2. *Proc. Natl. Acad. Sci. USA* 91:10,000-10,004.

Reed, J.R., D.R. Quilici, G.J. Blomquist, and R.C. Reitz. 1995. Proposed mechanism for the cytochrome P450 catalyzed conversion of aldehydes to hydrocarbons in the housefly, *Musca domestica*. *Biochemistry* 34:16,221-16,227.

Riddiford, L.M. (1991) Hormonal control of sequential gene expression in insect epidermis. In The Physiology of the Insect Epidermis. Ed. by A. Retnakaran, and K. Binnington, CSIRO, Melbourne. pp. 46-54.

Romer, F. 1980. Histochemical and biochemical investigations concerning the function of larval oenocytes of *Tenebrio molitor* L. (Coleoptera, Insecta). *Histochemistry* 69:69-84.

St. Leger, R.J. 1991. Integument as a barrier to microbial infections. In Physiology of the Insect Epidermis, Ed. by A. Retnakaran and K. Binnington, CSIRO, Melbourne. pp. 284-306.

Soroker, V., C. Vienne, and A. Hefetz. 1994. The postpharyngeal gland as a "Gestalt" organ for nestmade recognition in the ant *Cataglyphis niger*. *Naturwissenschaften*. 81:510-513.

Soroker, V., C. Vienne, and A. Hefetz. 1995. Hydrocarbon dynamics within and between nestmates in *Cataglyphis niger* (Hymenoptera: Formicidae). *J. Chem. Ecol.* 21:365-378.

Takahashi, S. and A. Gassa. 1995. Roles of cuticular hydrocarbons in intra- and interspecific recognition behavior of two Rhinotermidae species. *J. Chem. Ecol.* 21:1837-1845.

Tillman-Wall, J.A., D. Vanderwel, M.E. Kuenzli, R.C. Reitz, and G.J. Blomquist. 1992. Regulation of sex pheromone biosynthesis in the housefly, *Musca domestica*: Relative contribution of the elongation and reduction step. *Arch. Biochem. Biophys.* 299:92-99.

Vaz, A.H., R.A. Jurenka, G.J. Blomquist, and R.C. Reitz. 1988. Tissue and chain length specificity of the fatty acyl-CoA elongation system in the American cockroach. *Arch. Biochem. Biophys.* 267:551-557.

Vienne, C., V. Soroker, and A. Hefetz. 1995. Congruency of hydrocarbon patterns in heterospecific groups of ants: transfer and/or biosynthesis. *Insect Sociaux*. 42: 261-277.

Yoder, J. A., D.L. Denlinger, M.W. Dennis, and P.E. Kolattukudy. 1992. Enhancement of diapausing flesh fly puparia with additional hydrocarbons and evidence for alkane biosynthesis by a decarbonylation mechanism. *Insect Biochem. Molec. Biol.* 22:237-243.

Wang, D.L., J.W. Dillwith, R.O. Ryan, G.J Blomquist, and R.C. Reitz. 1982. Characterization of the acyl-CoA desaturase in the housefly, *Musca domestica* L. *Insect Biochem.* 12:545-551.

Wigglesworth, V.B. (1964) The Life of Insects. In The World Natural History. Ed. by R. Carrington. The New American Library, Inc., New York. 383 pp.

Wigglesworth, V.B. (1972) Principles of Insect Physiology. 7th Edition, Chapman and Hall, London. 827 pp.

Wigglesworth, V.B. (1991) Foreword. In The Physiology of the Insect Epidermis. Ed. by A. Retnakaran and K. Binnington. CSIRO, Melbourne. pp. xi-xii.

Willis, J.H. (1991) The epidermis and metamorphosis. In The Physiology of the Insect Epidermis. Ed. by A. Retnakaran, and K. Binnington, CSIRO, Melbourne. pp. 36-45.

PART TWO

Nestmate Recognition in Social Insects

3

Chemical Cues in Kin Recognition: Criteria for Identification, Experimental Approaches, and the Honey Bee as an Example

Michael D. Breed

Introduction

In this chapter I first discuss the rationale for studying recognition chemistry and then examine how the role of chemicals as recognition signals can be established experimentally. This leads to a brief evaluation of the available studies of recognition chemistry in social insects, and a classification of those studies according to the type of evidence presented. I then review studies conducted in my laboratory on the recognition chemistry of the honey bee, and evaluate the strengths and weaknesses of my approach to honey bee recognition. The final section of the paper contains suggestions for additional approaches that could be taken in order to broaden or amplify our knowledge of recognition chemistry.

Why Study Recognition Chemistry?

Kin discrimination is an extremely significant mechanism in animal behavior. The expression by an individual of differential behavior to kin and non-kin has profound evolutionary impacts, and by extension can substantially affect the behavioral ecology of a species. Typically, animals use olfaction or contact chemoreception to discriminate kin. Even humans, who have well developed visual abilities, can also make discriminations based on odors (Porter 1991). Chemically based discrimination has been demonstrated in anemones (Rinkevich et al. 1994), Crustacea (Caldwell 1992), social insects (Jaisson 1991, Smith and Breed 1995), Amphibia (Blaustein and Waldmann 1992), and a variety of other vertebrates (Halpin 1991).

Olfactory kin recognition usually relies on compounds that originally were probably present for other purposes. These include secretory compounds, such as surface hydrocarbons, which have a primary function as waterproofing, and excretory compounds, such as those found in the urine of mammals (Yamazaki et al. 1994). While it is possible that the abundance of certain compounds in secretions or excretions has either increased or decreased through selection affected by their use as recognition signals, there is no direct evidence to support this point in any animal.

Given the importance of kin discrimination in animal behavior it is not surprising that a number of investigators have attempted to determine which compounds are used by animals to make discriminations. The search for compounds faces the normal difficulties involved in pheromonal identifications, such as small quantities, lack of predictability of compound structure, and problems with bioassay development. While pheromone chemists are accustomed to dealing with multicomponent mixtures, the plethora of compounds that may be involved in discrimination is still exceptional, and adds an unusual layer of complexity.

Studies of recognition chemistry, then, provide an exceptional opportunity for investigating an evolutionarily important signaling system. The interest from a chemical point of view stems not from the novelty of the compounds involved, but from the potential complexity of the signal and from the establishment of linkages between chemistry and behavior.

Experimental Approaches to Recognition Chemistry

To establish that a compound serves as a recognition signal it is necessary to show that: 1) the compound is present on at least some members of the population, 2) that the concentration of the compound varies among members of the population, 3) the compound can be perceived, and 4) that manipulation of the concentration of the compound, with all other variables affecting chemoreception controlled, affects discrimination in an appropriate bioassay. Of course, if the fourth criterion is met, the third can be assumed. Three types of evidence have been used to support identifications of chemical kin recognition cues; these are correlational, removal and replacement, and chemical supplementation.

Correlational Studies

Correlational studies establish that compounds or types of compounds vary among individuals in association with discrimination ability. In most studies discrimination is tested in a bioassay. Cuticular hydrocarbons, for example, are correlated with colony differences in many wasps, ants, and the honey bee (see Table 3.2 for references). This is the weakest type of evidence, as no functional significance is attributed to any specific compound or compound class. While correlational studies could apply to any class or combination of classes of compound, they have been most frequently performed on cuticular hydrocarbons in social insects.

Cuticular hydrocarbon extracts are usually rich in hydrocarbons, which are readily separated into patterns of peaks by gas chromatography. The output is graph of a series of peaks that may correspond to single compounds. It is not unusual to find 30 to 40 peaks in such a chromatogram, and it is also not unusual to be able to visually discern intercolonial differences in chromatograms. The temptation to assume that the visualized differences have olfactory significance to the animal species in question is strong, but to make this assumption is completely unfounded. The leap from visualized or statistically analyzed differences to the assertion that cuticular hydrocarbons are recognition cues is insupportable, unless this information is supplemented with other types of experiments.

Nevertheless, correlational studies have substantial potential as a beginning step in the study of recognition chemistry. Such a study can establish which compounds are present or absent, and which classes of compounds vary among individuals, among colonies of social insects, between families, and among populations.

Removal and Replacement

In removal and replacement experiments compounds are selectively withheld or removed from individuals and then later may be applied or reapplied. Two experimental techniques are used in these experiments. The first is to manipulate exposure to cue sources by changing the nesting or foraging environment. The second is to wash or extract compounds from individuals or nesting materials. Removal and replacement experiments have the advantage of working with naturally occurring concentrations of compounds.

Manipulation of exposure to colonymates, nesting materials, or foraging resources is relatively straightforward. Bioassays can determine whether the manipulation has affected the recognition profile of the insects. If the removal eliminates discrimination and the replacement establishes discrimination then the manipulated materials are known to contain recognition cues.

The second method, treatment with solvents, has a number of pitfalls. While some selectivity of class or combination of classes of compound can be gained through solvent choice, the extracts contain many compounds and no evidence concerning the role of any specific compound is gained. Removal by solvents also requires that the biological materials be amenable to exposure to solvent (as are the paper nests of wasps) or that the experiment be designed so that solvent-sensitive living animals are not directly exposed to the solvent.

Removal and replacement methods have worked particularly well in social insect recognition systems that rely on nesting materials as intermediaries in the transmission of recognition compounds. In some species of social wasps the papery nest, which is derived from masticated plant material, is imbued with hydrocarbons by the wasps during nest construction. These hydrocarbons then are transferred from the nest to the wasps and at least some of the compounds in the hexane soluble fraction serve as recognition signals. This complex sequence of events was elucidated in an elegant series of experiments (Gamboa et al. 1986, Espelie and Hermann 1990, Singer and Espelie 1992). Similarly, in the honey bee, compounds in the comb wax, a glandular product of the bee, serve as recognition cues; this role was demonstrated by withholding wax from groups of bees and then exposing bees to wax (Breed et al. 1988b) and is reviewed in more detail below. Both of these demonstrations rely on the removal and replacement type of method.

Chemical Supplementation

In contrast, chemical supplementation experiments test for the effects of single compounds on recognition. In these experiments single compounds or limited sets of compounds are applied to living animals and then a bioassay is used to determine whether the compound(s) alter the recognition response. The main advantage of these tests is that hypotheses relate to the role of specific compounds, rather than classes of compounds. Unfortunately, compounds are usually applied in quantities that are much higher than natural concentrations. Chemical supplementation has only been tried extensively in the honey bee; these results are reviewed below.

Table 3.1 summarizes the information obtained from each type of analysis. Correlational analyses are assumed to include solvent extractions and GC or GC/MS analysis of the separated components.

TABLE 3.1. An analysis of how each of the three methods of assessing chemical communication in kin recognition addresses the criteria for establishing recognition activity.

Assessment Methods	Presence	Variation	Perception	Concent. Effect
Correlational	Yes	Yes	No	No
Removal and Repalcement	Yes /No*	No	Yes	No
Supplementation	No	No	Yes	Yes

*depends on choice of experimental conditions

Removal and replacement experiments give information on presence if a solvent is used or if the composition of the material being removed is known. Perception is assumed in removal and replacement as well as supplementation experiments because a positive bioassay result can be obtained only if the compound(s) are perceived. Direct evidence for perception at the receptor level can be obtained by electrophysiological recordings from sensillae (Getz and Smith 1987) or by recordings from intact antennae (Breed et al., in preparation), but these approaches do not link perception with a behavioral response.

It is outside the scope of this review to discuss in detail all of the available recognition chemistry studies. In Table 3.2 I have listed the studies of which I am aware. Correlational studies are, as noted above, the most common, with the other two approaches having been used less frequently.

Honey Bee Nestmate Discrimination

Honey bee discrimination mechanisms have been studied in a large number of laboratories. Most work has focused on behavioral mechanism or outcomes and not on chemistry; these studies are reviewed by Breed and Bennett (1987), Getz (1991) and Smith and Breed (1995). Here I review studies on the chemistry of the honey bee recognition system.

TABLE 3.2. Studies of social insect recognition chemistry, classified by the type of evidence presented. HC(s)=hydrocarbon(s). Studies that show subspecies, population, or colony differences in HCs or other compounds but which are primarily chemotaxonomic in purpose are not listed. A number of studies have used removal/replacement techniques involving queens and/or nesting materials but have not identified any specific compounds associated with these materials; these studies are also not included.

SPECIES	METHOD	RESULT	REFERENCE
CORRELATIONAL			
Reticulitermes flavipes, R. santonensis	Pentane wash	Intercolony variation in HC's	Bagnères et al. 1988, 1990
Zootermopsis angusticollis, Z. nevadensis, Z. laticeps	Tests among species with established HC phenotypes	Intercolony variation in cuticular HCs, correlated with recognition	Haverty et al. 1988, Haverty and Thorne 1989
Formica selysi, Manica rubida	Pentane wash, GC-MS	Intraspecific variation in cuticular HCs	Bagnères et al. 1991a
Camponotus vagus	Pentane wash	Interspecific variation in cuticular HCs	Bonavita-Cougourdan et al. 1987
Leptothorax lichtensteini	Hexane wash, GC-MS	Cuticular HC profiles of workers change over time	
Solenopsis invicta	Hexane wash, GC-MS	Nestmate recognition not correlated with five major cuticular HCs	Obin 1986
Solenopsis richteri	Hexane homogenate, GC-MS	Intracolony variation among foragers	Brill and Bertsch 1990
Solenopsis richteri	Headspace collection in hexane, GC-MS	Intercolony variation in HCs	Brill et al. 1986
Catagylyphis cursor	Pentane extract, GC-MS	Geographic variation in cuticular HC profiles	Nowbahari et al. 1990

Table 3.2 Continued

Table 3.2 Continued

Messor barbarus	Hexane extract, GC-MS	HC differences between mono- and polygnous colonies	Provost et al. 1994
Vespa crabro, Dolichovespula maculata, Vespula squamosa, Vespula maculifrons	Hexane extract, GC-MS	HC variation among species	Butts et al. 1991
Dolichovespula maculata	Hexane wash, GC-MS	Intercolony variability in HCs	Butts et al. 1993
Polistes exclamans	Hexane wash, GC-MS	Intercolony variability in HCs	Singer et al. 1992
Polistes metricus	Hexane wash, GC-MS	Intercolony variability in cuticular and paper HCs	Espelie et al. 1990, Layton et al. 1994
Polistes fuscatus	Hexane wash, GC-MS	Intercolony variability in cuticular HC's	Espelie et al. 1994
Polistes fuscatus	Hexane wash, GC-MS	Discriminant function identified candidate pheromones	Gamboa et al. 1996
Polistes atrimandibularis, Polistes biglumis	Pentane wash, GC-MS	Parasitic species matches unsaturated HC's of host	Bagneres et al. 1996
Lasioglossum zephyrum	GC-MS of Dufour's gland contents	Genetic correlation of variation with relatedness	Hefetz et al. 1986,
Lasioglossum zephyrum	GC of headspace samples in hexane	Related females covary in pheromone production	Smith and Wenzel 1988
Apis mellifera	Hexane wash, GC-MS	Intercolony variation in cuticular HCs	Page et al. 1991
Apis mellifera	Dichloromethane wash,GC-MS	Intercolony variation in comb wax HCs	Breed et al. 1995
Apis mellifera	Pentane wash, GC-MS	Intracolony variation in cuticular HCs	Arnold et al. 1996

Table 3.2 Continued

Table 3.2 Continued

REMOVAL AND REPLACEMENT			
Reticulitermes grassei, R. banyulensi	Pentane wash, coating lure with wash	Wash removes cues, coated lure = response	Bagneres et al. 1991
Solenopsis sp.,Myrmecaphodius excavaticollis	Hexane wash	This beetle acquires cuticular HCs from its host ant	Vander Meer and Wojcik 1982
Camponotus vagus	Pentane wash, coating on washed workers with extract	Wash removes cues, coating changes cues of coated ant	Bonavita-Cougourdan, et al. 1987
Camponotus floridanus	Hexane wash, coating workers with extract	Wash removes cues, coating changes cues of coated ant	Morel et al. 1988
Formica selysi, Manica rubida	Cross-fostering between species	Interspecific HC Induction or transfer	Bagnères et al. 1991a
Catagylyphis cursor	Cross-fostering between colonies	Cross-fostered ants have HC profiles intermediate - parent and host	Nowbahari et al. 1990
Polistes metricu	Hexane wash of nest, recoating nest with extract	Wash removes cues, recoating replaces them	Singer and Espelie 1992
Polistes metricus	Hexane wash of nest, coating nest material with extract	Wash removes cues, coating replaces them	Layton and Espelie 1995
Dolichovespula maculata	Hexane wash of nest, recoating	Negative result; no effect of extract on recognition	Butts and Espelie 1995
Apis mellifera	Exposure to comb	HC and organic acid cues transferred from comb to bees,cues are heritable	Breed et al. 1988a, b, Breed and Stiller 1992, Breed et al. 1995
SUPPLEMENTATION			
Apis mellifera	Treatment with putative cue	See table 3 for a listing of compounds tested and results	Breed and Stiller 1992, Breed et al 1992b

Chemical studies of wax and cuticular compounds are relevant to honey bee nestmate recognition, even though the authors of these studies may not have realized this significance at the time the studies were published. Early characterizations of honey bee hydrocarbons included those of Blomquist (1980), Francis et al. (1985, 1989), and Carlson (1988). This was followed by work directed at colonial differences that might be used in discriminations by Page et al. (1991). The most noteworthy general finding is that honey bee cuticular hydrocarbons are dominated by odd-chained n-alkanes, with the predominant compounds being C-25, C-27, C-29, C-31, C-33, and C-35. This is in rather sharp contrast to many of the ants and social wasps, in which either unsaturated or methyl-branched compounds are more prominent. More recently R.-K. Smith (1990) has identified unsaturated and methyl-branched compounds in hexane extracts of honey bees, but these compounds are present in minute amounts when compared with the alkanes.

Honey bees' wax is a complex mixture. In addition to the same alkanes as honey bee cuticular hydrocarbons, free fatty acids, and esters are found in significant quantities in comb wax (Tulloch 1980, Hepburn 1986). Wax is a product of glands on the sternal surface of the abdomen, but it is modified by mandibular gland products while it is being molded by the bees into comb. The acids, in particular, may be mandibular gland products, as they are characteristic of honey bee mandibular gland secretions (Crewe 1982).

Correlational Studies

Differences among honey bee species, subspecies, populations, and colonies have been found in mandibular gland esters (Crewe, 1982), cuticular hydrocarbons (Carlson and Bolton 1984, Page et al. 1991), and wax components (Carlson 1988, Breed et al, 1995b). Crewe (1982) found subspecies differences among populations of *Apis mellifera.* Page et al. (1991) showed that intercolonial differences within a population of *Apis mellifera* of cuticular hydrocarbons were adequate to allow statistical discrimination among colonies.

Breed et al. (1995b) found statistically significant differences in comb hydrocarbon concentrations among closely related colonies. These differences were present in 7 of 25 compounds analyzed. Differences in comb hydrocarbon concentrations may be adequate for colony discriminations; compound concentrations, however, were strongly intercorrelated, so that colonies could not be assigned to family groups using multivariate statistical analyses.

Removal and Replacement Studies

There is an interesting parallel between the classic removal and replacement studies of Espelie's group (Singer and Espelie, 1992, Layton and Espelie 1995) on paper wasps and studies of the role of comb wax in honey bee nestmate recognition. Experiments that are somewhat analogous to the wasp experiments are performed in the honey bee by removing bees from contact with wax, by changing the wax with which they have contact, or by exposing them to wax after a period without exposure.

Breed et al. 1995a showed that contact of bees with wax dramatically changes the recognitive profile of the bees. While family membership of bees without wax exposure can be discriminated under laboratory conditions (Breed 1983), acceptance of bees at a colony entrance in the field is dependent on exposure of bees to comb from that colony. Breed et al. (1995b) used gas chromatographic analyses to show that the surface chemistry of comb exposed bees is substantially different from those with no comb exposure. Some compounds increased in concentration due to exposure and others decreased. An analysis of trimethyl-silyl derivatives showed that acids, as well as alkanes, change concentration due to comb exposure. Acids may play a central role in honey bee nestmate recognition; this feature is discussed below, in the section on chemical supplementation.

Solvent extracts of comb wax were separated into fractions of different polarity using column chromatography (Breed et al. in prep). Using dichloromethane as a solvent, the recognition activity was captured in the solvent fraction; exposure to the insoluble fraction had no effect on bee's recognition characteristics. No single fraction showed as much recognition activity as the aggregate of the fractions, but the most active fraction was the one which contained the more polar compounds.

In a final variation on the theme of removal and replacement Breed et al. (submitted) collected pupae from colonies, separated emerging bees from their natal comb, and then exposed those bees to comb from the same colony or a different colony. Bees were tested for acceptability by guards at colony entrances. There were four possible treatments; 1) bees and comb from the same colony as the guards, 2) bees from the same colony as the guards, comb from different colony, 3) bees from a different colony, comb from the same colony, and 4) bees and comb both from a different colony than the guards. Only the first of these four treatments resulted in bees that were acceptable to the guards. This result shows that both cuticular cues and comb-derived cues are essential for nestmate recognition. Correlational evidence suggests that the cuticular cues could be hydrocarbons, but this conclusion is limited by the same factors that limit all correlational conclusions.

Chemical Supplementation Studies

A unique aspect of studies of the honey bee recognition system is the availability of chemical supplementation studies. To my knowledge, this technique has not been used in any other recognition system. It is the only experimental technique with gives information about both the action of a specific compound and the perception of that compound.

Despite the large advantages gained from chemical supplementation in obtaining information about perception and use of compounds, the technique has drawbacks. It distorts the naturally occurring concentration of the compound and changes the relative concentration of the compound with other compounds. The exposures necessary to yield an effect in the bioassays used in my laboratory cause the compounds under question to be present in much higher than natural concentrations.

The techniques used in the chemical supplementation experiments are relatively simple. Bees are collected in combs as pupae from field colonies and taken from their combs at emergence. The bees are then separated into groups of ten and maintained in the laboratory in paper cups with food and water *ad lib.* Bees in different cups are, of course, sisters and have a high degree of similarity in cuticular cues. Because they have not had adult exposure to comb they rely on cuticular cues for cupmate (analogous to nestmate) identification. If a bee is removed from one cup and placed in a cup containing sister bees she will usually be accepted (55 to 75% of the time) but will sometimes be bit or stung.

A compound is tested by placing a measured amount (usually 100 mg) of compound on a piece of filter paper. This paper is put in a cup containing bees and allowed to remain for five days. A bee is then removed and placed in a cup of sisters bees and the reaction of the sister bees is recorded. The rate of rejection over a number of replicates is then statistically compared with the rate of rejections in controls. If the rejection rate significantly differs from the control rate this indicates that the compound in question has activity in the recognition system.

The compounds yielding positive and negative results in chemical supplementation experiments with the honey bee are listed in Table 3.3. Of 38 compounds tested 22 yielded positive results. 11 of the 22 are esters produced in queen feces (Breed et al. 1992b); these compounds may serve in queen, rather than worker, recognition. Ignoring the queen fecal

TABLE 3.3. Compounds tested for recognition activity in the honey bee using the chemical supplementation technique. Sources: Breed et al. 1992a, 1992b and unpublished data

Alkanes & Alkenes - Positive Result	Alkanes & Alkenes - Negative Result
Hexadecane, octadecane, z-9-heneicosene, z-9-tricosene	Dodecane, tetradecane, pentadecane, heptadecane, eicosane, tricosane, pentacosane, nonacosane
Acids - Positive Result	**Acids - Negative Result**
palmitic acid, palmitoleic acid, oleic acid, linoleic acid, linolenic acid, tetracosaoic acid	tetradecanoic acid, stearic acid, octacosanoic acid
Esters - Positive Result	**Esters - Negative Result**
methyl docosanoate, hexyl decanoate*, hexyl dodecanoate*, decyl octanoate*, decyl decanoate*, octyl dodecanoate*, hexyl tetradecanoate*, decyl dodecanoate*, dodecyl decanoate*, tetradecyl decanoate*, tetradecyl dodecanoate*, hexadecyl decanoate*	methyl pentadecanoate, methyl octadecanoate, methyl eicosanoate, methyl triacontanoate, octyl tetradecanoate*

*queen fecal esters (Breed et al. 1992b)

compounds, 26 remain, of which 11 gave positive results. As a general pattern, alkanes are less likely to have recognition activity than acids or alkenes. There are fewer signals present than the number of compounds giving positive results would indicate, as bees do not distinguish between hexadecane and octadecane, between tetracosanoic acid and methyl docosanoate, or between linoleic and linolenic acids. Palmitic acid increases, rather than decreases the acceptability of bees to their sisters, suggesting a different role for this compound.

Discussion and Conclusions

The three methods discussed here--correlation, removal and replacement, and supplementation--provide powerful tools in the study of recognition chemistry. While the honey bee system has been more extensively studied than other systems--it is the only one in which all three methods have been applied--substantial advances have also been made on the recognition chemistry of termites, ants, and wasps. By using all three methods it should be possible to obtain at least a partial list of which compounds are responsible for recognitive processes in any given species.

Each method, however, has serious inherent difficulties. Correlational studies may lead the researcher down a completely incorrect path--the discovery of variation is only a very preliminary step in determining whether that variation has any significance to recognition. Removal and replacement studies are hampered by non-specificity, and investigators need to pay more attention to the limitations of solvent extraction methods. Chemical supplementation studies have focused on single compounds in complex blends and consequently may miss important contextual information.

Because of these difficulties, recognition chemistry is likely to remain a difficult area of inquiry. My view is that it is probable that a large number of compounds are used in recognition and that virtually any metabolite that makes its way to the surface of the animal could be a recognition compound. The large number of compounds means that the information content of any single compound is small, and that after a certain number of recognition-active compounds have been identified, a point of diminishing returns will be reached in the additional recognitive information added with new compound identifications. If all compounds have equal, or nearly equal effects, it may not be possible to develop a bioassay that is sensitive enough to detect the effects of each compound. I think it is therefore unlikely that we will ever be able to determine the complete catalogue of recognition compounds in a species.

The large number of correlative studies on social insects demonstrate that the cuticular surface of most, or perhaps all, social insects harbors adequate chemical variation to explain the ability of these insects to discriminate nestmates from non-nestmates. I am personally concerned, however, that a large number of the correlative studies extend their conclusions to the realm of causation without having the manipulative experiments in hand that would actually allow such extension. The survey of the literature on chemical recognition in social insects (Table 3.1) is certainly replete with studies that are overextended in this manner.

It is also inappropriate to focus solely on the alkanes, alkenes, and methyl-alkanes that predominate on the cuticular surfaces of insects (Blomquist et al. 1987, Lockey 1988, Hadley 1989, Howard 1993). While these compounds may have recognitive significance, alkanes, in particular, are devoid of the functional groups that typically stimulate chemoreceptors. In my studies of the honey bee, two alkanes do have recognition function, but these are atypically short-chained. The longer chained alkanes, which are more typical of insect cuticular hydrocarbon profiles, appear not to have recognitive function (Table 3.3). I suspect that the ease of extraction and visualization of alkanes, alkenes and methyl-alkanes has been a driving force in the focus on these compounds. Looking more broadly should be fruitful in finding additional recognition-active compounds. This is particularly true in the ants, as environmental influences (odors from food and nesting material) have been shown to have important effects in a number of species (Hölldobler and Wilson 1990).

The removal and replacement experiments are particularly important in helping to sort out the effects of nesting materials as transmitters of metabolic products versus truly environmentally derived odors. In the two systems in which removal and replacement has been extensively used--*Polistes* wasps and the honey bee--the results have pointed to nesting materials as an intermediary for metabolites. In ants there is strong evidence for true environmental odors, but chemical identifications need to be performed in order to confirm this.

The final level of experimental analysis is supplementation. Two important conclusions can be drawn from the supplementation experiments. First, it is inappropriate to assume that all of the compounds in a chromatogram have a recognitive function. This is a particularly important point relative to the correlative studies that have dominated the field. While these results certainly do not invalidate the data obtained in such studies, they do indicate that not all compounds represented in a chromatogram have equal value in recognition.

Second, the compounds active in recognition are scattered across a number of functional classes--alkanes, acids, alkenes and acid-alkenes. This indicates that functional class is not a predictor of recognition activity and that a broad net must be cast when choosing solvents to elute candidate recognition compounds, and that derivitization procedures to allow visualization of acids in gas chromatographs should be considered.

The combination of the three experimental approaches on the honey bee tells us quite a lot about this species' recognition chemistry. Correlative data indicate that there is adequate variation in both cuticular and comb compounds to allow nestmate discrimination in the honey bee. Removal and replacement experiments show that exposure to the comb plays a critical role in the system. Supplementation procedures implicate a variety of compounds as recognition signals, but indicate that the compounds used are a small subset of the available compounds.

We can look forward to substantial advances in recognition chemistry as more social insect systems are studied using combinations of the three approaches discussed in this chapter. Ultimately these studies will tell us much about the evolutionary roots of recognition. Additionally, we will be able to address the fascinating issue of how the recognition signatures of individuals are processed in the sensory system and how the information is used to shape behavior. The answers to these questions have been elusive, because of a lack of specificity of knowledge concerning the cues, and this deficiency may soon be eliminated.

Acknowledgment

This research was supported by a grant from the Animal Behavior program of the National Science Foundation, IBN 9408180.

Literature Cited

Arnold, G., Quenet, B. and Cornuet, J.-M., De Schepper, B, Estoup, A., Gasqui, P. 1996. Kin recognition in honeybees. Nature 379:498.

Bagnères, A,-G., Errard, C., Mulheim, C., Joulie, C., and Lange, C. 1991a. Induced mimicry of colony odors in ants. J. Chem. Ecol. 17:1641-1664.

Bagnères, A,-G., Killian, A., Clément, J.-L. and Lange, C. 1991b. Interspecific recognition among termites of the genus *Reticulitermes*: evidence for a role for the cuticular hydrcarbons. J. Chem. Ecol. 17:2397-2420.

Bagnères, A.-G., Clément, J.-L., Blum, M.S., Severson, R. F., Joulie, C. and Lange, C. 1990. Cuticular hydrocarbons and defensive compounds of *Reticulitermes flavipes* and *R. santonensis*: polymorphism and chemotaxomomy. J. chem. Ecol. 16:3213-3244.

Bagnères, A.-G., Lange, C., Clément, J.-L., and Joulie, C. 1988. Les hydrocarbures cuticulaires des *Reticulitermes* français: variations spécifiques et coloniales. Actes. Coll. Insect Soc. 4:34-42.

Bagnères, Anne-Genevieve, Killian, A., Clément, J.-L., and Lange, C. 1991b. Interspecific recognition among termites of the genus *Reticulitermes*. Evidence for a role for the cuticular hydrocarbons. J. chem. Ecol. 17:2397-2420.

Bagneres, A.-G., Lorenzi, M. C., Dusticier G., Turillazzi S. and Clement, J.-L. 1996. Chemical usurpation of a nest by paper wasp parasites. Science 272:889-892.

Blaustein, A. R. and Waldman, B. 1992. Kin recognition in anuran amphibians. Anim Behav. 44:207-221.

Blomquist, G. J., A. J. Chu, and S. Ramaley. 1980. Biosynthesis of wax in the honey bee, *Apis mellifera* L. Insect Biochem. 10:313-321.

Blomquist, G.J., Nelson, D.R., de Renobales, M. 1987. Chemistry, biochemistry, and physiology of insect cuticular lipids. Arch. Insect Biochem. Physiol. 6:227-265.

Blum, M.S. and Fales, H.M. 1988. Eclectic chemisociality of the honeybee: A wealth of behaviors, pheromones, and exocrine glands. J. Chem. Ecol. 14:2099-2107.

Blum, M.S., Fales, H.M., Jones, T.H., Rinderer, T.E. and Tucker, K.W. 1983. Caste-specific esters derived from the queen honey bee sting apparatus. Comp. Biochem. Physiol. 75B:237-238.

Bonavita-Cougourdan, A., Clément, J.-L., and Lange, C. 1987. Nestmate recognition: the role of cuticular hydrocarbons in the ant *Camponotus vagus*. J. ent. Sci. 22:1-10.

Bonavita-Cougourdan, A., Clément, J.-L., and Lange, C. 1989. The role of cuticular hydrocarbons in recognition of larvae by workers of the ant *Camponotus vagus*: changes in the chemical signature in response to social environment. Sociobiology 16:49-74.

Bonavita-Cougourdan, A., Clément, J.-L., and Povéda, A. 1990. Les hydrocarbures cuticulaires et les processus de reconnaissance chez les fourmis: le code d'information complexe de *Camponontus vagus*. Actes coll. Insect Soc. 6:273-280.

Breed, M.D. 1981. Individual recognition and learning of queen odors by worker honeybees (*Apis mellifera*). Proc. Natl. Acad. Sci. U.S.A. 78:2635-2637.

Breed, M. D. 1983. Nestmate recognition in honey bees. Anim. Behav. 31:86-91.

Breed, M. D. 1987. Multiple inputs in the nestmate discrimination system of the honey bee. In: The Chemistry and Biology of Social Insects (Ed. by J. Eder and H. Rembold), pp. 461-462. Munich: Verlag J. Peperny.

Breed, M. D. and Bennett, B. 1987. Kin recognition in highly eusocial insects. In: Kin recognition in animals (Ed. by D. J. C. Fletcher and C. D. Michener), pp. 243-285. Chichester:John Wiley.

Breed, M. D. and Stiller, T. M. 1992. Honey bee, *Apis mellifera*, nestmate discrimination: hydrocarbon effects and the evolutionary implications of comb choice. Anim. Behav. 43:875-883.

Breed, M.D., Butler, L., Stiller, T.M. 1985. Kin recognition by worker honey bees in genetically mixed groups. Proc. Natl. Acad. Sci. U.S.A. 82:3058-3061.

Breed, M. D., Fewell, J. H. and Williams, K. R. 1988a. Comb wax mediates the acquisition of nest-mate recognition cues in honey bees. Proc. natl. Acad. Sci. U.S.A., 85:8766-8769.

Breed, M. D., Stiller, T. M., and Moor, M. J., 1988b. The ontogeny of kin discrimination cues in the honey bee, *Apis mellifera*. Behav. Genet. 18:439-448.

Breed, M. D., T. M. Stiller, M. S. Blum, and R. E. Page, Jr. 1992. Honey bee nestmate recognition: effects of queen fecal pheromones. J. Chem. Ecol. 18:1633-1640.

Breed, M. D., Garry, M. F., Pearce, A. N., Bjostad, L., Hibbard, B., Page, R. E. 1995a. The role of wax comb in honey bee nestmate recognition: Genetic effects on comb discrimination, acquisition of comb cues by bees, and passage of cues among individuals. Anim. Behav. 50:489-496.

Breed, M. D., Page, R. E. Jr., Bjostad, L., Hibbard, B. 1995b. Genetic components of variation in comb wax hydrocarbons produced by honey bees. J. Chem. Ecol. 21:1329-1338.

Breed, M. D., Leger, E. A., Pearce, A. N., Wang, Y. J. submitted. How comb wax affects honey bee nestmate recognition.

Brill, J. H. and Bertsch, W. 1990. Comparison of cuticular hydrocarbon profiles of fire ants *Solenopsis richteri* from the same colony, using capillary column gas chromatography with pattern recognition. J. Chromatography. 517:95-101.

Brill, J. H., Mar, T., Mayfield, H.T., and Bertsch,W. 1986. Application of computer-based pattern recognition procedures inthe study of biological samples. Comparison of cuticular hydrocarbon profiles of different colonies of the black imported fire ant. Analyst 111:711-716.

Butler, C. G. and J. B. Free. 1952. The behaviour of worker honeybees at the hive entrance. Behaviour 4:262-292.

Butts, D.P., Espelie, K. E., and Hermann, H. R. 1991. Cuticular hydrocarbons of four species of social wasps in the subfamily Vespinae: *Vespa crabro, Dolichovespula maculata*, *Vespula squamosa*, and *Vespula maculifrons*. Comp. Biochem. Physiol. 99B:87-91.

Butts, D. P., Camann, M. A. and Espelie, K. E. 1993. Discriminant analysis of cuticular hydrocarbons of the baldfaced hornet, *Dolichovespula maculata*. Sociobiology 21:193-201.

Butts, D. P. and Espelie, K. E. 1995. Role of nest-paper hydrocarbons in nestmate recognition of *Dolichovespula maculata* workers. Ethology 100:39-49.

Carlin, N. F. and Holldobler, B. 1986. The kin recognition system of carpenter ants (*Camponotus spp.*) I. Hierarchical cues in small colonies. Behav. Ecol. Sociobiol. 19:123-134.

Caldwell R. L. 1992. Recognition signalling and reduced aggression between former mates in a stomatopod. Anim Behav. 44:11-19.

Carlson, D. A. 1988. Africanized and European honey-bee drones and comb-waxes: Analysis of hydrocarbon components for identification. pp. 264-274, in G. R. Needham, R. E. Page, M. Delfinado-Baker, and C. E. Bowman, eds. Africanized honey bees and bee mites. Ellis Horwood Ltd. Chichester.

Carlson, D. A. and Bolton, A. B.. 1984. Identification of Africanized and European honey bees using extracted hydrocarbons. Bull. Entomol. Soc. Am. 30:32-35.

Clément, J-L., Bonavita-Cougourdan, A. and Lange, C. 1987. Nestmate recognition and cuticular hydrocarbons in *Camponotus vagus*. In: The Chemistry and Biology of Social Insects (Ed. by J. Eder and H. Rembold), pp. 473-474. Munich: Verlag J. Peperny.

Crewe, R. M. 1982. Compositional variability, the key to the social signals produced by honey bee mandibular glands. In: The Biology of Social Insects (Ed. by M. D. Breed, C. D. Michener and H. E. Evans), pp. 318-322. Boulder, Colorado:Westview Press.

Crosland, M. W. J. 1989. Kin recognition in the ant *Rhytiponera confusa*. II. Gestalt odour. Anim. Behav. 37:920-926.

Espelie, K. E. and Hermann, H. R. 1990. Surface lipids of the social wasp *Polistes annularis* and its nest and nest pedicel. J. Chem. Ecol. 16:1841-1852.

Espelie, K. E., J. W. Wenzel, and Chang, G. Surface lipids of social wasp *Polistes metricus* and its nest and nest pedicel in relation to nestmate recognition. J. chem Ecol. 16:2229-2241.

Espelie, K. E., Butz, V. M., and Dietz, A. 1990. Decyl-decanoate - A major component of the tergite glands of honeybee queens. J. Apic. Res. 29:15-19.

Espelie, K. E., Gamboa, G. J., Grudzien, T. A., Bura, E. A. 1994. Cuticular hydrocarbons of the paper wasp, *Polistes fuscatus*: a search for recognition pheromones. J. Chem. Ecol. 20:1677-1687.

Ferguson, I. D., Gamboa, G. J. and Jones, J. K. 1987. Discrimination between natal and non-natal nests by the social wasps *Dolichovespula maculata* and *Polistes fuscatus*. J. Kans. Entomol. Soc. 60:65-69.

Fletcher, D. J. C. and Michener, C. D. eds. 1987. Kin recognition in animals. Chichester: John Wiley and Sons.

Francis, B. R., Blanton, W. E., and Nunamaker, R. A. 1985. Extractable surface hydrocarbons of workers and drones of the genus *Apis*. J. Apic. Res. 24:13-26.

Francis, B. R., Blanton, W. E., Littlefield, J. L. and Nunamaker, R. A. 1989. Hydrocarbons of the cuticle and hemolymph of the adult honey bee. Ann. Entomol. Soc. Am. 82:486-494.

Franks, N. R., Blum, M., Smith, R.-K. and Allies, A. B. 1990. Behavior and chemical disguise of cuckoo ant *Leptothorax kutteri* in relation to its host *Leptothorax acervorum*.. J. Chem. Ecol. 16:1431-1444.

Frumhoff, P. C. 1991. The effects of the cordovan marker on apparent kin discrimination among nestmate honey bees. Anim. Behav. 42:854-856.

Gamboa, G. J., Foster, R. L. and Richards, K. W. 1987. Intraspecific nest and brood recognition by queens of the bumble bee, *Bombus occidentalis*. Can. J. Zool. 65: 2893-2897.

Gamboa, G. J., Reeve, H. K. and Pfennig, D. W. 1986. The evolution and ontogeny of nestmate recognition in social wasps. Ann. Rev. Entomol. 31: 431-454.

Gamboa, G. J., Reeve, H. K., Ferguson, I. and Wacker, T. L. 1986. Nestmate recognition in social wasps: the origin and acquisition of recognition odours. Anim. Behav. 34:685-695.

Gamboa, G. J., Grudzien, T. A., Espelie, K. A. and Bura, E. A. 1996. Kin recognition pheromones in social wasps: combining chemical and behavioural evidence. Anim. Behav. 51:625-629.

Getz, W. M. and Page, R. E. 1991. Chemosensory kin–communication systems and kin recognition in honey bees. Ethol. 87: 298–315.

Getz, W. M. and Smith, K. B. 1986. Honeybee kin recognition: learning self and nestmate phenotypes. Anim. Behav. 34:1617-1626.

Getz, W. M. and Smith, K. B. 1987. Olfactory sensitivity and discrimination of mixtures in the honey bee *Apis mellifera*. J. Comp Physiol. 160:239-245.

Getz, W. M., Brückner, D. and Smith, K. B. 1989. The ontogeny of cuticular chemosensory cues in worker honey bees *Apis mellifera*. Apidologie 20:105-113.

Getz, W. M. 1991. The honey bee as a model kin recognition system. In Hepper, P. G. Kin recognition. Cambridge University Press: Cambridge. pp. 358-412.

Hadley, N. F. 1989. Lipid water barriers in biological systems. Prog. Lipid. Res. 28:1-33.

Halpin, Z. T. 1991. Kin recognition cues of vertebrates. In Hepper, P. G., ed. Kin recognition. Cambridge University Press:Cambridge. pp. 220-258.

Haverty, M. I., Page, M., Nelson, L. J., and Blomquist, G. J. 1988. Cuticular hydrocarbons of dampwood termites, *Zootermopsis*: intra- and inter-colony variation and potential as taxonomic characters. J. Chem. Ecol. 14:1035-1058.

Haverty, M. I. and Thorne, B. L. 1989. Agonistic behavior correlated with hydrocarbon phenotypes in dampwood termites, *Zootermopsis*. J. Insect Behav. 2:523-543.

Hepburn, H. R. 1986. Honey Bees and Wax. Berlin: Springer-Verlag.

Hefetz, A., Bergström, G., and Tengö, J. 1986. Species, individual, and kin specific blends in Dufour's gland secretions of halictine bees--Chemical evidence. J. Chem. Ecol. 12:97-208.

Hepper, P. G. 1991. Kin recognition. Cambridge University Press: Cambridge.

Hölldobler, B. and Michener, C. D. 1980. Mechanisms of identification and discrimination in social Hymenoptera. In: Evolution of Social Behavior: Hypotheses and Empirical Tests. (Ed. by H. Markl), pp. 35-58. Weinheim: Verlag Chemie.

Hölldobler, B. and Wilson, E. O. 1990. The ants. The Belknap Press of Harvard University Press, Cambridge Massachusetts.

Howard, R. W., McDaniel, C. A., and Bloquist, G. J. 1980. Chemical mimicry as an integrating mechanism: cuticular hydrocarbons of a termitophile and its host. Science 210:431-433.

Howard, R. W., McDaniel, C. A., and Blomquist, G. J. 1982. Chemical mimicry as in integrating mechanism for three termitophiles associated with *Reticulitermes virginicus*. Psyche 89:157-167.

Howard, R. W., McDaniel, C. A., Nelson, D. R., Blomquist, G. J., Gelbaum, L. T., Zalkow, L. H. 1982. Cuticular hydrocarbons of *Reticulitermes virginicus* and their role as potential species- and caste-recognition cues. J. chem. Ecol. 8:1227-1239.

Howard, R. W., Stanley-Samuelson, D. W., and Akre, R. D. 1990. Biosyntheis and chemical mimicry of cuticular hydrocarbons from the obligate predator, *Microdon albicomatus* and its ant prey, *Myrmica incompleta*. J. Kans. Entomol. Soc. 63:437-443.

Howard, R. W., Akre, R. D., and Garnett, W. B. 1990. Chemical mimicry in an obligate predator of carpenter ants. Ann. Entomol. Soc. Am. 83:607-616.

Howard, R. W. 1993. Cuticular hydrocarbons and communication. In Insect lipids: Chemistry, Biochemistry, and Biology. Stanley-Samuelson, D. W. and Nelson , D. R., eds. Univ. Nebraska Press, Lincoln pp. 179-226.

Jaisson, P. 1991. Kinship and fellowship in ants and social wasps. In Hepper, P. G. Kin recognition. Cambridge University Press: Cambridge. pp. 60-93.

Kalmus, H. and Ribbands, C. R. 1952. The origin of odors by which honey bees distinguish their companions. Proc. Royal Soc. (B) 140:50-59.

Layton, J. M. Camann, M. A., and Espelie, K. E. 1994. Cuticular lipid profiles of queens, workers, and males of the social wasp, Polistes metricus are colony-specific. J. Chem. Ecol. 20:2307-2321.

Layton, J. M. and Espelie, K. E. 1995. Effects of nest paper hydrocarbons on nest and nestmate recognition in colonies of *Polistes metricus*. J. Insect Behav. 8:103-113.

Lenoir, A. Clément, J.-L., Nowbahari, M. and C. Lange. 1987. Les hydrocarbures cuticulaires de la fourmi *Cataglyphis cursor*: variations géographigues et rôle dan la reconaissance coloniale. Actes Coll. Ins. Sc. 4:71-77.

Lockey, K. H. 1988. Lipids of the insect cuticle. Comp. Biochem. Physiol. 89B:595-645.

Lorenzi, M. C. 1992. Epicuticular hydrocarbons of *Polistes biglumis bimaculatus*: preliminary results. Evol. Ecol. Ethol. 2:61-63.

Mar, T., Brill, J., Bertsch, W, Fletcher, D. J. C., Crewe, R. 1987. Investigation of cuticular hydrocarbons with selected honey bee populations by gas chromatography with pattern recognition. J. chromatography. 399:277-290.

Markow, T. A. and E. C. Toolson 1990. Temperature effects on epicuticualr hydrocarbons an sexual isolation in *Drosophila mojavensis*. in Ecological and evolutionary genetics of *Drosophila*, J. S. F. Barker, W. T. Starmer, and R. J. MacIntyre, eds. Plenum: New York. pp.315-331.

McDonald, C. A. 1990. Cuticular hydrocarbons of the Formosan termite *Coptotermes formosanus*. Sociobiology 16:265-273.

Michener, C. D. and Smith, B. H. 1987. Kin recognition in primitively social insects. In: Kin recognition in animals (Ed. by D. J. C. Fletcher and C. D. Michener), pp. 209-242. Chichester:John Wiley.

Mintzer, A.C., Williams, H. J., and Vinson, S. B. 1987. Identity and variation of hexane soluble cuticular components produced by the acacia ant *Pseudomyrmex ferruginea*. Comp. Biochem. Physiol. 86B:27-30.

Morel, L. and Vander Meer, R. K. 1987. Nestmate recognition in *Camponotus floridianus*: Behavioral and chemical evidence for the role of age and social experience. In Chemistry and Biology of Social Insects (Ed. by J. Eder and H. Rembold), pp. 471-472. Munich: Verlag J. Peperny.

Morel, L., Vander Meer, R. K. and Lavine, B. K. 1988. Ontogeny of nestmate recognition cues in the red carpenter ant (*Camponotus floridanus*): behavioral and chemical evidence for the role of age and social experience. Behav. Ecol. Sociobiol. 22:175-183.

Moritz, R. F. A. and Crewe, R. M. 1988. Chemical signals of queens in kin recognition of honeybees, *Apis mellifera* L. J. Comp. Physiol. A 164:83-89.

Moritz, R. F.A. and Southwick, E. E. 1987. Metabolic test of volativel odor labels as kin recognition cues in honey bees (*Apis mellifera*). J. exp. Zool. 243:503-507.

Nowbahari, E., Lenoir, A., Clément, J.-L., Lange, C., Bagnères, A.-G. and Joulie, C. 1990. Individual, geographical and experimental variation of cuticular hydrocarbons of the ant *Cataglyphis cursor*: their use in nest and subspecies recognition. Biochem. syst. Ecol. 18:63-73.

Obin, M. S. 1986. Nestmate recognition cues in laboratory and field colonies of *Solenopsis invicta* Buren: Effect of environment and role of cuticular hydrocarbons. J. Chem. Ecol. 12:1965-1975.

Obin, M. S. and VanderMeer, R. K. 1988. Sources of nestmate recognition cues in the imported fire ant *Solenopsis invicta* Buren. Anim. Behav. 36:1361-1371.

Oldroyd, B. P., T. E. Rinderer, and S. M. Buco. 1991. Honey bees dance with their super-sisters. Anim. Behav. 42:121-129.

Page, R. E. and Erickson, E.. 1986. Kin recognition and virgin queen acceptance by worker honey bees (*Apis mellifera* L.). Anim. Behav. 34:1061-1069.

Page, R. E., R. A. Metcalf, R. L. Metcalf, E. H. Erickson, and R. L. Lampman. Extractable hydrocarbons and kin recognition in honeybee (*Apis mellifera* L.). J. Chem. Ecol. 17:745-756.

Page, R. E., Robinson, G. E. and Fondrk, K. 1989. Genetic specialists, kin recognition, and nepotism. Nature 338:576-579.

Pfennig, D. W., Gamboa, G. J., Reeve, H. K., Shellman-Reeve, J. and Ferguson, I. D. 1983. The mechanism of nestmate discrimination in social wasps (*Polistes*, Hymenoptera, Vespidae). Behav. Ecol. Sociobiol. 13:299-305.

Porter, R. H. 1991. Mutual mother-infant recognition in humans. In Hepper, P. G., ed. Kin recognition. Cambridge University Press:Cambridge. pp. 413-432.

Provost, E., Riviere, G., Roux, M., Morgan, E. D., Bagnères, A.-G. 1993. Change in the chemical signature of the ant *Leptothorax lichtensteini* with time. Insect Biochem. Molec. Biol. 8:945-957.

Provost, E., Riviere, G., Roux, M., A.-G. Bagnères, Clément, J.-L. 1994. Cuticular hydrocarbons whereby *Messor barbarus* ant workers putatively discriminate between monogynous and polygynous colonies. Are workers labeled by queens? J. Chem. Ecol. 20:2985-3003.

Rinkevich, B., Frank, U., Bak, R. P. M., and Müller, W. E. G. 1994. Alloimmune responses between Acropora hemprichi conspecifics: nontransitive patterns of overgrowth and delayed cytocytoxicity. Marine Biology 118:731-737.

Ross, K. G. , Vander Meer, R. K., Fletcher, D.J.C., and Vargo, E. L. 1987. Biochemical, phenotypic and genetic studies of two introduced fire ants and their hybrid. Evolution 41:280-293.

Singer, T. L. and Espelie, K. E. 1992. Social wasps use nest paper hydrocarbons for nestmate recognition. Anim. Behav. 44:63-68.

Singer, T. L., Camann, M. A., and Espelie, K. E. 1992. Discriminant analysis of cuticular hydrocarbons of social wasps *Polistes exclamans* and surface hydrocarbons of its nest and pedicel. J. chem. Ecol. 18:785-797.

Smith, B. H. 1983. Recognition of female kin by male bees through olfactory signals. Proc. natl. Acad. Sci. U.S.A. 80:4551-4553.

Smith, B. H. and Wenzel, J. W. 1988. Pheromonal covariation and kinship in social bee *Lasioglossum zephyrum*. J. Chem. Ecol. 14:87-94.

Smith, B. H. and Breed, M. D. 1995. The chemical basis for nestmate recognition and mate discrimination in social insects. in Chemical Ecology of Insects II, R. T. Carde and W. J. Bell, eds. Chapman and Hall: New York. pp. 287-317.

Smith, R.-K. 1990. Chemotaxonomy of honey bees. Part I. European and African workers. Bee Science 1:23-32.

Tulloch, A. P. 1980. Beeswax--composition and analysis. Bee World 61:47-62.

Vander Meer, R. K. , Saliwanchik, D. and Lavine, B. 1989. Temporal changes in colony cuticular hydrocarbon patterns of *Solenopsis invicta*. Implications for nestmate recognition. J. chem. Ecol. 15:2115-2125.

Vander Meer, R. K. and Wojcik, D. P 1982. Chemical mimicry in the myrmecophilous beetle *Myrmecaphodius excavaticollis*. Science 218:806-808.

Vander Meer, R. K. Jouvenaz, D. P. and Wojcik, D. P. 1989. Chemical mimicry in a parasitoid of fire ants. J. chem Ecol. 15:2247-2261.

Vander Meer, R.K. 1988. Behavioral and biochemical variation in the fire ant, *Solenopsis invicta*. In Interindividual Behavioral Variation in Social Insects, R. L. Jeanne, ed. Westview Press: Boulder. pp. 223-255.

Yamazaki, K. Beauchamp, G. K., Shen, F.-W., Bard J. and Boyse, E. A. 1994. Discrimination of odortypes determined by the major histocompatibility complex among outbred mice. Proc. Natl. Acad. Sci. (USA). 91:3735-3738.

4

Nestmate Recognition in Ants

Robert K. Vander Meer and Laurence Morel

Kin/Nestmate Recognition

All ants are highly eusocial, which means 1) individuals care for the young, 2) there are castes (reproductive division of labor), and 3) there is an overlap of at least two generations in which workers (normally sterile) assist their mother in rearing sisters and brothers. This altruistic (or nepotistic) behavior is thought to characterize kin selection (Hamilton 1964), defined by Wilson (1987) as "Differential survival or reproductivity that changes the proportion of genes through time due to the circumstance that individuals favor or disfavor relatives other than direct offspring." To achieve kin selection there must be kin recognition. Herein lies much of the driving force for the large volume of research over the past decade concerned with kin and nestmate recognition (e.g. Fletcher and Michener 1987). There are more pragmatic reasons to understand nestmate recognition.

Ants have developed a formidable array of active and/or passive semiochemical and non-chemical defenses (Hermann and Blum 1981). The many active defensive behaviors are initiated when an intruder is recognized as non-nestmate by resident workers. Thus, nestmate recognition represents the first line of defense for a colony. The chemistry and associated behaviors of kin/nestmate recognition are intrinsically interesting. Beyond that, knowledge of nestmate recognition is essential for a comprehensive understanding of both ant defenses and the organisms (symphiles) that have broken the recognition code and are able to infiltrate ant colonies and exploit colony resources (Kistner 1979). These areas of research may provide the basis of innovative control strategies for pest ant species.

What Are the Possible Recognition Scenarios?

The simplest ant colony situation is one where there is a single queen (monogyne), inseminated by a single male (monoandrous), the colony resides in a single nest (monodomous), and workers from each colony defend a territory (intolerant of workers from adjacent colonies). This scenario is found in nature, but so are polyandry (insemination by more then one male), polygyny (more then one queen), and polydomy (more then one nest). The latter conditions are common and are often accompanied by a lack of or diminished territoriality. This has profound effects on recognition possibilities.

Individuals within a colony may recognize each of the other individuals in their colony (**individual recognition**; Figure 4.1, Situation #1). This appears to be unlikely except where a colony is composed of only a few individuals. In small and large colonies **kin**

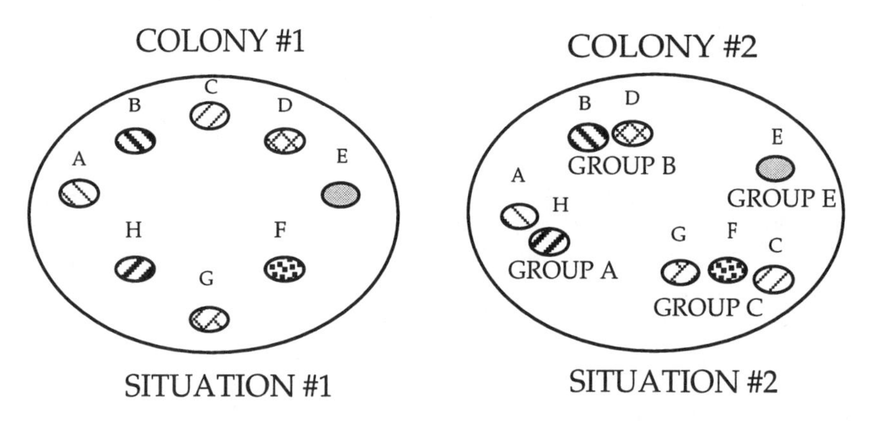

FIGURE 4.1. The diagrammatic representation of worker - worker recognition possibilities within a colony. Workers are represented by ovals filled with different patterns if recognized as different or the same pattern if not recognized as different. Situation #1: The colony is composed of eight workers, each of which recognizes all of the others as different, yet they also recognize them as nestmates. In large colonies this situation becomes difficult. Situation #2: Segregation of workers may occur based on worker recognition of a common mother or father in species that exhibit colony polygyny or polyandry. It is possible for members of each polygynous or polyandrous group to recognize each other as different, members of the other groups as more different, yet all members of the colony as nestmates. Since recognition is made up of chemical cues and neural templates, these diagrams can also be used to visualize the cue and template possibilities within colonies.

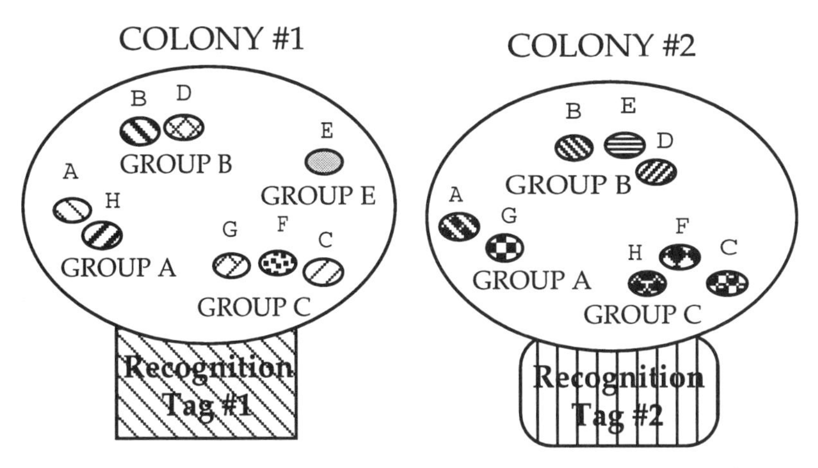

FIGURE 4.2. Diagrammatic representation of nestmate recognition for two polygynous or polyandrous colonies whose workers exhibit class recognition (Groups). The behaviors associated with class recognition are thought to be subtle in nature. Superimposed on class recognition is nestmate recognition - each colony member recognizes other colony members as nestmates. The commonality is represented by the colony TAG. Members of other colonies have different colony TAGs and are recognized as non-nestmate. Nestmate recognition is characterized by aggressive interactions.

recognition or **class (group) recognition** may occur. This category can be broken down into several subsections. Individual groups within a colony may be segregated to some degree based on recognition of a common maternal (**matriline recognition**) or a common paternal (**patriline recognition**) phenotype or both if polyandry and polygyny occur (Figure 4.1. Situation #2). At the next level is recognition of belonging to an individual nest (**nestmate recognition**; Figure 4.2). As indicated in Figure 4.2, it is possible to have intra-nest class recognition occur simultaneously with nestmate recognition. Class recognition may manifest itself in subtle interactions, such as preferential feeding of a worker's closest kin; whereas aggressive behaviors are elicited from resident workers when con-specific (in territorial species) or hetero-specific intruders are recognized. In theory, nestmate recognition also can involve subtle interactions, but these have not been observed. Nestmate recognition consolidates the colony against con- or hetero-specific ant competitors; whereas, kin recognition represents the optimization of gene flow of their nearest relative, the end result is presumably, **kin selection.**

Nestmate recognition is functionally equivalent to kin recognition in the special case where the species is monogynous and monoandrous, exemplified by the fire ant, *Solenopsis invicta*. Nestmate recognition may also promote fitness through kin selection if queens of a polygyne colony are more related to each other then are queens from other colonies. In all other situations of polygyny and/or polyandry, kin recognition may occur within the nest and inter-nest interactions are considered separately as nestmate recognition.

Most of the above discussion of kin recognition has been dogma for the last two decades with researchers working toward validation of the hypotheses. Recently, some interesting alternative views to the kin recognition story have surfaced in the literature. Carlin (1989), cautions about ascribing adaptive significance to results obtained under artificial laboratory conditions. He suggests that inter-colony discrimination may be based on any random odor idiosyncrasies and thus is not under selection. This is especially relevant for species demonstrated to use environmental cues in recognition. Also, within colony discrimination (kin recognition) may be a nonadaptive side-effect of clearly adaptive between colony discrimination (nestmate recognition). Grafen (1990) expands on this theme, stating that purported demonstrations of kin recognition are too weak to be effective and are again better explained simply as a side-effect of species recognition systems. We will proceed with these intriguing hypotheses in mind along with the realization that virtually all recognition studies on ants involve nestmate recognition rather than kin recognition.

The Recognition Process: Cues, Template, and Response

For recognition to occur each individual must have a label or set of cues (conceptually analogous to a bar code), a mechanism for detecting and interpreting the label of another individual (bar-code reader and central processing unit -- CPU), and a learned template of cues stored in memory (computer storage device that, as we shall see later, must have, read/write capabilities) with which to compare the incoming signal. The recognition process is completed with an appropriate behavioral response, which need not be overt.

The Cues: Where Do They Come From?

It is generally accepted that nestmate recognition cues are chemical in nature and are detected by a mere sweep of ant A's antennae across the cuticle of ant B (Wilson 1971). Recognition, especially inter-specific, may be elicited over a short distance (1-2cm) without antennal contact. As we shall see in detail later, nothing is known definitively about the chemical nature of nestmate recognition cues, in spite of the many publications implicating cuticular hydrocarbons. Thus, there are no chemical class / nestmate recognition associations to examine and the possibilities are open.

We define colony odor as all odors associated with a particular colony. Nestmate recognition cues are a subset of colony odor (Vander Meer 1988). The all-encompassing colony odor is composed of two main categories, odors derived from the environment and those from physiological sources. Environmental odors may be from the air, soil, food, or any other source not produced or modified by members of the colony. Physiological sources are derived from cuticular lipids, exocrine gland, excretory and regurgatory products, as well as other odors modified by the individual or group and released to the colony. Literature examples of nestmate recognition illustrate the importance of environmental cues, e.g. *Acromyrmex octospinosus* (Jutsum et al. 1979); physiological cues, e.g. *Psuedomyrmex feruginea* (Mintzer and Vinson 1985); and a combination of environmental and physiological cues, e.g. *Solenopsis invicta* (Obin and Vander Meer 1988). Thus, all combinations of the two broad categories are possible.

Cue Detection

The ant's antennae are analogous to our nose -- detecting volatile compounds in the air, but perhaps also capable of detecting "non-volatile" compounds by contact. The signal received is transduced and processed by the central nervous system, which then may trigger a behavioral response. The situation is similar to our personal experiences, where our nose detects many different odors but not all elicit a behavioral response. And similarly for insects, electroantennagram studies reveal that an insect's antennae may detect a compound, but observable behaviors may not be triggered. Antennectomized *Formica lugubris* workers were unable to recognize nestmates (LeMoli et al., 1983). After detection of an intruder's cue profile (a sub-set of the overall colony odor), the profile must be compared with the resident worker's neural **template** or memory pattern of its colony's recognition cues.

The Template

The hypothesis that the template associated with heritable cues is genetically determined and directly linked to the cues has not been supported experimentally. Thus, the following discussion only considers the development of a learned template. In all ant species nestmate recognition discriminators and the neural template are derived to some extent from all possible sources -- the environment, the individual, a class of worker, or queen from within the colony, or collectively from all individuals in a colony. The relative importance of each discriminator type may vary from species to species. Hölldobler and Carlin (1986) proposed just such a hierarchy for *Camponotus* species, involving queen, worker and environment derived discriminators.

Depending on the polyandrous and/or polygynous status of a colony, there may be many possible discriminator groups (also known as classes or referents; see Figure 4.1, Situation #2). It is easy to visualize a worker having to learn **multiple templates** (Bennett and Breed 1987), perhaps thousands, especially in polyandrous, and/or polygynous colonies, in order to assess the relatedness of nestmates. This presents an extremely complicated learning situation for individual workers. To simplify the task for both the ants and their human investigators the concept of **multiple mean templates** was proposed so that, for example, a worker from a polyandrous, polygynous colony could theoretically have mean templates for each queen and her offspring regardless of worker patrilineage. This is still a difficult experimental model and has been expanded further to a colony having a **single mean template**. The latter approach is experimentally indistinguishable from the "gestalt" model of Crozier and Dix (1979), which assumes that colony odors (nestmate recognition cues) are exchanged among colony workers through normal social interactions, such that all workers possess the same odor profile. Thus, genetically varied nestmates are anonymous to each other (Hölldobler and Carlin 1987) within the colony, but not between colonies. This relates also to the "fellowship" concept of Jaisson (1987). The "gestalt" model has been demonstrated in several ant species (*Pristomyrmex pungens*, Tsuji, 1990; *Camponotus* species, Carlin and Hölldobler, 1987). In reductionist terms, multiple templates, multiple mean templates and single mean template become an **individualistic model, multiple "gestalt" template model** and **colony "gestalt" model**, respectively. All of the above may occur; however, from an experimental point of view, devising an unambiguous experiment that measures subtle discrimination of non-kin via multiple templates or multiple mean templates has not yet been accomplished. So, we are left with experiments centered around the concept of a single mean template or colony "gestalt", which measures nestmate recognition rather than kin recognition, except the previously mentioned special case of monogynous, monoandrous species where kin and nestmate recognition are synonymous in that they both promote kin selection.

Are Cues and Templates Static or Dynamic?

We know that recognition cues can be derived from environmental and/or heritable sources. Environmental cues dynamically change over time, as food sources vary and the surrounding habitat changes with the seasons. Therefore, the part of the neural template based on environmental cues must also be dynamically changing. What about heritable recognition cues? One's first reaction is that heritable cues are genetically controlled, therefore, they should be static. It is well known that component ratios of lepidopteran sex pheromones are necessarily static, in order to maintain species separation. On the other hand, ant alarm pheromone components from *Crematogaster* species have been reported to vary considerably from individual to individual within the same nest (Brand, and Pretorius 1986), and head derived pheromones of *Tetramorium caespitum* vary with caste (Pasteels, et al. 1980), as do the mandibular gland products of *Atta sexdens rubropilosa* (Do Nascimentoi, et al. 1993). Pheromone variation within a colony led the former authors to suggest that alarm pheromones would not be useful nestmate recognition cues. What about the chemicals associated with an ant's cuticle? After all, recognition occurs with a simple sweep of the antennae across the cuticle of the intruder (Wilson 1971). Where quantitation has been reported, hydrocarbons are the major class of chemicals found on ant cuticle (Lok et al. 1975), but as should be expected, other lipid classes are also present. Cuticular hydrocarbons are under genetic control, as evidenced by high levels of concordance between the identification of *S. invicta* and *S. richteri* and hybrids based on hydrocarbon patterns and isozyme analyses (Vander Meer et al. 1985; Ross et al. 1987). Cuticular hydrocarbons are readily analyzed by gas chromatograph and have been used as a tool to study variability in the heritable component of colony odor. Vander Meer et al. (1989) used pattern recognition analyses to determine that at a given time, within colony cuticular hydrocarbon variation was less than colony to colony variation and nine *S. invicta* colonies sampled were readily distinguished. In addition, and perhaps most importantly, they found that colony hydrocarbon patterns varied over time. This phenomenon has been confirmed in *Leptothorax* species (Provost, et al. 1993). If the premise that cuticular hydrocarbons are representative of the heritable component of colony odor, and thus are heritable recognition cues, is correct, then heritable cues are not static but dynamically changing with time along with cues derived from the environment. This has profound effects on how we view the other half of the recognition process - the template. The neural imprint or template of colony recognition cues must also be dynamic rather than static and track changes in the cue profile. Therefore, a colony's nestmate recognition cue profile cannot be learned by newly eclosed workers (callows) as a fixed pattern but must be continuously updated through a process of iterative learning (Vander Meer 1988). Wallis (1963) proposed a similar scheme based on observations of *Formica* species. He suggested that each worker was probably "continually habituating to slight variations in the odor of its nestmates".

A generalized cue/template model is shown in top half of Figure 4.3 for territorial (monogyne) species -- they recognize con-specific workers from other colonies as different. Environmental cues are expected to vary quantitatively and qualitatively, whereas heritable cues will vary only in the relative intensity of the compounds involved (quantitative change only). Each colony in Figure 4.3 is designated by a letter. If colonies A to D were analyzed for recognition cues at some point in time they would each have a distinct profile derived from environmental and/or heritable sources, as shown in Figure 4.3. The template for each colony reflects the cues of that colony at that point in time. If these same colonies were sampled at another time, they would again have distinct cue profiles, but different from the previous sampling. For the system to work the template must reflect the cue changes. Hetero-colonial intruders have cues that do not match the template of the resident colony at any given time, resulting in aggressive behaviors.

Polygyne species often do not defend a territory and do not recognize members of other nests as different. There is free flow of environmentally derived odors from a larger area than expected for monogyne colonies and with free flow of workers from many matrilines and patrilines, workers experience a broad range of heritable cues. However, at any given point in time each polygyne worker has a distinct cue profile just as in the monogyne situation. The difference between monogyne and polygyne nestmate recognition lies in the template (Figure 4.3, bottom half), which is broader and less distinct than their monogyne counterpart. The fire ant, *S. invicta*, provides an excellent example of both extremes in a single species.

How do *S. invicta* polygyne and monogyne aggression bioassay data (Morel et al. 1990) conform to the model presented above? Polygyne residents do not recognize monogyne or polygyne intruders as different; however, monogyne residents are very aggressive toward both types of intruders (Figure 4.4). Thus, the cues of polygyne and monogyne intruders fit within the polygyne template. The response of polygyne or monogyne intruders introduced into monogyne colonies is small because the intruders are attacked so quickly they have no chance to respond

MONOGYNE NESTMATE RECOGNITION

CUES

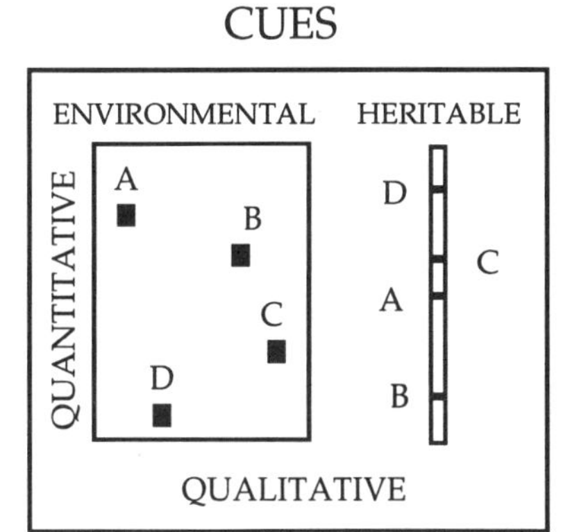

TEMPLATE

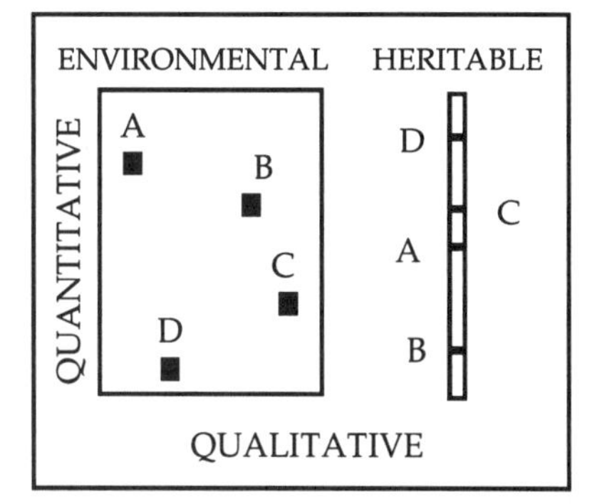

POLYGYNE NESTMATE RECOGNITION

CUES

ENVIRONMENTAL HERITABLE

QUANTITATIVE

X Y W Z

W X Z Y

QUALITATIVE

TEMPLATE

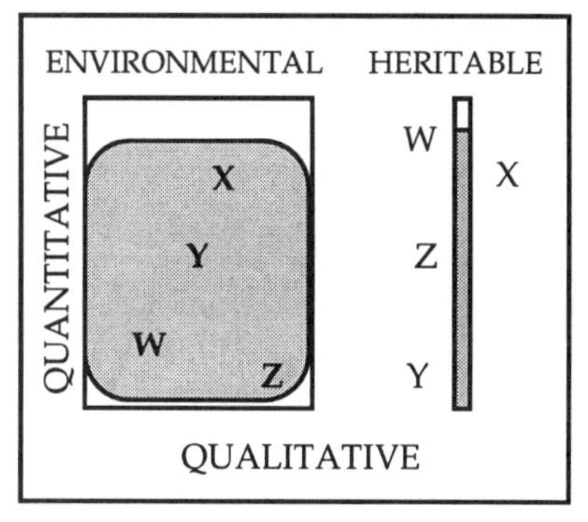

FIGURE 4.3. Diagrammatic comparison of the cue/template nestmate recognition system for monogyne and polygyne intraspecific populations. The upper cue/template pair are the same as seen in Figure 5. and represent the monogyne situation. The bottom cue/template pair represents the polygyne situation. For both types the cues part of the process is identical -- at any point in time each worker from each colony has a distinct cue profile. Differences in the two population types lie in the template. Multiple matrilines, patrilines, and lack of territoriality lead to a broader template that accepts intruders with a wider variety of cue profiles. Thus, polygyne and monogyne intruders are accepted into polygyne colonies, but polygyne intruders are not accepted into monogyne colonies.

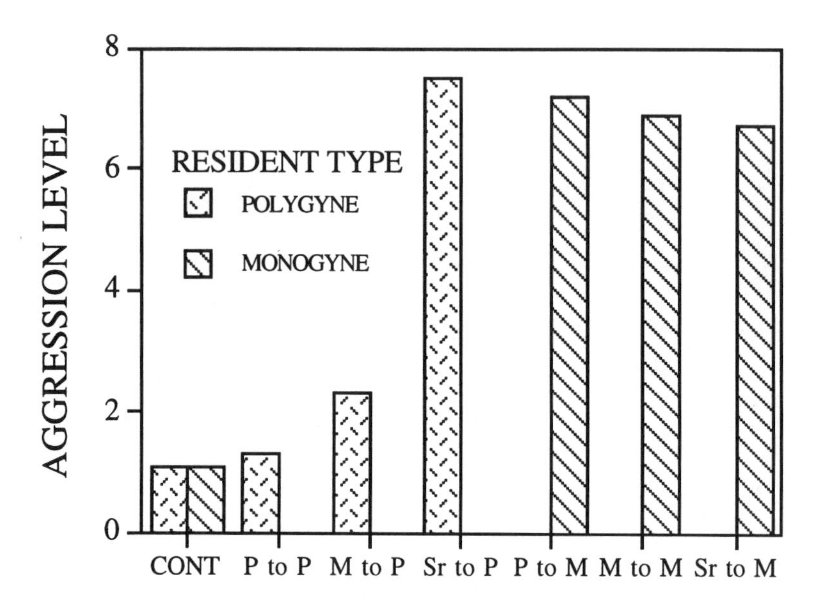

FIGURE 4.4. Aggressive responses of polygyne and monogyne *Solenopsis invicta* residents to intruders. M = monogyne; P = polygyne; Sr = *Solenopsis richteri;* cont = control; M>P = monogyne worker introduced into a polygyne colony. Level 1-3 = investigative; 4-6 = challenge; 7-9 = attack.

(Figure 4.5). For polygyne workers introduced into polygyne colonies there is acceptance in both directions and no aggression is elicited. However, monogyne intruders are not attacked by polygyne residents and have a chance to respond in encounters with residents whose cues do not match their templates. Heterospecific workers are readily recognized by both polygyne and monogyne *S. invicta* residents because the heritable component of their cues is qualitatively distinct. The model represented in Figure 4.3 accommodates the bioassay data. Further, queenright polygyne *S. Invicta* colonies brought into the laboratory for > three months, thus isolated from the polygyne population, developed acute intraspecific discrimination capabilities (Obin et al. 1993). Having isolated multiple queen colonies does not provide the necessary breadth of cue sources. Does this fit with the ontogeny of nestmate recognition?

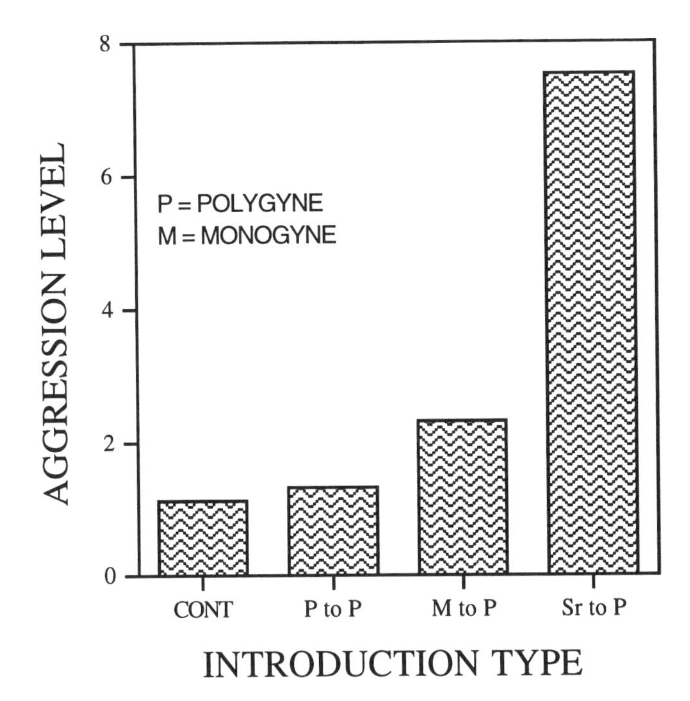

FIGURE 4.5 Aggressive responses of polygyne and monogyne *Solenopsis invicta* intruders to residents. M = monogyne; P = polygyne; cont = control; M to P = monogyne worker introduced into a polygyne colony; Sr = *S. Richteri*. Level 1-3 = investigative; 4-6 = challenge; 7-9 = attack.

The Ontogeny of Nestmate Recognition

Newly eclosed ant workers are called callows. Jaisson (1985) states the callow worker's situation succinctly: "...the newly emerged individual appears in an environment which is by definition hostile to anything alien, and must itself recognize its own colony as being an exception to this general hostility rule". Jaisson (1985) lists five callow worker attributes that help to explain the integration of young workers into the group: (1) The absence of aggressive behavior; (2) The presence of compounds that inhibit adult aggression; (3) Attractiveness for adults; (4) The potential for absorbing the colony odor; (5) Reduced mobility.

Colonies composed of two different species can be formed through the successful adoption of foreign brood or callow workers (Fielde 1903; Plateaux 1960 and Jaisson 1971). Thus, the callow worker characteristics listed above that account for colony integration, are non-specific, at least to a certain degree. The absence of aggressive behaviors and reduced mobility lend themselves to lessened aggression by mature resident workers, since it does take two to tangle. This allows the potential adoptee time in which to acquire the colony odor (nestmate recognition cues) of its new colony. Mechanisms that allow intruders (parasites, predators, etc.) to survive long enough to acquire colony odor will surface again in discussions of myrmecophiles and brood pheromones. Hölldobler and Michener (1980) suggest that successful callow worker adoptions can be explained by pupal pheromones contained in part in exuvial fluids. Jaisson (1985) more specifically believes the substances involved are only produced at the time of worker eclosion, because assistance from nurse workers is often necessary for successful eclosion. LaMon and Topoff (1985) give the following account of fire ant eclosion. At the initiation of eclosion, gross pupal movements occur, which loosen the pupal cuticle, often tearing at the joints. "At the same time the surface of the cuticle becomes covered with a sticky, viscous exudate. When adult workers are present they remove the eclosing pupae from the brood pile and avidly strip away the pupal skin..." Workers often consume the shreds of cuticle they tear off the eclosing pupae. This description of the social facilitation of eclosion in fire ants supports the contention of Jaisson (1985). Although potent phagostimulants appear to be present, it has not been demonstrated and it may not be necessary that attractants are involved. Vigorous movement in previously quiescent pupae may be adequate to get a brood tending worker's attention (LeMasne 1953). The intense interactions of adult workers with eclosing pupae and the resulting callow should accelerate the absorption of the colony's odor (Jaisson 1985). We think this along with reduced callow worker mobility and aggressive behaviors, but without invoking the presence of attractive compounds (pheromones), is enough to allow integration of the callow worker into its colony or adopted colony.

The net result of the above is that early social experience plays a role in colony integration of callow workers (Le Moli and Mori 1984, 1985; Morel and Blum 1988; Errard 1984, 1986) and in the acquisition of the colony label (Morel et al. 1988). In addition, a sensitive period can be defined, after which workers can no longer be successfully transferred to another colony. The mature worker's fully developed exocrine gland system and repertoire of aggressive behaviors precludes it from surviving long enough in another colony to acquire that colony's distinct odor. How does the above fit our scheme as presented in Figure 4.3? A newly eclosed callow worker has a distinct cue profile that differs from adult workers from the same colony, consequently, it should be attacked. Indeed, low levels of resident aggression toward callows are reported (Morel et al. 1988); however, through lack of reciprocal behaviors and immobility, callows gain time to acquire the constantly changing colony odor. Concomitant with the acquisition of colony odors, its repertoire of behaviors and mobility increase, such that successful adoption of this worker into another colony becomes increasingly difficult because it is losing the capacity to survive long enough to acquire a new colony's odor. Thus, the ontogeny of nestmate recognition fits very well into the general model shown in Figure 4.3, and further provides important insights into mechanisms of parasite and predator integration into ant colonies.

Cue -Template Matching Possibilities or How Are Recognition Decisions Made?

Cue similarity versus discrete odor matching models were investigated in *S. invicta* by Obin and Vander Meer (1989), using readily manipulated environmental cues (diet) to simulate the possible models described above. The data support a cue similarity model rather than models requiring discrete odor-matching (as in Getz 1982). Provost (1991) reported support for the graded response model, as non-related *Leptothorax lichtensteini* intruders induced a wide range of agonistic behaviors from resident workers. Provost (1991) also found 100% acceptance of related intruders and suggests this may be evidence in support of a threshold model. It seems to us that a threshold model must work in both directions -- acceptance and rejection. The Obin and Vander Meer (1989) paper did not address the question of graded response versus threshold response, however, there is ample evidence in the literature of ant nestmate recognition that graded responses occur in the real world (see the behavior section below).

Behavior Elicited

To study nestmate recognition and its associated chemistry an appropriate bioassay(s) must be developed. Assays have generally focused on the aggressive behaviors displayed by resident workers toward intruders that are recognized as non-nestmates. Interactions can be classified into two broad behavioral categories: non-aggressive (licking and trophallaxis) and aggressive (mandible opening, grasping, and flexing of the gaster with release of venom) (Morel, and Blum 1988, for *Camponotus floridanus*; Henderson, et al. 1990, for *Formica montana*). Aggressive behaviors in turn can be ranked in a hierarchical aggression scale (Table 2; Carlin, and Hölldobler 1986, for *Camponotus* spp.; Obin, and Vander Meer 1989, for *Solenopsis invicta*). The use of hierarchical behavioral units to score nestmate recognition is itself evidence against the threshold hypothesis for cue/template matching (Getz and Chapman, 1987) and evidence for a graded behavioral response depending on the degree of cue and template similarity. The caveat is that a worker introduced into a foreign con-specific colony may elicit a wide range of behavioral responses from encountered resident workers. Colony and individual variation in aggressive worker responses has been demonstrated for *Rhytidoponera confusa* (Crosland, 1990). Does this mean that each resident has a template different enough to promote the different behavioral responses to the intruder (support for the graded response) or are resident workers in different physiological modes that affect their behavior to the intruder (the effect is a graded response, but the cause is independent of the cue / template system)? To avoid this problem Obin and Vander Meer (1988) scored aggression assays using the most aggressive response shown out of 20 interactions. Bioassays have also been developed that utilize a surrogate worker, which makes it possible to address the chemistry of nestmate recognition.

Habituation Versus Active Recognition

It is possible that resident workers are habituated to their colony recognition cues and recognize non-nestmates rather than their nestmates, thus instead of nestmate recognition we would be dealing with **non-nestmate recognition**. Arguing against this proposition are examples where intra-specific, inter-nest recognition is dominated by heritable cue differences (Obin and Vander Meer 1988). In this situation only quantitative variation is expected for intra-specific heritable cues (as opposed to qualitative and quantitative variation for environmentally derived cues), thus in a habituated situation, small quantitative cue variation would not be expected to elicit a behavioral response. Therefore, recognition of nestmates will be assumed in this review.

The Chemical Cues

Some Problems with Nestmate Recognition Chemistry

Nestmate recognition in ants has been under intense investigation for well over a decade, yet, little is known about the chemical make-up of the recognition cues. There are many reasons for this. From our experience with the fire ant, *Solenopsis invicta,* it has been impossible to devise an appropriate nestmate recognition bioassay for this highly aggressive and sensitive species, although other species are more amenable (Bonavita-Cougourdan et al., 1987). Another problem, usually unrecognized, is that a solvent rinse of an ant's cuticle contains exocrine gland products, regurgatory and/or excretory products, as well as the expected cuticular lipids. The fire ant is a good example of this, because it produces easily analyzed alkaloids in its poison gland. Large amounts of these alkaloids are released when the ants are placed, dead or alive, in hexane. These compounds have many behavioral and physiological properties (Obin and Vander Meer 1985). In this case the surface lipids are grossly contaminated. For other studies we have made many unsuccessful attempts at obtaining alkaloid free fire ant body parts. The same problems probably apply to other less readily detectable behavior modifying exocrine gland products that can not be analyzed by gas chromatography.

What Is the Nature of Cue Chemistry?

Intra-Specific Examples. The surface lipids of most ants investigated thus far contain a large proportion of hydrocarbons -- straight chain, methyl, dimethyl, trimethyl branched and unsaturated. Early papers on the integration of myrmecophiles and termitophiles into their host colony showed a correlation between host hydrocarbon pattern and integration. These papers led to the "hydrocarbon bandwagon effect", which has delayed our coming to grips with the difficult nestmate recognition problem. Howard et al. (1980) demonstrated that the termitophile, *Trichopsenius frosti*, biosynthesized a cuticular hydrocarbon mixture qualitatively equal to that of its host, *Reticulitermes flavipes*. Without behavioral evidence the authors stated that cuticular hydrocarbons probably serve as the primary integration mechanism for this termitophile. Later Howard et al. (1982a) showed an asymmetric behavioral response from one of two sympatric termite species presented with critical-point dried *R. virginicus* worker surrogates treated with cuticular hydrocarbons from the two species. Other cuticular lipids were not tested to determined if a similar response would have been obtained. Howard et al. (1982b) reported three other termitophiles that appeared to use cuticular hydrocarbons as an integrating mechanism. There were no associated bioassays. Vander Meer and Wojcik (1982) got the wagon rolling further in their paper on the integration of a myrmecophile into fire ant colonies. Unfortunately, they wrote about hydrocarbon patterns and beetle integration in the same sentence and only at the end did they state that integration involves a passive defense that enables the beetle to survive long enough to acquire the species and environmental components of colony odor. As we will illustrate in the next paragraphs, a preoccupation with hydrocarbons as nestmate recognition cues has been a major handicap faced by researchers in this area.

Obin (1986) demonstrated that *Solenopsis invicta* nestmate recognition cues were both environmentally and heritably derived. Thus, aggression between colonies reared in the laboratory under identical conditions was reduced, compared to aggression between field colony workers. He also demonstrated differences between the patterns of the five major cuticular hydrocarbons found on workers from laboratory and field colonies; however, a correlation between similarity of hydrocarbon pattern and aggression was not obtained. Although this was an informative study regarding the source of nestmate recognition cues in *S. invicta*, no definitive conclusions could be reached regarding the role of cuticular hydrocarbons.

Bonavita-Cougourdan et al. (1987) studied the chemistry of nestmate recognition in *Camponotus vagus*. They washed worker heads and thoraces in pentane, analyzed the washes by GC, GC/MS and argentation TLC. As stated earlier, hydrocarbons dominate the chemical classes in cuticular washes, but they are not the only chemical class present. Bonavita-Cougourdon et al. (1987) found readily analyzed hydrocarbons in their GC and GC/MS analyses, and concluded that the total cuticular rinse contained only saturated hydrocarbons. Bioassays of the total rinse demonstrated differences in aggressive behavior between lures treated with resident worker versus alien worker rinses. They conclude that in *C. vagus*, nestmate recognition cues are composed of cuticular hydrocarbons. The lack of direct behavioral assays with the isolated hydrocarbons detracts from the authors' claims. However, they were able to devise a lure (surrogate worker) and bioassay that is amenable to testing the chemistry of nestmate recognition cues.

Bonavita-Cougourdan et al. (1989) showed that in choice experiments *C. vagus* workers preferentially retrieve larvae from their own colony rather than foreign conspecific larvae. They also determined that foreign larvae kept for 20 days with test colony sisters are no longer recognized as different and that the cuticular GC profile of the foreign larvae becomes more like that of test colony larvae. This is a good example of using hydrocarbons as a tool to study transfer and/or changes in the surface chemistry of individuals. However, the authors assume that the readily analyzed cuticular hydrocarbons are the responsible agents for worker recognition of larvae, as well as, worker-worker recognition and species recognition. Similarly, Bonavita-Cougourdan et al. (1990; 1993) found from principle component analysis of cuticular hydrocarbon profiles that dimethylalkanes are responsible for colonial chemical signature and the larvae, workers, sexual males or females and the queen possess characteristic chemical signatures composed from the n-alkanes and monomethylalkanes. They hypothesize that cuticular hydrocarbons are the cue that allows foragers to differentiate brood-tender and forager sub-castes. It is interesting that statisitical methodologies are capable of making these distinctions; however, until the behavioral bioassays are executed to demonstate a behavioral correlation, the "chemical signatures" are only useful to the human observers.

Henderson et al. (1990) studied internest aggression and nestmate recognition in the polygynous ant, *Formica montana*. They used pentane, hexane, ether, isopropanol, and water as cuticular rinses. Aggression bioassay results for pentane, hexane, and ether washes were not different for either nestmate or non-nestmate interactions, but were different from water and isopropanol washes. These authors assumed that pentane, hexane, and ether only remove hydrocarbons from the cuticular surface, and they hypothesize that hydrocarbons are used by this species in nestmate recognition. Pentane, hexane, and ether will dissolve all classes of non-polar lipids, thus no conclusions can be made about their hypothesis.

Nowbahari et al. (1990), investigated the individual, geographical and experimental variation of cuticular hydrocarbons in the ant, *Cataglyphis cursor*, and discussed their use in nestmate recognition. They cite the previous work, sometimes inappropriately, e.g. to support their contention that hydrocarbons are involved in nestmate recognition. Then they repeated the now classic mistake of assuming that if GC and GC/MS analysis of a worker rinse only detects what can be interpreted as hydrocarbons, then this rinse must contain only hydrocarbons. They conclude that intercolony aggressiveness in *C. cursor* is regulated, at least in part, by cuticular hydrocarbons. "In part", because application of the total rinse did not induce full levels of aggression.

Morel et al. (1988) made hexane soaks of *Camponotus floridanus,* and carried out chemical and pattern recognition analyses on the hydrocarbons that dominated the GC profile. However, they recognized that the ant rinse contains more than hydrocarbons. Thus they state "...although hydrocarbons are the major class of GC detectable compounds in *C. floridanus* soaks, we cannot say that they are responsible for nestmate recognition." Here the hydrocarbons and their GC profiles were used as a model for heritable cues known to dominate nestmate recognition in this species (Carlin and Hölldobler, 1986).

Vander Meer et al. (1989) studied the temporal changes in colony cuticular hydrocarbons of *S. invicta.* These authors did not state that cuticular hydrocarbons were involved in nestmate recognition but used them as a general model for heritable cues. Multivariate analyses were used to determine that at any point in time individual colonies are distinguishable through their hydrocarbon pattern; however, these patterns change over time for a particular colony (see the template section for details).

Interspecific Mixed Colonies. GC analyzed solvent rinses of workers from mixed colonies of *Formica selysi* and *Manica rubida* showed that hydrocarbon profiles had characteristics of both species (Errard and Jaisson 1991; Bagneres, et al. 1991). They assumed that these hydrocarbon changes permitted the two species to inhabit the same nest. Most interesting is the suggestion that heterospecific adoptees may be capable of switching epicuticular hydrocarbon biosynthesis to match that of the host colony. This certainly is an intriguing hypothesis that requires further investigation.

Habersetzer and Bonavita-Courgourdan (1993) investigated the cuticular components from mixed colonies of the slave-making ant, *Polyergus rufescens* and the slave species, *Formica rufibarbis.* In contrast to the above, the GC profiles of enslaved *F. rufibarbis* workers did not take on the characteristics of the *Polyergus* slave-maker. However, enslaved *F. Rufibarbis* workers lost their colony characteristic components. These authors assumed that the observed GC components they were hydrocarbons, based on retention time range, use of a non-polar solvent and a non-polar GC column. Unfortunately, these criteria are inadequate to support the hydrocarbon assumption. They conclude that one should bear in mind that chemicals other than the cuticular hydrocarbons may be involved in the recognition process.

All of the studies that have included chemical analyses have focused on hydrocarbons. Cuticular hydrocarbons no doubt function to prevent insect dessication, and they can have other functions, e.g. sex, alarm, and thermoregulation pheromones; defense; alarm; and as kairomones (see review Howard and Blomquist, 1982). More recently, Blum (1987) reviewed and speculated on the use of epicuticular hydrocarbons in social insects. He pointed out that epicuticular hydrocarbons exhibit enough variation between species and colonies to "qualify admirably" as species-specific signals and colonial signatures (of course other lipid classes would also qualify). These same epicuticular hydrocarbons (plus other lipids) should readily absorb exocrine gland products (and environmental odors), acting as a slow release matrix and adding information to the already chemically diverse epicuticular surface. Indeed, cuticular hydrocarbons may be involved in nestmate recognition, but the critical bioassays directly linking the chemistry and behavior have yet to be done.

Alarm Pheromones as Recognition Cues?

Jaffe and Marcuse (1983) found that *Odontomachus bauri* workers use volatile compounds as nestmate recognition cues. An aggression bioassay gave positive results for all body parts. Volatiles were detected by gas chromatography from the gaster and head but not the thorax. The proportions of volatile compounds differed from colony to colony, but no behavioral assays were done to test whether or not these compounds were a part of the recognition cue system. Jaffe and Sanchez (1984) investigated nestmate recognition in the ant *Camponotus rufipes*. They determined that introduction of worker heads but not thoraces or gasters were bitten more often by non-nestmates than nestmates. Freeze-dried workers were not effective in the bioassay, thus the authors conclude that cephalic odors, most likely alarm pheromones, are responsible for nestmate recognition. Similarly, alarm pheromones were suggested to be nestmate recognition cues for *Solenopsis geminata* (Jaffe 1986).

Jaffe (1987) brought together his ideas about nestmate recognition and alarm pheromones. He classifies nestmate recognition as advanced, if a specific chemical signal is used, e.g. alarm pheromone, and less advanced if non-specific signals are used, e.g. environmental odors, cuticular hydrocarbons, and/or more than one exocrine gland blend. According to Jaffe (1987) the gradation from non-advanced to advanced tracks the evolutionary advancement of the species considered. Thus, primitive Ponerinae, such as *Odontomachus bauri,* use non-specific signals, whereas the more complex *Ectatomma ruidum* specifically uses its alarm pheromone for nestmate recognition. In view of what we know now, there are problems with this scheme. "Advanced" species do utilize what Jaffe would call non-specific nestmate recognition signals, most notably *S. invicta*. We have already cited references to the intracolonial variation of alarm pheromones (Pasteels et al. 1980; Brand and Pretorius 1986; Do Nascimentoi, et al. 1993), which detracts from their use as recognition cues. It is difficult to understand the utility of alarm pheromones as nestmate recognition cues. A resident worker may release alarm pheromone in response to an intruder, but release must come after the recognition process occurs. Since recognition assays measure aggression, it is easy to visualize how presentation of an alarm pheromone could result in heightened activity and responsiveness to subsequently encountered cues, such as recognition cues.

For alarm pheromones to be operable as nestmate recognition cues they would have to be present on the surface of worker ants at levels below the alarm response threshold, otherwise alarm behavior would occur all the time. Jaffe (1987) astutely brings territoriality into the nestmate recognition picture, by postulating that an intruder experiencing a foreign territory will release alarm pheromone, thus initiating the recognition process. If this is true, recognition of being in the wrong place must still occur prior to release of the pheromone. Alarm pheromones are intimately linked with nestmate recognition, but in nature their release most probably is a result of recognition rather than the cause of recognition.

Interspecific adopted workers present an intriguing problem if alarm pheromones are involved in nestmate recognition. There is no question that the interspecific adoptee has an alarm pheromone profile that is qualitatively distinct from members of the host colony. How can it survive and what happens when the adoptee is in a situation that induces release of its alarm pheromone? It should be perceived as different by the normal workers in the colony and one would expect aggression to ensue.

The Bottom Line

Much has been theorized and learned about nestmate recognition in the past two decades. It is an essential ingredient in the recipe for eusociality and as such will continue to be studied extensively in the decades to come. An area of considerable interest to us is the unambiguous elucidation of the chemistry of nestmate recognition. Usually chemical techniques are far ahead of the development of suitable bioassays, but not in this case. The time is right for us to develop a clearer understanding of the nature of nestmate recognition cues, which will lead to a better understanding of the entire process. It is evident that more and more social insect researchers are developing expertise in molecular techniques, which can be brought to bear on many social insect questions, among them -- the issues centered around the elusive kin recognition problem. Ants are as diverse a group of insects as any, providing exceptions to every rule, thus generalizations can be dangerous. There are many examples of workers isolated during eclosion being attacked when returned to their mother colony (they did not acquire the colony-specific odor). However, Stuart (1987) reports that for *Leptothorax curvispinosus*, isolated workers produce persistent, colony-specific recognition cues after eclosion and are accepted when returned to their mother colony. Another divergence example, queen discriminators play a dominant role in nestmate recognition for

Camponotus spp. (Carlin, and Hölldobler, 1983, 1986, 1987), whereas they do not in *S. invicta* (Obin, and Vander Meer 1989) or *Rhytidoponera confusca* (Crosland 1990). These results are not contradictory (Carlin and Hölldobler 1991), but highlight the diversity of mechanisms and dangers of generalizations. Ant nestmate recognition is complex enough to keep the reductionist happy and to remind us of the sign in RVM's office, "Heisenberg may have slept here". As we more on to the next level of nestmate recognition understanding, it is clear that the future of this area of research has to be as exciting as the past has been.

References Cited

Blum, M. S. 1987. The bases and evolutionary significance of recognitive olfactory acuity in insect societies. Behavior in Social Insects, Experientia Supplementum. 54: 277-293.

Bonavita-Cougourdan, A., J. L. Clément, and C. Lange. 1987. Nestmate recognition: The role of cuticular hydrocarbons in the ant *Camponotus vagus* Scop. J. Entomol. Sci. 22: 1-10.

Bonavita-Cougourdan, A., J. L. Clément, and C. Lange. 1989. The role of cuticular hydrocarbons in recognition of larvae by workers of the ant *Camponotus vagus*: Changes in the chemical signature in response to social environment (Hymenoptera: Formicidae). Sociobiology. 16: 49-74.

Bonavita-Cougourdan, A., J. L. Clément, and Proveda, A. 1990. Les hydrocarbures cuticulaires et les processus de reconnaissance chez les fourmis: le code d'information complexe de *Camponotus vagus*. Actes Coll. Insectes Sociaux, 6: 273-280.

Bonavita-Cougourdan, A., J. L. Clément, and C. Lange. 1993. Functional subcaste discrimination (foragers and brood-tenders) in the ant *Camponotus vagus* Scop.: Polymorphism of cuticular hydrocarbon patterns. J. Chem. Ecol. 19: 1461-1477.

Brand, J. M., and V. Pretorius. 1986. Individual variation in the major alarm pheromone components of two *Crematogaster* species. Biochem. Syst. and Ecol. 14: 341-343.

Breed, M. D., and B. Bennett. 1987. Kin recognition in highly eusocial insects. In D. J. C. Fletcher, and C. D. Michener, eds., Kin Recognition in Animals, pp. 243-285. New York, N. Y.: John Wiley & Sons.

Carlin, N. F. 1989. Discrimination between and within colonies of social insects: Two null hypotheses. Netherlands Journal of Zoology. 39: 86-100.

Carlin, N. F., and B. Hölldobler. 1983. Nestmate and kin recognition in interspecific mixed colonies of ants. Science. 222: 1027-1029.

Carlin, N. F., and B. Hölldobler. 1986. The kin recognition system of carpenter ants (*Camponotus* spp.) I. Hierarchical cues in small colonies. Behav. Ecol. Sociobiol. 19:123-134.

Carlin, N. F., and B. Hölldobler. 1987. The kin recognition system of carpenter ants (*Camponotus* spp.) II. Larger colonies. Behav. Ecol. Sociobiol. 20: 209-217.

Carlin, N. F., and B. Hölldobler. 1991. The role of the queen in ant nestmate recognition: reply to Crosland. Anim. Behav. 41: 525-527.

Crosland, M. W. J. 1990. Variation in ant aggression and kin discrimination ability within and between colonies. J. Insect Behav. 3: 359-379.

Crosland, M. W. J. 1990. The influence of the queen, colony size and worker ovarian development on nestmate recognition in the ant *Rhytidoponera confusa*. Anim. Behav. 39: 413-425.

Crozier, R. H. and M. W. Dix. 1979. Analysis of two genetic models for the innate components of colony odor in social Hymenoptera. Behav. Ecol. Sociobiol. 18: 105-115

Errard, C. 1984. Evolution en fonction de l'age des relations sociales dans les colonies mixtes pluris specifiques chez les fourmis des genres *Camponotus* et *Pseudomyrmex*. Insectes Soc. 31: 185-194.

Errard, C. 1986. Role of early experience in mixed-colony odor recognition in the ants *Manica rubida* and *Formica selysi*. Ethology. 72: 243-249.

Errard, C., and P. Jaisson. 1991. Les premières étapes de la reconnaissance interspécifique chez les fourmis *Manica rubida* et *Formica selysi* (Hymenoptera, Formicidae) élevées en colonies mixtes. C.R. Acad. Sci. Paris, Series III, 313: 73-80.

Espelie, K. E., J. W. Wenzel, and G. Chang. 1990. Surface lipids of social wasp *Polistes metricus* Say and its nest and nest pedicel and their relation to nestmate recognition. J. Chem. Ecol. 16: 2229-2241.

Fielde, A. M. 1903. Artificial mixed nests of ants. Biol. Bull. 5: 320-325.

Fletcher, D. J. C. and C. D. Michener. 1987. Kin Recognition in Animals. pp. 465, New York, John Wiley & Sons.

Getz, W. M. 1982. An analysis of learned kin recognition in social Hymenoptera. J. Theor. Biol. 99: 585-597.

Getz, W. H., and R. S. Chapman. 1987. An odor discrimination model with application to kin recognition in social insects. Internat. J. Neurosci. 32: 963-978.

Grafen, A. 1990. Do animals really recognize kin? Anim. Behav. 39: 42-54.

Habersetzer, C. and A. Bonavita-Courgordan. 1993. Cuticular spectra in the slave-making ant *Polyergus rufescens* and the slave species *Formica rufibarbis*, Physiological Entomol. 18: 160-166.

Hamilton, W. D. 1964. The genetical evolution of social behavior. I, II. J. Theor. Biol. 7: 1-52.

Henderson, G. J. F., Anderen, J. K. Phillips, and R. L. Jeanne. 1990. Internest aggression and identification of possible nestmate discrimination pheromones in polygnous ant *Formica montana*. J. Chem. Ecol. 16: 2217-2228.

Hermann, H.R. and M.S. Blum. 1981. Defensive mechanisms in the social Hymenoptera. In H.R. Hermann, ed. Social Insects, Vol.II. pp. 78-197. New York, Academic Press.

Hölldobler, B. and N. Carlin. 1986. The kin recognition system of carpenter ants (*Camponotus* spp.). I. Hierarchical cues in small colonies. Behav. Ecol. Sociobiol. 20: 219-227.

Hölldobler, B. and N. Carlin. 1987. Anonymity and specificity in the chemical communication signals of social insects. J. Comp. Physiol. A. 161: 567-581.

Hölldobler, B., and C. D. Michener. 1980. Mechanisms of identification and discrimination in social Hymenoptera. In H. Markl, ed., Evolution of Social Behavior: Hypotheses and Emperical Tests, pp. 35-58. Weinheim: Verlag Chemie GmbH.

Howard, R. W., and G. J. Blomquist. 1982. chemical ecology and biochemistry of insect hydrocarbons. Ann. Rev. Entomol. 27: 149-72.

Howard, R. W., C. A. McDaniel, and G. J. Blomquist. 1980. Chemical mimicry as an integrating mechanism: Cuticular hydrocarbons of a termitophile and its host. Science. 210: 431-433.

Howard, R. W., C. A. McDaniel, and G. J. Blomquist. 1982b. Chemical mimicry as an integrating mechanism for three termiophiles associated with *Reticulitermes virginicus* (Banks). Psyche. 89: 157-167.

Howard, R. W., C. A. McDaniel, D. R. Nelson, G. J. Blomquist, L. T. Gelbaum, and L. H. Zakow. 1982a. Cuticular hydrocarbons of *Reticlitermes virginicus* (Banks) and their role as potential species- and caste-recognition cues. J. Chem. Ecol. 8: 1227-1239.

Jaffe, K. 1986. Nestmate recognition and territorial marking in *Solenopsis geminata* and in some Attini. In C. S. Lofgren and R. K. Vander Meer, eds. Fire Ants and Leaf-cutting Ants: Biology and Management. pp. 211-222. Boulder, CO. Westview Press.

Jaffe, K. 1987. Evolution of territoriality and nestmate recognition in ants. Experientia Supplementum. 54: 295-311.

Jaffe, K., and M. Marcuse. 1983. Nestmate recognition and territorial behaviour in the ant Odontomachus bauri Emery (Formicidae:Ponerinae). Insectes Sociaux. 30: 466-481

Jaffe, K., and C. Sanchez. 1984. On the nestmate recognition system and territorial marking behaviour in the ant *Camponotus rufipes*. Insectes Sociaux. 31: 302-315.

Jaisson, P. 1971. Expériences sur l'agressivité chez les fourmis. C. R. Acad. Sci. Paris, ser. D 273: 2320-2323.

Jaisson, P. 1985. Social behaviour. In, G. A. Kerkut and L. I. Gilbert eds., Comprehensive Insect Physiology, Biochemistry and Pharmacology, pp. 673-694. Oxford, Pergamon Press.

Jaisson, P. 1987. The construction of fellowship between nestmates in social Hymenoptera. Behavior in Social Insects, Experientia Supplementum. 54: 313-331.

Jutsum, A. T., T. Saunders, and J. Cherrett. 1979. Intraspecific aggression in the leaf-cutting ant *Acromyrmex octospinosus*. Anim. Behav. 27: 839-844.

Kistner, D.H. 1979. Social and evolutionary significance of social insect symbionts. In H.R. Hermann, ed. Social Insects, Vol.I. pp. 339-413. New York, Academic Press.

LaMon, B, and H. Topoff. 1985. Social facilitation of eclosion in the fire ant, *Solenopsis invicta*. Develop. Psychobiol. 18: 367-374.

Le Moli, F., and A. Mori. 1984. The effect of early experience on the development of "aggressive" behaviour in *Formica lugubris* Zett. (Hymenoptera: Formicidae). Z. Tierpsychol. 65:241-249.

Le Moli, F., and A. Mori. 1985. The influence of early experience of workers on enslavement. Anim. Behav. 33: 1384-1387.

Le Moli, F., A. Mori, and S. Parmigiani. 1983. The effect of antennalectomy on attack behaviour of *Formica lugubris* Zett. (Hymenoptera: Formicidae). Boll. Zool. 50: 201-206.

Lok, J. G., E. W. Cupp, and G. J. Blomquist, 1975. Cuticular lipids of the imported fire ants, *Solenopsis invicta* and *richteri*. Insect Biochem. 5: 821-829.

Mintzer, A., and S. B. Vinson. 1985. Kinship and incompatibility between colonies of the acacia ant *Pseudomyrmex ferruginea*. Behav. Ecol. Sociobiol. 17: 75-78.

Morel, L., and M. S. Blum. 1988. Nestmate recognition in *Camponotus floridanus* callow worker ants: are sisters or nestmates recognized. Anim. Behav. 36: 718-725.

Morel, L., R. K. Vander Meer, and C. S. Lofgren. 1990. Comparison of nestmate recognition between monogyne and polygyne populations of *Solenopsis invicta* (Hymenoptera: Formicidae). Ann. Entomol. Soc. Am. 83: 642-647.

Morel, L., R. K. Vander Meer, and Lavine, B. K. 1988. Ontogeny of nestmate recognition cues in the red carpenter ant (*Camponotus floridanus*): Behavioral and chemical evidence for the role of age and social experience. Behav. Ecol. Sociobiol. 22: 175-183.

Do Nascimentoi, R.R., Morgan, E.D., Billen, J., Schoeters, E., Della Lucia, T.M.C., and Bento, J.M.S. 1993. Variation with caste of the mandibular gland secretion in the leaf-cutting ant *Atta sextens rubropilosa*. J. Chem. Ecol., 19: 907-918.

Nowbahari, E., A. Lenoir, J. L. Clément, C. Lange, A. G. Bagneres, and C. Joulie. 1990. Individual, geographical and experimental variation of cuticular hydrocarbons of the ant Cataglyphis cursor (Hymenoptera: Formicidae): Their use in nestmate and subspecies recognition. Biochem. System. and Ecol. 18: 63-73.

Obin, M. S. 1986. Nestmate recognition cues in laboratory and field colonies of *Solenopsis invicta* Buren (Hymenoptera: Formicidae). J. Chem. Ecol. 12: 1965-1975.

Obin, M. S., and R. K. Vander Meer, 1985. Gaster flagging by fire ants (Solenopsis spp.): Functional significance of venom dispersal behaviour. J. Chem. Ecol. 11: 1757-1768.

Obin, M. S., and R. K. Vander Meer, 1988. Sources of nestmate recognition cues in the imported fire ant *Solenopsis invicta* Buren (Hymenoptera: Formicidae). Anim. Behav. 36: 1361-1370.

Obin, M. S., and R. K. Vander Meer. 1989. Mechanism of template-label matching in fire ant, *Solenopsis invicta* Buren, nestmate recognition. Anim. Behav. 38: 430-435.

Obin, M. S., and R. K. Vander Meer. 1989. Nestmate recognition in fire ants (*Solenopsis invicta* Buren). Do queens label workers. Ethology. 80: 255-264.

Obin, M.S., Morel, M., and Vander Meer, R.K. 1993. Unexpected and well-developed nestmate recognition in laboratory colonies of polygyne imported fire ants (Hymenoptera: Formicidae). J. Insect Behav. 6: 579-589.

Pasteels, J. M., J. C. Verhaeghe, J. C. Braekman, D. Daloze, and B. Tursch. 1980. Caste-dependent pheromones in the head of the ant *Tetramorium caespitum*. J. Chem. Ecol. 6: 467-472.

Plateaux, L. 1960. Adoptions expérimentales de larves entre des fourmis de genres différents: *Leptothorax nylanderi* Först. et *Solenopsis fugax* Latr. Insectes Sociaux. 7: 163-170.

Provost, E. 1991. Nonnestmate kin recognition in the ant *Leptothorax lichtensteini*: Evidence that genetic factors regulate colony recognition. Behav. Genetics. 21: 151-167.

Provost, E., E., Riviere, G., Roux, M., Morgan, E.D., and, Bagneres, A.G. 1993. Chamge in the chemical signature of the ant *Leptothorax lichtensteini* Bondroit with time. Insect Biochem. Mol. Biol. 23: 945-957.

Ross, K. G., R. K. Vander Meer, D. J. C. Fletcher, and E. L. Vargo. 1987. Biochemical phenotypic and genetic studies of two introduced fire ants and their hybrid (Hymenoptera: Formicidae). Evolution 41: 280-293.

Stuart, R. J. 1987. Individual workers produce colony-specific nestmate recognition cues in the ant, *Leptothorax curvispinosus*. Anim. Behav. 35: 1062-1069.

Stuart R.J., 1991. Kin recognition as a functional concept, Anim. Behav. 41: 1093-1094.

Stuart, R. 1992. Nestmate recognition and the ontogeny of acceptability in the ant, *Leptothorax curvispinosus*. Behav. Ecol. Sociobiol., 30: 403-408.

Tsuji, K. 1990. Kin recognition in *Pristmyrmex pungens* (Hymenoptera: Formicidae): asymmetrical change in acceptance and rejection due to odour transfer. Anim. Behav. 40: 306-312.

Vander Meer, R. K. 1988. Behavioral and biochemical variation in the fire ant, *Solenopsis invicta*. In R. L. Jeanne, ed., Interindividual Behavioral Variability in Social Insects. pp. 223-255. Boulder, CO. Westview Press.

Vander Meer, R. K., C. S. Lofgren, and F. M. Alvarez. 1985. Biochemical evidence for hybridization in fire ants. Florida Entomol. 68: 501-506.

Vander Meer, R. K., D. Saliwanchik, and B. Lavine. 1989. Temporal changes in colony cuticular hydrocarbon patterns of *Solenopsis invicta:* Implications for nestmate recognition. J. Chem. Ecol. 15: 2115-2125.

Vander Meer, R. K., and D. P. Wojcik. 1982. Chemical mimicry in the myrmecophilous beetle *Myrmecaphodius excavaticollis*. Science. 218: 806-808.

Wallis, D. I. 1963. A comparison of the response to aggressive behavior in two species of ants, *Formica fusca* and *Formica sanguinea*. Animal Behav. 11: 164-171.

Wilson, E. O. 1971. The insect societies. Cambridge, MA. Harvard University.

Wilson, E. O. 1987. Kin Recognition: An Introductory synopsis. In D. J. C. Fletcher, and C. D. Michener, eds., Kin Recognition in Animals, pp. 7-18. New York, N.Y., John Wiley & Sons.

5

Nest and Nestmate Discrimination in Independent-Founding Paper Wasps

Theresa L. Singer, Karl E. Espelie, and George J. Gamboa

Introduction

Social wasps have served as model systems for investigating numerous sociobiological questions, including those associated with kin-selection theory. The various levels of sociality and the phenotypic plasticity exhibited by social wasps make them ideal model organisms for studies of behavioral evolution. The knowledge we acquire from studies of wasps contributes to our understanding of other, more complex, social animals including vertebrates (West-Eberhard, 1991). In this chapter, we review the research on kin recognition ability and the ontogeny of nest and nestmate recognition in two subfamilies of Vespidae, the primitively eusocial polistine wasps and the highly eusocial vespine wasps.

Many social wasps have independently-founded colonies that undergo four colony phases (Greene, 1991; Matsuura, 1991; Reeve, 1991). The first is the founding phase (or preemergence phase) in which new nests are initiated. The second phase is the worker phase, which begins with the eclosion of the first workers and ends at the onset of the emergence of reproductives. The third phase is the reproductive phase, during which gynes and males eclose (excluding "early" males that mate with workers, not gynes). The last phase is the intermediate phase in which the colony declines and the sexuals disperse from the natal nest. The last phase continues until the initiation of a new nest the following spring (Reeve, 1991). The ability of *Polistes* to recognize their nest and nestmates has been examined in all four phases of the colony cycle (Gamboa, 1996). The behavioral response of resident females to nestmates and non-nestmates appears to depend, to some extent, on the caste of the resident females (Fishwild and Gamboa, 1992), the context of recognition, and the phase of the colony (Gamboa et al., 1991).

Most of the published research on kin and nestmate recognition in wasps has been on species from temperate zones. This is undoubtedly due to the fact that the vast majority of wasp researchers live in temperate areas. In many studies of the mechanism of recognition, wasps have been reared and maintained in the laboratory with attempts to control environmental influences or, in some cases, to provide the most natural environment possible. Both approaches have been important and necessary in order to decipher the recognition system. Some of the more recent investigations of recognition in social wasps have been field studies. Field tests of recognition can be more difficult than laboratory tests because of the vulnerability of wasp colonies to predators, conspecific usurpers, parasites, and parasitoids. Additionally, the manipulation of wasps and their nests outside of the laboratory may be problematic because of social wasps' aggressive behavior. On the other hand, field assays of recognition are probably much more sensitive than lab assays for detecting recognition ability when it is present. Consistent with this statement is the fact that, for *P. fuscatus*, the differential tolerance displayed toward nestmates and non-nestmates (and toward kin and non-kin) is much greater in the field than in the laboratory (Gamboa et al., 1991).

Foundresses, gynes, workers, and males of several wasp species have been examined to determine if they are able to discriminate nestmates from non-nestmates, and natal nests from foreign nests (Gamboa et al., 1986b; Espelie et al., 1990; Gamboa, 1996). A number of studies have investigated the mechanisms of recognition including the nature, identity, and origin of recognition cues. In addition, several studies have probed the evolution and adaptiveness of kin recognition and the role of recognition in nest usurpation and social parasitism (Gamboa, 1996). In this chapter, we discuss the pertinent research that has led to our current understanding of nest and nestmate recognition in social wasps. We concentrate our discussion on the mechanism of recognition, particularly the chemical nature of recognition cues.

Evidence for Nestmate and Kin Discrimination

One of the first laboratory studies of nestmate discrimination in social wasps documented that *Polistes metricus* foundresses preferentially associated with nestmates over non-nestmates, and that foundresses had significantly more tolerant interactions (and significantly fewer aggressive interactions) with their nestmates than with non-nestmates (Ross and Gamboa, 1981). Laboratory foundresses of *P. fuscatus* also associated more frequently with nestmates than non-nestmates and preferentially initiated nests with nestmates (Bornais et al., 1983), regardless of whether they overwintered with them (Post and Jeanne, 1982). It was clear from these studies that females of both species detect individually borne cues, likely odors, that may be acquired by the wasps from the environment (exogenous cues) or may be produced by the wasps themselves (endogenous cues), or both (Gamboa et al., 1986a,b).

Shellman and Gamboa (1982) reported that the isolation of *P. fuscatus* gynes from their nest and nestmates at eclosion disrupted the ontogeny of recognition ability. From this study, they concluded that exposure to the natal nest was required in order for gynes to later discriminate their nestmates from non-nestmates. These results suggested that gynes learn and/or acquire recognition odors from the natal comb. In studies of *P. fuscatus* and *P. carolina*, Pfennig et al. (1983a,b) determined that recognition cues are, indeed, odors, and that they are learned (and possibly remembered for life) in the adult stage within a few hours after emergence. Pfennig et al. (1983a) documented that exposure to the natal nest was both necessary and sufficient for the development of recognition ability in *P. fuscatus*. It was also determined from an experiment that involved exposing wasps to foreign combs that recognition odors could be both learned and acquired from the nest (Pfennig et al., 1983a). Similarly, Venkataraman et al. (1988) found that females of *Ropalidia marginata* discriminate nestmates from non-nestmates, but only after prior exposure to their nest and nestmates. As with *P. fuscatus* (Shellman and Gamboa, 1982), *R. marginata* females that had been isolated immediately upon emergence later failed to discriminate nestmates from non-nestmates.

Ryan et al. (1985) found that gynes of the baldfaced hornet, *Dolichovespula maculata*, are significantly more tolerant of their female nestmates than they are of non-nestmate gynes. Nestmate preference was exhibited by gynes regardless of whether they had been isolated at emergence, had been exposed only to their female nestmates, or had been exposed previously to both their female nestmates and natal nest. Exposure to the natal nest did not seem to be required in order for *D. maculata* gynes to discriminate nestmates from non-nestmates although it did significantly affect the variances in tolerance. This appears to contrast with the apparent requirement of exposure to the nest by *Polistes* for nestmate recognition. However,

Ryan et al. (1985) observed a marked difference between the behavior of *D. maculata* and *P. fuscatus* gynes as they emerged from their pupal cells. *P. fuscatus* gynes usually emerge within 15 minutes after the pupal cap is removed, while *D. maculata* gynes remain in their uncapped cells for as long as 3 hours. During this time in the uncapped cell, gynes of *D. maculata* may learn recognition odors from the natal nest. Therefore, the nest may, in fact, be important in the development of nestmate recognition ability in *D. maculata* as it is in *Polistes* (Ryan et al., 1985).

In laboratory and field studies, Gamboa and his students discovered that gynes of *P. fuscatus* treat non-nestmate aunts and nieces with the same tolerance as they treat their nestmate sisters, but are as intolerant of non-nestmate cousins as they are of unrelated non-nestmate females (Gamboa et al., 1987; Gamboa, 1988; Bura and Gamboa, 1994). These were the first reports that social wasps are capable of recognizing non-nestmate kin. In a related study with *P. exclamans*, Pfennig (1990) reported that wasps were more tolerant of non-nestmates from nearby nests than they were of non-nestmates from distant nests. Because *Polistes* females are philopatric, *P. exclamans* workers on nearby nests are probably more closely related to one another than they are to conspecifics from distant nests. This is especially likely since *P. exclamans* commonly build satellite nests near the natal nest (Strassmann, 1981). The results of these studies of *P. fuscatus* and *P. exclamans* strongly suggest that the recognition odor of *Polistes* has a genetic component.

The ability of nestmates of the opposite sex to recognize each other may be important in the context of mating. Ryan and Gamboa (1986) found that *P. fuscatus* males and nestmate virgin gynes were significantly more tolerant of each other than were unrelated, non-nestmate males and gynes, and that males copulated more frequently with non-nestmate virgin gynes. Conversely, Ross (1983) found in a study of eastern yellowjackets, *Vespula maculifrons*, that virgin females mated preferentially with nestmate males. It is not clear why males and gynes of *P. fuscatus* preferentially mated with non-nestmates while males and gynes of *V. maculifrons* preferentially mated with nestmates in the laboratory. Although laboratory mating preferences are suggestive, there is presently no evidence that social wasps use their recognition abilities to preferentially outbreed or inbreed in the field.

Males of *P. fuscatus* can recognize their male nestmates, although the age of males appears to affect the manifestation of recognition. Older *P. fuscatus* males (average age of 45 days) did not show a spatial preference for their male nestmates in the laboratory (Ryan et al.,

1984). However, in a subsequent laboratory study, it was found that younger males (average age of 2 days) spent significantly more time in close proximity to nestmate males than non-nestmate males (Shellman-Reeve and Gamboa, 1985). This is apparently the only documentation of male "brother" recognition ability in social insects, although its adaptive significance is unclear.

The work described above clearly demonstrates that wasps are capable of discriminating between nestmates and non-nestmates and, to some extent, between non-nestmate kin and non-kin. This discrimination ability is displayed by all castes or classes of wasps: foundresses, workers, gynes and males.

Relevance of Cuticular and Nest Surface Hydrocarbons to Recognition Ability

The recognition cues of wasps may be of genetic origin, or they may be obtained environmentally from nest-building materials or food sources, or they may be a combination of both. The behavioral evidence indicates that a combination of both components is the most likely source of recognition odors (Gamboa et al., 1986a,b; Bura and Gamboa, 1994). However, the evidence for a genetic component is considerably stronger than the evidence for an environmental component (Gamboa, 1996). The cuticular hydrocarbons of several social wasp species have been examined in order to obtain preliminary information on the chemicals suspected to be recognition odors.

Gas chromatography/mass spectrometry (GC/MS) revealed that adult females, adult males, larvae, and eggs of *P. annularis* have similar cuticular hydrocarbon compositions (Espelie and Hermann, 1990). The same complex mixture of *n*-alkanes, monomethylalkanes, and dimethylalkanes was also found on the paper and pedicel of *P. annularis* nests.

Colonies of *P. metricus* also were shown to have species-specific cuticular hydrocarbon compositions (Espelie et al., 1990). Again, the hydrocarbon profiles of the female wasps were similar to those of the nest paper and pedicel. Principle components analysis of the cuticular lipid profiles of *P. metricus* females and nests revealed that the hydrocarbons of nestmates were more similar than the hydrocarbons of non-nestmates. Additionally, the hydrocarbon profiles of the wasps matched those of their natal nest paper. These results indicated that adults of *P. metricus* have colony-specific hydrocarbon profiles (Espelie et al., 1990). The implication is that the wasps, deliberately or passively, spread these hydrocarbons on the surface of their nest. The source of the hydrocarbons may be sternal secretions that the wasps, especially the foundresses, rub onto the nest surface and pedicel with their gasters (Hermann and Dirks, 1974; Post et al., 1984; Dani et al., 1992).

Polistes exclamans workers also had species-specific as well as colony-specific hydrocarbon profiles (Singer et al., 1992). In a discriminant function analysis of the cuticular hydrocarbon profiles of *P. exclamans* workers, the wasps were placed into separate groups according to their colonies. More interestingly, wasps that were from colonies located close to one another on the same building had similar, yet colony-specific, hydrocarbon profiles. It is likely that these nearby colonies were inhabited by related individuals. Spring foundresses of temperate *Polistes* tend to be philopatric, i.e. they initiate new nests near their natal nest (Reeve, 1991). Thus, foundresses build nests in close proximity to the nest of a sister. *P. exclamans* queens and mated workers also build satellite nests near their natal nest, possibly as a refuge for the colony if the natal nest is destroyed or parasitized (Strassmann, 1981).

The hydrocarbon profiles of *P. fuscatus* females resemble those of other *Polistes* species, but differ in that a greater proportion of the hydrocarbons are *n*-alkanes (Espelie et al., 1994). Stepwise discriminant function analysis identified three methylbranched compounds that may serve as recognition pheromones (Figure 5.1). These three compounds met the postulated criteria for recognition pheromones: 1) they were the most important hydrocarbons for correctly assigning wasps to their own colony; 2) they appeared to be heritable since they were ranked highly among the hydrocarbons that correctly assigned wasps to sister groups; 3) they were similar for foundresses and workers, which is expected for compounds that proclaim colony identity and; 4) the three compounds had a distinctive stereochemistry, which may make them easier for wasps to recognize (Butts et al., 1993; Espelie et al., 1994).

Factor analysis of correspondences and principle components analysis of the cuticular hydrocarbons of *P. biglumis bimaculata* and *P. dominulus* (Bonavita-Cougourdan et al., 1991; Lorenzi, 1992) revealed species- and colony-specific profiles. *P. dominulus* foundresses and their female offspring had noticeably different hydrocarbon profiles (Bonavita-Cougourdan et al., 1991). These differences in composition may allow wasps on the nest to recognize dominant females or the caste of a female.

In a discriminant analysis of the cuticular hydrocarbons of queens, workers and males of *P. metricus*, Layton et al. (1994) found differences between workers and queens, and between males and females. Queens, workers, and males could also be distinguished according to colony by

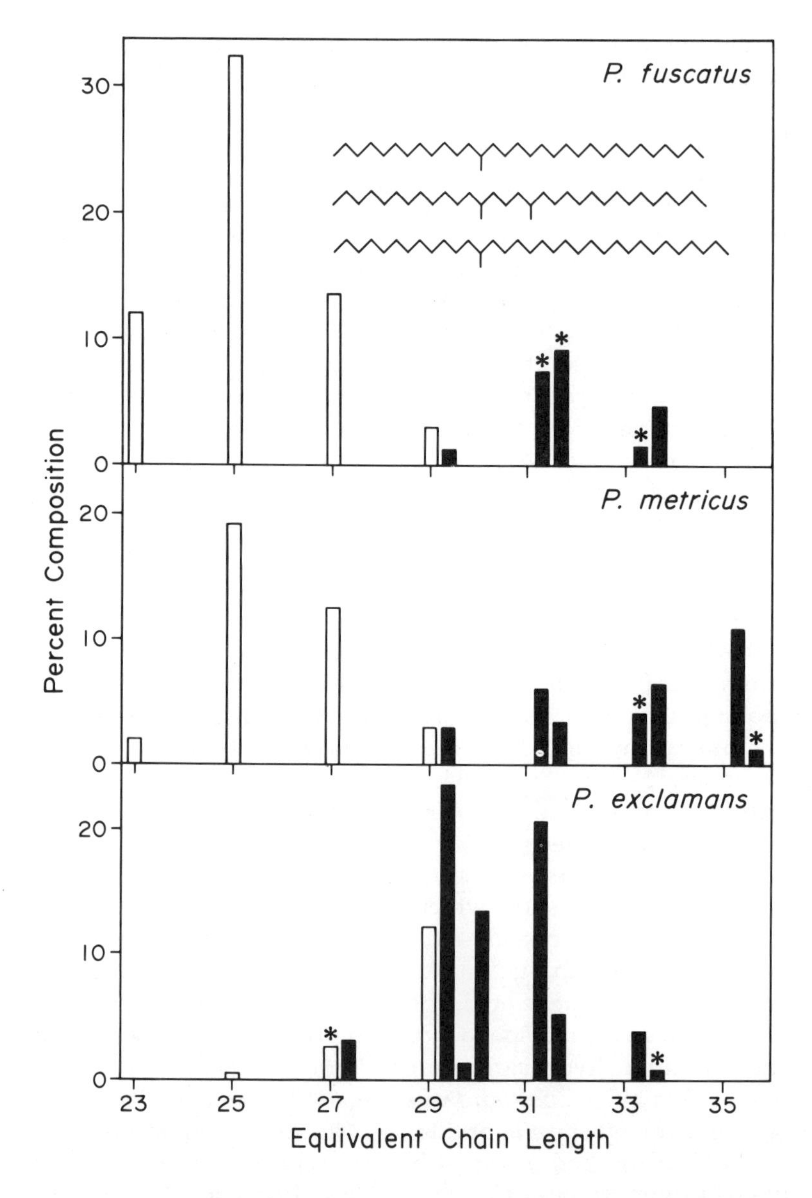

FIGURE 5.1. Cuticular hydrocarbon profiles of: *Polistes fuscatus* (Espelie et al., 1994); *P. metricus* (Espelie et al., 1990; Layton et al., 1994); and *P. exclamans* (Singer et al., 1992). Open bars are *n*-alkanes and solid bars are methylbranched hydrocarbons. Components marked with an asterisk have been shown by statistical analysis to be important in providing colony specificity. Structures of the three hydrocarbons that provide colony specificity for *P. fuscatus* are shown in the top panel.

their hydrocarbon profiles. In these analyses, different hydrocarbon components separated the wasps according to colony, caste, and sex, suggesting that specific cuticular hydrocarbons may provide different information about the identities of wasps in a colony.

Hydrocarbons of several species of wasps in the subfamily Vespinae also have been characterized. The European hornet, *Vespa crabro*, the baldfaced hornet, *Dolichovespula maculata*, the southern yellowjacket, *Vespula squamosa*, the eastern yellowjacket, *Vespula maculifrons*, and the German wasp, *V. germanica*, were shown to have species-specific and caste-specific hydrocarbon profiles (Brown et al., 1991; Butts et al., 1991). There was a strong similarity between the hydrocarbon profiles of *V. maculifrons* and *V. squamosa* which belong to different subgenera. Interestingly, *V. squamosa* is a social parasite of *V. maculifrons* (Greene, 1991). Chemical mimicry of the cuticular hydrocarbons of *V. maculifrons* may be part of the evolutionary process that facilitates nest usurpation by *V. squamosa*. More recently, Butts et al. (1993, 1995) showed that *D. maculata* and *V. crabro* have colony-specific hydrocarbon profiles.

The cuticular hydrocarbons have now been characterized for a number of social wasp species. The compositions are species-specific, and, in each case, the cuticular hydrocarbons are comprised of multiple classes of hydrocarbons (Table 5.1). The surface lipids of *D. maculata* workers, for example, include alkenes, *n*-alkanes, midchain-methylbranched alkanes, terminal-methylbranched alkanes, and dimethylbranched alkanes (Butts et al., 1991). In addition to the variation in the proportions of the different classes of cuticular hydrocarbons, there are variations in the chain lengths of the compounds within each of these classes. The most dominant hydrocarbons usually have a backbone of 25, 27, 29, 31, or 33 carbons. Thus, there is sufficient variation in cuticular hydrocarbon profiles to provide for species-specificity *and* for colony-specificity. It may be easier for a wasp to distinguish between hydrocarbons that differ in their methyl-branching pattern than between hydrocarbons that differ slightly in chain length. For example, a wasp may be able to discriminate between 3-methylnonacosane and 13-methylnonacosane more readily than she is able to differentiate between *n*-nonacosane (n-C_{29}) and *n*-hentriacontane (n-C_{31}). In fact, statistical analyses indicate that methylbranched and dimethylbranched alkanes are the most important classes of cuticular hydrocarbons for providing social wasps with colony-specificity (Figure 5.1).

TABLE 5.1. Percent Composition by Class of Cuticular Hydrocarbons for Workers of Six Species of Social Wasps[1]

	Polistinae					Vespinae
	Polybiini		Polistini			
Hydro-carbon Class	*P. aztecus*	*P. occidentalis*	*P. fuscatus*	*P. metricus*	*P. exclamans*	*D. maculata*
Straight chain						
Alkenes	66.3	0.0	4.0	2.3	0.7	8.9
n-Alkanes	29.0	25.1	67.1	42.7	18.9	36.8
Methyl branch						
Midchain	0.0	27.3	9.9	14.4	49.4	26.3
Terminal	0.0	15.4	0.0	0.4	18.1	7.1
Dimethyl	0.0	16.5	13.4	20.2	7.1	7.9

[1]*Parachartergus aztecus* (Espelie and Hermann, 1988); *Polybia occidentalis* (K. E. Espelie and R. L. Jeanne, unpublished results); *Polistes fuscatus* (Espelie et al., 1994); *Polistes metricus* (Layton et al., 1994); *Polistes exclamans* (Singer et al., 1992); *Dolichovespula maculata* (Butts et al., 1991).

The Role of Nest Surface Hydrocarbons in Nest Recognition

The presence of colony-specific hydrocarbons on the surface of wasp nests suggested that these compounds serve as nest recognition cues. In one investigation, gynes of *P. fuscatus* and *D. maculata* exhibited the ability to discriminate between a fragment of their own nest and a fragment of a foreign nest (Ferguson et al., 1987). Each gyne was released into a large box containing a fragment of her natal nest and a fragment of a foreign nest. In some experiments the nest fragments contained brood, while in others the comb fragment lacked brood. The length of time that the nest fragments of *P. fuscatus* were isolated from brood or adults was varied from several hours to many days. Regardless of the condition or history of the nest fragments, the wasps were able to distinguish their own nest fragment from a foreign one, indicating that the recognition cues are detectable and persistent on the natal comb.

Espelie et al. (1990) demonstrated in the laboratory that the presence of hydrocarbons on the nest surface was important for *P.*

metricus females to be able to choose their own nest from a foreign nest when presented with the two nests in a large Plexiglas box. When the hydrocarbons were present on the nests, the wasps sat on their own nest more often than on the foreign nest. But when hydrocarbons were extracted from the nests with hexane, wasps no longer showed a preference for their own nest. When the hydrocarbons were reapplied to the respective nests, the wasps again selected their own nest.

P. exclamans workers are also capable of discriminating between fragments of their own nest and fragments of another nest, and between fragments of a nearby nest and a distant nest (Pfennig, 1990). To determine the preference of wasps for either of two nest fragments, females were introduced one at a time into a box that was divided into three "zones" (two end zones each containing a nest fragment and a middle zone). The locations of the wasps were recorded intermittently. When introduced into a box with a nest fragment from their own nest and a fragment from a foreign nest, wasps were found more often in the zone with their own nest fragment. Given a choice between fragments of a nearby-foreign nest and a distant-foreign nest, the wasps were observed more often in the zone of the nearby nest fragment.

A different result was found when workers of *P. exclamans* were tested in a nest recognition bioassay in which nest preference was based on actual contact with the nest, rather than presence in a zone (Singer et al., 1992). *P. exclamans* workers were placed into a Plexiglas box with their own nest and a foreign nest. The presence of the wasps on their own nest, the foreign nest, or no nest was recorded intermittently. Workers sat on their own nest only 9% of the time and on the foreign nest only 3% of the time. It is not clear why these wasps showed little interest in either nest. *P. exclamans* are easily excitable and are less successfully maintained in the laboratory than *P. metricus* (personal observations). In the field, *P. exclamans* are known to switch between nests and to make satellite nests (Strassmann, 1981; Willer and Hermann, 1989), and, therefore, they may exhibit less affinity for their natal nest than other species of *Polistes*. It would be interesting to conduct a bioassay utilizing monitoring zones as did Pfennig (1990), and then remove the hydrocarbon layer of nests in order to determine the importance of surface hydrocarbons in nest recognition.

Little is known about the recognition ability of tropical *Polistes* species. Singer (unpublished) examined the nest recognition ability of *Polistes instabilis* in Costa Rica. Brood were removed from nests, and the nests were then cut into 3 or 4 fragments. One fragment from each nest was extracted in hexane to remove surface hydrocarbons. Individual female wasps were exposed to either: 1) two fragments from their own nest (control)(N = 54); 2) one fragment from their own nest and one from a distant nest (N = 53); or 3) two fragments from their own nest, one of which had been extracted in hexane (N = 48). Each wasp was left in the box with the nest fragments for 30 minutes, and the location of the wasp was recorded every 5 minutes. *P. instabilis* failed to show nest discrimination in this bioassay. The wasps did not sit on their own nest fragment more often than on the foreign nest fragment, nor did they sit on an untreated fragment of their own nest more often than on an extracted fragment. However, negative results do not necessarily indicate the lack of recognition ability (Gamboa et al., 1991) nor do they preclude the possibility that surface hydrocarbons play a role in nest recognition. The significance of the presence or absence of nest recognition ability in *P. instabilis* may be more amenable to interpretation when we have more information about its nestmate recognition ability, mode of colony-founding, use of satellite nests, and frequency of nest-switching, predation, and parasitism.

Recent work indicates that *P. metricus* workers use nest paper hydrocarbons to recognize their natal nest (Singer and Espelie, 1997). Pupae were excised from laboratory nests, placed in gelatin capsules, and then stored individually in plastic jars in one of the following conditions: 1) with an untreated, natal nest fragment (N = 65); 2) with a natal nest fragment that had the surface hydrocarbons removed with hexane (N = 61); or 3) with no nest fragment (N = 43). Once a wasp had emerged and spent four days in her jar, she was placed in a Plexiglas box containing a nest fragment from her own nest and a nest fragment from a foreign nest. For 30 minutes, the location of the wasp was recorded continuously, and then 24 hours later the location of the wasp was recorded for an additional 30 minutes. On the second day of observation, workers that had been exposed to the untreated nest fragments subsequently chose their own nest fragment significantly more often than they chose the foreign nest fragment. Wasps that had emerged on an extracted nest fragment and wasps that had emerged in complete isolation did not choose their own nest fragment more often than the foreign nest fragment. Thus, it appears that exposure of a newly emerged wasp to her nest surface hydrocarbons is essential for her to subsequently recognize the natal nest.

It is obvious that temperate social wasps have the ability to discriminate their natal nest from other nests, and that nest surface hydrocarbons play an important role in this process. Wasps apparently learn the colony-specific hydrocarbon profile (i.e., the colony odor) on the nest surface shortly after they emerge from their cells, which subsequently enables them to recognize both their natal nest and their nestmates. This implies that the hydrocarbon profile of the nest is very similar to that of the colony's adult females.

The Role of Nest Surface Hydrocarbons in Nestmate Recognition

The ability of wasps to recognize their nests via chemoreception may be an ancestral characteristic in solitary wasps that served as a preadaptation for the evolution of nestmate recognition in social wasps (Gamboa et al., 1986b). The nest may also play a pivotal role in the evolutionary history of nestmate recognition in other social insects. For example, Breed et al. (1995) recently suggested that the evolutionary derivation of nestmate recognition in honey bees is based on nest recognition in solitary ancestors. Singer and Espelie (1992) proposed that newly emerged wasps of *P. metricus* learn and/or acquire the colony-specific hydrocarbon profile from their nest surface which enables them to not only recognize the natal nest, but also to discriminate between nestmates and non-nestmates. In order to test this hypothesis, *P. metricus* pupae were removed from laboratory nests. Each pupa was put in a gelatin capsule, and then placed in a plastic jar in one of four conditions: 1) with the untreated natal nest (N = 50); 2) with a nest that had been extracted in hexane to remove surface hydrocarbons (N = 52); 3) with an extracted nest to which the hydrocarbons had been reapplied (N = 35); or 4) in complete isolation with no nest (N = 73)(Singer and Espelie, 1996). Four days after the worker emerged, she was assayed for recognition ability by using a "triplet test" with a nestmate and a non-nestmate (Pfennig et al., 1983a,b; Singer and Espelie, 1992). The nestmate and the non-nestmate had been previously exposed to both their natal nest and their nestmates. The interactions of all three wasps were recorded for one hour.

Singer and Espelie (1996) found that newly emerged *P. metricus* workers were able to discriminate nestmates from non-nestmates only if they had been exposed to the nest with its surface hydrocarbons present. Wasps that had been exposed to an extracted nest or had not been exposed to a nest did not display a preference for either nestmates or non-nestmates. Moreover, the newly emerged wasps were accepted more often by nestmates than by non-nestmates only when the newly emerged wasps had previously been exposed to their nest surface hydrocarbons. Similarly, adults in laboratory colonies of *P. metricus* were more likely to accept newly emerged nestmates that had been exposed to a fragment of the natal nest with the surface hydrocarbons present than nestmates exposed to an extracted natal nest fragment (Layton and Espelie, 1995). These results suggested that newly emerged wasps had acquired recognition odors from their natal nest, which were recognized by experienced resident nestmates. Another possibility is that the prior exposure of the young wasps to nest surface hydrocarbons, and their subsequent ability to recognize the natal nest and nestmates, caused them to behave in a manner that made their nestmates more likely to accept them.

Ryan et al. (1985) reported that *D. maculata* gynes had the ability to recognize nestmate females. They found that although exposure to the nest may not be necessary for the development of nestmate recognition ability, prior exposure to the natal nest significantly affected the development of this ability. Butts and Espelie (1995) also examined the nestmate recognition ability of *D. maculata*. Pupae were excised from field nests, and wasps were allowed to emerge onto nest fragments with and without surface hydrocarbons. Exposure to nest surface hydrocarbons did not affect the ability of the wasps to discriminate between a nestmate and a non-nestmate in subsequent triplet tests (Butts and Espelie, 1995). It is possible that the recognition assay employed by Butts and Espelie (1995) may not be sensitive to detecting nestmate recognition ability in *D. maculata*. The behavior of *D. maculata* appears to be affected by confinement in a laboratory setting. Adults are very aggressive and hyperactive in the laboratory, and this undoubtedly affects laboratory bioassays of recognition (personal observations). An improved bioassay, such as using laboratory-reared *D. maculata* whose behavior may be less affected by being confined in the laboratory, may provide more meaningful results.

Aunts of *P. fuscatus* display roughly similar frequencies of acceptance and rejection of their non-nestmate nieces in the field (Bura and Gamboa, 1994). Gamboa et al. (1996) reasoned that aunt-niece interactants would be the class of conspecifics that would most likely reveal which hydrocarbons are most important in kin recognition. Gamboa et al. (1996) combined the results of behavioral interactions between *P. fuscatus* aunts and nieces with statistical analyses of the differences in cuticular hydrocarbon profiles of interacting wasps. The field bioassays of aunt-niece recognition involved flushing several resident females from the nest. While these wasps were away from their nest, a nest from a colony founded by a queen that was a sister of the original queen was switched with the original nest. When the flushed wasps (nieces) returned to the replacement nest, the treatment of these wasps by their aunt was recorded and scored as an acceptance or rejection. The wasps involved in this bioassay were collected, and their cuticular hydrocarbons analyzed. The null hypothesis that the hydrocarbon profile matches of the interactants was independent of their treatment of each other (i.e., whether they accepted or rejected each other) was rejected at a significance level of 0.0015. Thus, differences in the hydrocarbon profiles of interacting wasps were linked to whether they accepted or rejected each other. This is the first evidence of a linkage between the chemical profiles of interactants and their tolerance of each other for any social insect (Gamboa et al., 1996).

A stepwise discriminant function analysis was employed by Gamboa et al. (1996) to predict the acceptance or rejection of nieces by aunts on the basis of differences between the niece's cuticular hydrocarbon profile and the colony profile of their aunt. The analysis yielded a 100% correct assignment of all wasps to either the accepted or rejected groups. Five of the highest ranked cuticular hydrocarbons for colony- and sister-group specificity that were identified in a previous analysis (Espelie et al., 1994) were also highly ranked in the discriminant function analysis. There were, however, additional hydrocarbons that ranked highly as predictors of acceptance or rejection but had relatively low colony-specificity. Thus, *P. fuscatus* females appear to use a number of different cuticular hydrocarbons for recognition, some of which have relatively low colony- and sister-group specificity.

The studies described above demonstrate convincingly that cuticular and nest surface hydrocarbons are important and necessary in the recognition system of social wasps. The proximate and ultimate origin of these compounds, however, remains less certain. Present evidence indicates that recognition pheromones are both genetically and environmentally derived, although the evidence for genetic derivation is the more compelling. In temperate *Polistes* the vast majority of adults in a colony are closely related to the queen, and most are offspring of the queen. Prior to the emergence of other adults, the queen may coat the nest with her own cuticular hydrocarbons by rubbing her gaster on the nest surface. Once she has done this, the nest surface would have the chemical signature of the queen as well as any chemicals that may be derived from environmental sources such as nest-building material. Upon eclosion, the newly emerged adults are exposed to the surface of the natal nest from which they learn the colony odor. In a single-foundress colony, half of the alleles of female offspring are from the queen, and all of the alleles of male offspring are from the queen. It would be reasonable to assume that the hydrocarbon profiles produced by offspring are very similar to the profile of their queen. Since a newly emerged wasp does not use her own surface chemistry to form a recognition template (Gamboa et al., 1986a,b), the young wasp recognizes nestmates via the odors she learned from the nest surface. Evidence from several studies indicates that the young wasp also acquires a nest odor through direct contact with the nest surface (e.g. Pfennig et al., 1983a). As more wasps emerge, passive or deliberate contact of colony members with the nest surface may change the profile of hydrocarbons on the nest. However, since colony members are related, these changes may be relatively minor. This scenario can be extended to include recognition by resident females of non-nestmate kin such as aunts, nieces, and first cousins. These non-nestmate kin may have cuticular hydrocarbon profiles that sufficiently match those of the residents' nest to allow these non-nestmate wasps to be accepted into the residents' colony.

Exploiting the Recognition System

As in almost every group of living organisms, there are exploiters or "cheaters" who use an alternative strategy to enhance their reproductive success. In social wasps, the exploiters include conspecific usurpers and obligate social parasites.

Conspecific usurpations are common in temperate *Polistes* and especially in *P. fuscatus*: Gamboa et al. (1992) reported that, on average, a colony experiences one usurpation attempt per day during the founding phase of the colony. Conspecific usurpations of *P. fuscatus* colonies are usually conducted by lone foundresses who were probably displaced from their own nests. Usurpers are almost always unrelated to the colonies they usurp (Klahn, 1988). The targeting of unrelated nests for usurpation may be mediated through the detection of kin-specific chemical cues, although this hypothesis has not been tested.

Nest-stroking behavior may contribute to the ability of wasps to usurp or parasitize a nest. It has been hypothesized that the stroking of the gaster against the surface of the nest is performed by a dominant female in order to communicate her status (West-Eberhard, 1982). It was observed that the alpha foundress of *P. dominulus* colonies performed more nest-rubbing behavior than a subordinant foundress (Dani et al., 1992). When the alpha female was removed, the rate of nest-stroking by the subordinant female increased significantly. When the alpha female was reintroduced to the nest, the stroking rate of the subordinate remained high, but the stroking activity of the alpha female increased to a rate that was higher than before she was removed from the nest. Although these results are consistent with the hypothesis that nest-stroking behavior communicates dominance, they suggest that this behavior functions to apply colony recognition pheromones to the nest.

Nest-stroking may be employed by social parasites and conspecifics to usurp the host nest. *P. atrimandibularis* and *P. sulcifer* are social parasites of *P. biglumis bimaculata* and *P. dominulus*, respectively (Cervo et al., 1990; Turillazzi et al., 1990). During the process of usurpation, both the *P. biglumis bimaculata* foundress and the *P. atrimandibularis* intruder were observed stroking the nest surface with their abdomens. When the parasite was alone on the nest, she continued this stroking behavior along with inspection of the surface and cells of the nest (Cervo et al., 1990). Turillazzi et al. (1990) also observed abdominal stroking by *Polistes sulcifer* intruders on nests of *P. dominulus*. Covering a nest with one's own chemical odor may be necessary in order for a usurper to become the dominant individual on the nest *and* to be accepted by subsequently emerging wasps.

Conclusions

The various studies discussed in this chapter demonstrate that nest and nestmate recognition ability are widespread in temperate social wasps. The ability to recognize one's own nest may be an ancestral preadaptation of solitary wasps that subserves the ability of social wasps to recognize and discriminate nestmates from non-nestmates (Gamboa et al., 1986b). The evidence that the chemical profile of the nest surface is very similar to that of the surface chemistry of the adults from the colony supports this evolutionary scenario. It is apparent that wasps learn and can acquire a nest odor after they emerge onto the nest surface. The odor that they learn from the nest serves as a template that they match with the chemical phenotypes of conspecifics. Cuticular and nest surface hydrocarbons play critical roles in the recognition system. The hydrocarbons on the cuticles of the wasps in a colony consist of colony-specific, heritable compounds that identify wasps as members of the colony. The colony-specific odor is also found on the nest, and perhaps this odor can be enhanced or changed over the colony cycle.

The nest is the center of social life for wasps (Starr, 1991). It serves as the place where the brood is maintained and defended against enemies. The nest provides a location for the wasps to rest, find refuge, and communicate. It is a physical boundary within which non-colony members are accepted or rejected, and it is the domain in which the queen can be recognized and regarded as the dominant individual (Starr, 1991). Using the durable, structural component of a colony, i.e. the nest, to harbor the colony recognition cues seems logical. All colony members are exposed to the nest surface upon emergence and may contribute to the colony odor by deliberate or passive contact with the nest. Hydrocarbons present on the cuticle of all insects are stable, hydrophobic compounds that are non-volatile and water insoluble (see Blomquist et al., Chapter 2). Therefore, the transfer of these compounds onto the nest provides a convenient, efficient, and fairly permanent method of communicating the recognition cues of the colony to colony members.

The similarity of the cuticular hydrocarbons of an individual wasp and the template formed from the hydrocarbon profile of the natal nest determines if a wasp will be accepted or rejected as a colony member. In the cue similarity threshold model, assessment for colony acceptance is based on whether or not the wasp falls above or below the acceptance threshold. In this system there is no gradient in the treatment of a wasp that is a function of the degree of relatedness between interactants (Gamboa, 1996). Thus, wasps are treated very tolerantly, like nestmate sisters, or very intolerantly, like unrelated non-nestmates. The acceptance threshold appears to be more restrictive in queens than in workers, and may become more restrictive in different parts of the colony cycle (Gamboa et al., 1991; Fishwild and Gamboa, 1992). It has been shown that certain components of the cuticular hydrocarbons of wasps from the same colony can be used to identify wasps as members or relatives of a colony. These particular components may be applied to the nest surface and subsequently learned by emerging adults as the colony odor. The learned template is then used by a wasp to match with the cuticular hydrocarbon profile of other wasps that it encounters. The treatment that an encountered wasp receives is dependent upon whether the match between the template and perceived odor falls above or below an acceptance threshold. In field studies examining tolerance displayed toward non-nestmate kin, the aunts, nieces, and first cousins are either accepted or rejected, but are not treated intermediately in tolerance as a function of their degree of relatedness (Gamboa, 1996).

Further studies are necessary for a more complete understanding of the factors involved in nest and nestmate discrimination in social wasps. More work is needed to determine if the nest surface hydrocarbons are derived solely from the wasps that inhabit the nest or whether a portion of these lipids are obtained from the environment. It would be interesting to supply different wasp colonies with various nest-building materials and a variety of caterpillar species (in the field or in the laboratory), and then to determine the effects of these manipulations on the composition of the nest and cuticular hydrocarbons and on the ability of the wasps to recognize their nest and nestmates.

Further studies are also required to determine if the wasps can update their learning of recognition cues. Almost certainly the odors associated with the comb change over the colony cycle, and one would assume that wasps would update their template periodically in order to minimize the probability of rejecting their own nestmates.

The sensory mechanism of how wasps detect and differentiate specific hydrocarbons has not yet been explored. Undoubtedly, the antennae are the primary receptors of chemical stimuli, but how antennal receptors transmit their information to ganglia that process it and subsequently trigger behavioral responses is unknown. Our knowledge of the matching process that involves comparing the template and chemical profile of the encountered wasp is particularly deficient.

There is very little information concerning the recognition abilities of social wasps from the tropical regions of the world. In the tropics, wasps are not forced into a winter diapause, neighboring colonies may not be in the same developmental phase, and predation pressure from ants and other animals may be greater than in temperate regions. Consequently, these wasps may possess dramatically different recognition abilities and mechanisms of recognition. The ecological pressures of conspecific usurpations and brood theft as well as the severity of competition for food and nest sites almost certainly differ between temperate and tropical social wasps. Thus, the selection pressures affecting nest and nestmate recognition probably differ between temperate and tropical social wasps. In fact, it is conceivable that for some tropical social wasps, there may be little or no advantage for recognizing one's nest or nestmates. Interestingly, Gastreich et al. (1990) were unable to demonstrate nestmate recognition in the tropical social wasp, *Parachartergus colobopterus*, even though they carried out both laboratory and field recognition tests.

Despite the numerous studies of nest and nestmate recognition in temperate social wasps, our knowledge of their recognition system remains imperfect. Not only are there gaps in our understanding of the mechanism of recognition, but our search for an apprehension of the adaptiveness and evolutionary history of recognition has just begun. As in any scientific endeavor, each finding generates new questions and hypotheses. Although less than perfect, our knowledge of the recognition system of temperate social wasps is as complete as that of any other animal. As such, it is well-suited for comparisons with the recognition systems of other animals. Understanding the differences and similarities between the recognition systems of paper wasps and other social animals will provide insight into the selective pressures that have shaped the evolution of kin recognition.

References Cited

Bonavita-Cougourdan, A., G. Theraulaz, A.G. Bagnères, M. Roux, M. Pratte, E. Provost and J.L. Clément, 1991. Cuticular hydrocarbons, social organization and ovarian development in a polistine wasp: *Polistes dominulus* Christ. *Comp. Biochem. Physiol. 100B*:667-680.

Bornais, K.M., C.M. Larch, G.J. Gamboa, and R.B. Daily, 1983. Nestmate discrimination among laboratory overwintered foundresses of the paper wasp, *Polistes fuscatus* (Hymenoptera: Vespidae). *Can. Entomol. 115*:655-658.

Breed, M.D., M.F. Garry, A.N. Pearce, B.E. Hibbard, L.B. Bjostad, and R.E. Page, Jr., 1995. The role of wax comb in honey bee nestmate recognition. *Anim. Behav. 50*: 489-496.

Brown, W.V., J.P. Spradbery and M.J. Lacey, 1991. Changes in the cuticular hydrocarbon composition during development of the social wasp, *Vespula germanica* (F.) (Hymenoptera: Vespidae). *Comp. Biochem. Physiol. 99B*:553-562.

Bura, E.A. and G.J. Gamboa, 1994. Kin recognition by social wasps: asymmetric tolerance between aunts and nieces. *Anim. Behav. 47*:977-979.

Butts, D.P., M.A. Camann and K.E. Espelie, 1993. Discriminant analysis of cuticular hydrocarbons of the baldfaced hornet, *Dolichovespula maculata* (Hymenoptera: Vespidae). *Sociobiology 21*:193-201.

Butts, D.P., M.A. Camann and K.E. Espelie, 1995. Workers and queens of the European hornet *Vespa crabro* L. have colony-specific cuticular hydrocarbon profiles (Hymenoptera: Vespidae). *Ins. Soc. 42*:45-55.

Butts, D.P. and K.E. Espelie, 1995. Role of nest-paper hydrocarbons in nestmate recognition of *Dolichovespula maculata* (L.) workers (Hymenoptera: Vespidae). *Ethology 100*:39-49.

Butts, D.P., K.E. Espelie and H.R. Hermann, 1991. Cuticular hydrocarbons of four species of social wasps in the subfamily Vespinae: *Vespa crabro* L., *Dolichovespula maculata* (L.), *Vespula squamosa* (Drury), and *Vespula maculifrons* (Buysson). *Comp. Biochem. Physiol. 99B*:87-91.

Cervo, R., M.C. Lorenzi and S. Turillazzi, 1990. Nonaggressive usurpation of the nest of *Polistes biglumis bimaculata* by the social parasite *Sulcopolistes atrimandibularis* (Hymenoptera Vespidae). *Ins. Soc. 37*:333-347.

Dani, F.R., R. Cervo and S. Turillazzi, 1992. Abdomen stroking behaviour and its possible functions in *Polistes dominulus* (Christ) (Hymenoptera, Vespidae). *Behav. Proc. 28*:51-58.

Espelie, K.E., G.J. Gamboa, T.A. Grudzien and E.A. Bura, 1994. Cuticular hydrocarbons of the paper wasp, *Polistes fuscatus*: a search for recognition pheromones. *J. Chem. Ecol. 20*:1677-1687.

Espelie, K.E. and H.R. Hermann, 1990. Surface lipids of social wasp *Polistes annularis* (L.) and its nest paper and pedicel. *J. Chem. Ecol. 16*:1841-1852.

Espelie, K.E., J.W. Wenzel and G. Chang, 1990. Surface lipids of social wasp *Polistes metricus* Say and its nest and nest pedicel and their relation to nestmate recognition. *J. Chem. Ecol. 16*:2229-2241.

Ferguson, I.D., G.J. Gamboa and J.K. Jones, 1987. Discrimination between natal and non-natal nests by the social wasps *Dolichovespula maculata* and *Polistes fuscatus* (Hymenoptera: Vespidae). *J. Kansas Entomol. Soc. 60*:65-69.

Fishwild, T.G. and G.J. Gamboa, 1992. Colony defense against conspecifics: caste-specific differences in kin recognition by paper wasps, *Polistes fuscatus*. *Anim. Behav.* 43:95-102.

Gamboa, G.J., 1988. Sister, aunt-niece, and cousin recognition by social wasps. *Behav. Gen. 18:*409-423.

Gamboa, G.J., 1996. Kin recognition in social wasps. In: *Natural History and Evolution of Paper Wasps* (S. Turillazzi and M.J. West-Eberhard (Eds.)), Oxford University Press, Oxford, UK, pp. 161-177.

Gamboa, G.J., R.L. Foster, J.A. Scope and A.M. Bitterman, 1991. Effects of stage of colony cycle, context, and intercolony distance on conspecific tolerance by paper wasps (*Polistes fuscatus*). *Behav. Ecol. Sociobiol. 29:*87-94.

Gamboa, G.J., T.A. Grudzien, K.E. Espelie and E.A. Bura, 1996. Kin recognition pheromones in social wasps: combining chemical and behavioural evidence. *Anim. Behav. 51:*625-629.

Gamboa, G.J., J.E. Klahn, A.O. Parman and R.E. Ryan, 1987. Discrimination between nestmate and non-nestmate kin by social wasps (*Polistes fuscatus,* Hymenoptera: Vespidae). *Behav. Ecol. Sociobiol. 21:*125-128.

Gamboa, G.J., H.K. Reeve, I.D. Ferguson and T.L. Wacker, 1986a. Nestmate recognition in social wasps: the origin and acquisition of recognition odours. *Anim. Behav. 34:*685-695.

Gamboa, G.J., H.K. Reeve and W.G. Holmes, 1991. Conceptual issues and methodology in kin recognition research: a critical discussion. *Ethology 88:*109-127.

Gamboa, G.J., H.K. Reeve and D.W. Pfennig, 1986b. The evolution and ontogeny of nestmate recognition in social wasps. *Annu. Rev. Entomol. 31:*431-454.

Gamboa, G.J., T.L. Wacker, K.G. Duffy, S.W. Dobson and T.G. Fishwild, 1992. Defence against intraspecific usurpation by paper wasp cofoundresses (*Polistes fuscatus*, Hymenoptera: Vespidae). *Can. J. Zool. 70:*2369-2371.

Gastreich, K.R., D.C. Queller, C.R. Hughes and J.E. Strassmann, 1990. Kin discrimination in the tropical swarm-founding wasp, *Parachartergus colobopterus. Anim. Behav. 40:*598-601.

Greene, A., 1991. *Dolichovespula* and *Vespula*. In: *The Social Biology of Wasps* (K.G. Ross and R.W. Matthews (Eds.)), Cornell University Press, Ithaca, New York. pp. 263-305.

Hermann, H.R. and T.F. Dirks, 1974. Sternal glands in polistine wasps: morphology and associated behavior. *J. Georgia Entomol. Soc. 9:*1-8.

Klahn, J., 1988. Intraspecific comb usurpation in the social wasp *Polistes fuscatus. Behav. Ecol. Sociobiol. 23:*1-8.

Layton, J.M., M.A. Camann and K.E. Espelie, 1994. Cuticular lipid profiles of queens, workers, and males of social wasp *Polistes metricus* Say are colony-specific. *J. Chem. Ecol. 20:*2307-2321.

Layton, J.M. and K.E. Espelie, 1995. Effects of nest paper hydrocarbons on nest and nestmate recognition in colonies of *Polistes metricus* Say. *J. Insect Behav. 8:*103-113.

Lorenzi, M.C., 1992. Epicuticular hydrocarbons of *Polistes biglumis bimaculatus* (Hymenoptera Vespidae): preliminary results. *Ethol. Ecol. Evol. 2:*61-63.

Matsuura, M., 1991. *Vespa* and *Provespa*. In: *The Social Biology of Wasps* (K.G. Ross and R.W. Matthews (Eds.)), Cornell University Press, Ithaca, New York. pp. 232-262.

Pfennig, D.W., 1990. Nest and nestmate discrimination among workers from neighboring colonies of social wasps *Polistes exclamans*. *Can. J. Zool. 63*:268-271.

Pfennig, D.W., G.J. Gamboa, H.K. Reeve, J.S. Reeve and I.D. Ferguson, 1983a. The mechanism of nestmate discrimination in social wasps (*Polistes*, Hymenoptera: Vespidae). *Behav. Ecol. Sociobiol. 13*:299-305.

Pfennig, D.W., H.K. Reeve and J.S. Shellman, 1983b. Learned component of nestmate discrimination in workers of a social wasp, *Polistes fuscatus* (Hymenoptera: Vespidae). *Anim. Behav. 31*:412-416.

Post, D.C. and R.L. Jeanne, 1982. Recognition of former nestmates during colony founding by the social wasp *Polistes fuscatus* (Hymenoptera: Vespidae). *Behav. Ecol. Sociobiol. 11*:238-285.

Post, D.C., M.A. Mohamed, H.C. Coppel and R.L. Jeanne, 1984. Identification of ant repellent by social wasp *Polistes fuscatus* (Hymenoptera: Vespidae). *J. Chem. Ecol. 10*:1799-1807.

Reeve, H.K., 1991. *Polistes*. In: *The Social Biology of Wasps* (K.G. Ross and R.W. Matthews (Eds.)), Cornell University Press, Ithaca, New York. pp. 99-148.

Ross, K.G., 1983. Laboratory studies of the mating biology of the eastern yellowjacket, *Vespula maculifrons* (Hymenoptera: Vespidae). *J. Kansas Entomol. Soc. 56*:523-537.

Ross, N.M. and G.J. Gamboa, 1981. Nestmate discrimination in social wasps (*Polistes metricus*, Hymenoptera: Vespidae). *Behav. Ecol. Sociobiol. 9*:163-165.

Ryan, R.E., T.J. Cornell and G.J. Gamboa, 1985. Nestmate recognition in the bald-faced hornet, *Dolichovespula maculata* (Hymenoptera: Vespidae). *Z. Tierpsychol. 69*:19-26.

Ryan, R.E., G.C. Forbes and G.J. Gamboa, 1984. Male social wasps fail to recognize their brothers (*Polistes fuscatus*, Hymenoptera: Vespidae). *J. Kansas Entomol. Soc. 57*:105-110.

Ryan, R.E. and G.J. Gamboa, 1986. Nestmate recognition between males and gynes of the social wasp *Polistes fuscatus* (Hymenoptera: Vespidae). *Ann. Entomol. Soc. Am. 79*:572-575.

Shellman, J.S. and G.J. Gamboa, 1982. Nestmate discrimination in social wasps: the role of exposure to nest and nestmates (*Polistes fuscatus*, Hymenoptera: Vespidae). *Behav. Ecol. Sociobiol. 11*:51-53.

Shellman-Reeve, J.S. and G.J. Gamboa, 1985. Male social wasps (*Polistes fuscatus*, Hymenoptera: Vespidae) recognize their male nestmates. *Anim. Behav. 33*:331-332.

Singer, T.L., M.A. Camann and K.E. Espelie, 1992. Discriminant analysis of cuticular hydrocarbons of social wasp *Polistes exclamans* Viereck and nest surface hydrocarbons of its nest paper and pedicel. *J. Chem. Ecol. 18*:785-797.

Singer, T.L. and K.E. Espelie, 1992. Social wasps use nest paper hydrocarbons for nestmate recognition. *Anim. Behav. 44*:63-68.

Singer, T.L. and K.E. Espelie, 1996. Nest surface hydrocarbons facilitate nestmate recognition for *Polistes metricus* Say. *J. Insect Behav.* 9: 857-870.

Singer, T.L. and K.E. Espelie, 1997. Exposure to nest paper hydrocarbons is important for nest recognition by a social wasp, *Polistes metricus* Say (Hymenoptera: Vespidae). *Ins. Soc.* (submitted).

Starr, C.K., 1991. The nest as the locus of social life. In: *The Social Biology of Wasps* (K.G. Ross and R.W. Matthews (Eds.)), Cornell University Press, Ithaca, New York. pp. 520-539.

Strassmann, J.E., 1981. Kin selection and satellite nests in *Polistes exclamans*. In: *Natural Selection and Social Behavior: Research and Theory*. (R.D. Alexander and D.W. Tinkle (Eds.)), Chiron Press, New York. pp. 18-44.

Turillazzi, S., R. Cervo and I. Cavallari, 1990. Invasion of the nest of *Polistes dominulus* by the social parasite *Sulcopolistes sulcifer* (Hymenoptera: Vespidae). *Ethology 84*:47-59.

Venkataraman, A.B., V.B. Swarnalatha, P. Nair, C. Vinutha and R. Gadagkar, 1988. The mechanism of nestmate discrimination in the tropical social wasp *Ropalidia marginata* and its implications for the evolution of sociality. *Behav. Ecol. Sociobiol. 23*:271-279.

West-Eberhard, M.J., 1982. The nature and evolution of swarming in tropical social wasps (Vespidae, Polistinae, Polybiini). In: *Social Insects in the Tropics, Vol. 1* (P. Jaisson (Ed.)), Université de Paris-Nord, Paris. pp. 97-128.

West-Eberhard, M.J., 1991. Introduction. In: *The Social Biology of Wasps* (K.G. Ross and R.W. Matthews (Eds.)), Cornell University Press, Ithaca, New York. pp. 1-4.

Willer, D.E. and H.R. Hermann, 1989. Multiple foundress associations and nest switching among females of *Polistes exclamans* (Hymenoptera: Vespidae). *Sociobiology 16:*197-216.

6

Nestmate Recognition in Termites

Jean-Luc Clément and Anne-Geniève Bagnères

Introduction

The earliest studies on termites were performed in tropical and equatorial species nesting in mounds (Emerson, 1933). For many years thereafter it was thought that all termites formed closed familial colonies made up of the descendants of one couple of sexual alates, and that they usually exhibited highly aggressive intra- and interspecific behavior (Pickens, 1934; Wallis, 1964; Nel, 1968; Andrew, 1991). Later studies revealed the existence of supplementary functional reproductives (neotenics) in many species (Noirot, 1956, 1969) and showed that there was a great deal of variation in agonistic behavior between species and even within the same population. These observations led researchers to study the complex phenomenon of nestmate recognition. The purpose of this chapter is to review the main findings in this field.

The study of nestmate recognition requires a multidisciplinary approach. Genetic studies are needed to define the genetic structure of the society: is it a family, an inbred tribe, or a population? Ethological studies are needed to document the response of each caste of the society to individuals from other colonies or genetically different parts of the population during each season. Chemical studies are needed to identify signals and to determine the glands that secret them. Chemical lure experiments using natural molecules are finally needed to confirm the existence of a chemical recognition system.

Sufficient data for meaningful evaluation are available in only a few termite genera. The best known are *Reticulitermes* ((Howard et al., 1978, 1980, 1982a,b); (Clément, 1977a,b, 1978a,b, 1979a,b, 1980,

1981a,b,c, 1982a,b,c,d, 1984; Clément et al., 1981, 1985, 1986, 1988, 1989); (Bagnères, 1989; Bagnères et al., 1988, 1990, 1991); (Parton et al., 1981); (Haverty et al., 1988, 1989, 1990, 1991)), *Zootermopsis* (Haverty et al., 1988, 1989; Blomquist et al., 1979; Korman et al., 1991), *Coptotermes* (Howick et al, 1980; Su et al., 1988, 1991; Brown et al., 1990; Haverty et al., 1990; Korman et al., 1991; Strong et al., 1993) and *Nasutitermes* (Nel, 1968; Thorne, 1982a,b; Levings et al., 1984; Gush et al., 1985; Howard et al., 1988; Everaerts et al., 1988a,b; Thorne et al., 1991). Most of the evidence presented in the following text pertains to these four genera.

Genetic Structure of Termite Societies

When studying social insects, the first questions that must be answered concern the genetic structure of the nest. There are a wide range of possibilities. If the colony was founded by a pair of primary winged sexuals, the descendants form one family and their genome is subject to Mendelian heredity laws. If the initial founders have been replaced by supplementary functional reproductives from the initial family, the nest can be considered as an inbred tribe if the colony is closed. In the open colony, however, there may be exchange not only of workers, but also of primary and supplementary functional reproductives (neotenics). In this case each sample of workers corresponds to a sample of the population responding to the Hardy-Weinberg law (distribution of genotypes according to the frequence of alleles in panmictic reproduction, Mayr, 1963). The only way to determine if colonies are inbred or not is to perform molecular genetics studies.

Termites have two dispersal strategies (swarming and budding) corresponding with the two types of reproduction. Swarming requires winged individuals which fly away from the nest and reproduce by panmixy at a certain time of the year, and budding requires neotenics. Not all termites use these two strategies alike and these differences have repercussions on the genetic structure of the colonies (inbreeding or not). Aggressive intercolonial behavior and therefore the colonial cohesion related to perception of signals, determines whether a colony is open or closed.

Reproduction by Winged Sexuals

Termites of the *Termitinae* family generally disperse by swarming. The time of swarming is genetically preset to the hour of the day. As a result of this mode of dispersal, colonies are isolated. In the simplest cases (e.g. *Bellicositermes* and *Macrotermes*), colonies contain a single couple and members are all brothers and sisters (Noirot, 1956, 1969; Nutting, 1969; Miller, 1969). These colonies are closed, and when the mound is dissected a royal couple is found in a special chamber from which they normally never emerge. This type of colony constitutes a family in the Mendelian sense. Some colonies are co-founded by groups of winged sexuals. In these cases it is difficult to know whether or not the founders are genetically close or not, i.e. whether they are brothers or sisters insuring reproduction with, or in place, of the founding couple (Harms, 1927; Coaton, 1949; Kaiser, 1956; Noirot, 1956, 1969; Roy-Noel, 1974). In *Nasutitermes corniger* (Dudley and Beaumont, 1889a, b, 1890; Dietz and Snyder, 1923; Thorne, 1982), some nests have many laying queens that coexist in the nest without aggressivity. The other nests have only one founding couple. B. Thorne (1982) has hypothesized that the formation of polygynous colonies results from "budding" and fusion. In some Rhinotermitidae (e.g. *Reticulitermes*) winged sexuals are rarely found. They are certainly replaced by neotenics in few years.

Nests containing a number of winged supplementary functional reproductives have been observed in *Macrotermes natalensis* (Bequaert, 1921) and *Odontotermes obesus* (Roonwal and Gupta, 1952; Roonwal and Chotani, 1962). If the winged sexuals come from different colonies, they form non-inbred tribes. If they come from the same, the tribes are inbred. The only way of determining if tribes are inbred or not is to perform molecular genetics studies.

Reproduction by Neotenics

Neotenics are common in *Rhinotermitidae*, *Kalotermitidae*, and *Hodotermitidae* (Buchli, 1950, 1958; Miller, 1969; Nutting, 1969) and are often the only reproductives found in the colonies. Supplementary functional reproductives are also observed in *Termitidae* (Noirot, 1956, 1969; Nutting, 1969). Weyer (1930a,b) discovered over 100 neotenics in one colony of *Microtermes*. They can even be found coexisting with winged sexuals (Skaife, 1954 a,b, 1955; Nutting, 1970). When the society is closed year round, it is possible to have inbred tribes, but there is no genetic data to support this hypothesis. In case of an open colony a panmictic population can be formed. A molecular genetic study is needed in this case, also.

Molecular Analysis of the Colony Genetic Structure

The genetic structure of the population can be studied by assessing the frequency of alleles in polymorphic genes identified in individual workers. Two techniques are available for the purpose, i.e., analysis of enzymatic and DNA polymorphism. Broughton and Kistner (1991) showed that study of phenotypes by hybrization of DNA was possible in *Zootermopsis* but they did not extend their study to detection of the genetic structure of the colony.

The only useful data comes from assessment of enzymatic polymorphism. Strong and Grace (1993) studied 13 *Coptotermes* colonies and showed no polymorphism in 29 loci. Korman and Pashley (1991) found that the rate of gene polymorphism in *Coptotermes* was 16.7% (3 of 18 loci). However the fact that *Coptotermes formosanus* used in these studies were imported, and probably came from neighboring colonies, reduces the value of these results. Higher rates of gene polymorphism have been observed in *Zootermopsis angusticollis*, *Z. laticeps*, and *Z. nevadensis* with Korman et al. (1991) reporting 11 polymorphic loci, but the authors did not go on to describe the genetic structure of the colonies. On the contrary, the rate of polymorphic loci in the European *Reticulitermes* was much much closer to that reported in other insects (52%). Some loci had up to 8 alleles (Clément, 1981c). Conversely, in the United States an excess of homozygotous phenotypes was observed in a population of *Reticulitermes flavipes* studied by Reilly (1987), indicating a high rate of inbreeding. However the latter study did not cover the whole area of distribution.

The most complete study of genetic structure was performed in four species of *Reticulitermes* from several hundred natural colonies obtained throughout Europe (Clément, 1981c). This study showed that the situation is in fact extremely complex, with great differences, not only from one species to another, but also within a given population (Clément, 1981c). The study of enzymatic polymorphism also permitted calculation of genetic distances between species, populations and nests (Clément, 1981a,c, 1982d).

In *Reticulitermes santonensis* (Clément, 1981c) from southwestern France (Charente-Maritime), diagnostic alleles of polymorphic genes (32% heterozygotes) are characteristic of the species (Clément, 1984). Forty-three percent of colonies present phenotype frequencies identical to those predicted using Mendel's laws, thus demonstrating that these colonies may be considered as true families founded by winged alate sexuals, or neotenics coming from various nests. Fifty-seven percent of colonies exhibited an excess of homozygotes, and can therefore be considered to be inbred. A high number of neotenics (sometimes 10 to 20 together) are observed in these natural colonies. The high degree of relatedness (or low genetic distance) between the colonies implies that they are open, a hypothesis that is supported by ethological studies. The facility with which neotenics can appear is thus not compensated by the open nature of the colonies. These neotenic brothers and sisters are developed from nymphs with short pterothecae.

The species *Reticulitermes (lucifugus) grassei* (Clément, 1981c) from western Europe (southwest France, western Spain, Portugal and southern Spain) presents genetic structures of colonies corresponding to the variety of the different ecosystems in this area. In Atlantic zones, where dead wood is abundant, 60% of colonies can be considered as populations since the incidence of genotypes in the colonies are consistent with the law of Hardy-Weinberg. The genetic distance between nests is very small. We can consider the zone between Bordeaux in France and Santiago de Compostella (Galicia, Spain) as a single panmictic population. This is a unique case in social insects. At each extremity of this zone, (Charente-Maritime, France, and South of Spain) 82% to 100% of the colonies are familial societies. In the North, at the boundary of the distribution area, the conditions do not allow a homogeneous distribution. In the South the soil is too dry to allow budding.

The species *R. (lucifugus) banyulensis* (Clément, 1981c) which lives on the northeast Mediterranean coastline of Spain and Southeast of France (Pyrénées orientales, Roussillon) is composed of 50 to 60% of familial colonies, and 20 % to 25% of small populations. The remaining colonies present an excess of heterozygotes or homozygotes. The genetic distances between colonies and populations are high. The conditions in these areas, where the ground is dry and dead wood is scarce, are not conducive to budding.

In Italy, *R. (lucifugus) lucifugus* is composed of 50% of familial colonies, 20% of population colonies, 20% of colonies with an excess of heterozygotes, and 10% of inbred colonies. A decrease in the incidence of polymorphous alleles is observed from the north to south and there is a high degree of relatedness in colonies suggesting an open structure with many exchanges between populations.

Thus, it appears that the situation is very complex in the same genus and even in the same species: monogamous open or closed colony, polygamous open society resembling a population structure. It should be pointed out that the proportion of inbred colonies is low everywhere, except for one very prolific species (*R. santonensis*) which exhibits an open structure and a particular developmental mode that favors transformation of nymphs into neotenics. In the super-species *lucifugus* the incidence of inbred colonies (excess of homozygotes) is low in each area: 9%, 0%, 5%, 12%, and 10%, respectively. The open society is a way of limiting inbreeding due to neotenics and allows rapid dispersal through the soil (Clément, 1981c; Reilly, 1987).

Evidence for a Colonial Recognition System

The study of recognition requires data from four analyses: A. Analysis of intercolonial behavior to determine if there is a communication system, the influence of that system on whether the colony is open or closed in relation with the aforesaid genetic studies and experimental proof of its existence; B. Analysis of the nature of the signals: chemical, acoustic signals, etc, and its origine (glands...); C. Study of the variation of the chemical signals between castes, colonies, and populations; and D. Study of the regulation of the chemical signature.

Ethological Evidence of Colonial Recognition

Thorne and Haverty (1991) published an exhaustive review of intracolonial, intraspecific, and interspecific agression in termites. An outstanding feature of this excellent report was that it analyzed colonial recognition in terms of aggression rather than avoidance which is an economic reaction that is difficult to measure if not by the genetic consequences. Most studies describe aggressive responses during encounters between individuals from different colonies of the same species.

Intraspecific recognition has been extensively studied in the *Reticulitermes* (Pickens, 1934, Clément, 1978a, 1980, 1981a, 1986; Bagnères, 1989; Bagnères et al., 1990,1991). Clément (1978a) has proposed a scale for a better quantification of aggression. This study of *Reticulitermes* showed that intraspecific aggression varies greatly from one species to another and within the colonies. These reactions can vary depending on the season (Table 6.1).

Ethological experiments (Dudley and Beaumont, 1889a,b; Andrew, 1911, Thorne, 1982b, Levine and Adams, 1984, Traniello and Beshers, 1985) showed that individuals of *Nasutitermes* species generaly attacked individuals from other nests. Some data are also available in higher termites of the genus *Macrotermes* (Kettler and Leuthold, 1995) or *Microtermes* (Adams, 1991). Data also come from ethological tests performed in the laboratory in *Coptotermes* (Howick and Creffield, 1980; Su and Scheffrahn, 1988).

Observations have also documented cases where aggression is absent. Encounters can be followed by licking. Grassé (1986) indicated that there was no aggression between individuals from different *Bellicositermes bellicosus* nests. Haverty and Thorne (1989) reported the same results for different species of *Zootermopsis*. Su and Haverty (1991), and Su and Scheffrahn (1988), made the same observation in

TABLE 6.1. Mean of intraspecific agression index (Ag) with the confidence interval for $\alpha = 0.05$. The index (Ag) between two colonies is calculated by the formula (M + (m/2))x2.5, where M is the average number of dead workers, m is the average number of injured workers. Each confrontation between two colonies included five replicates using 20 workers from each colony in a petri dish of 50mm diameter for 24 hours. The number of confrontations is indicated between parentheses.

		WINTER		SUMMER
R. santonensis	(4)	1.55 ± 0.35	(15)	2.17 ±1.75
R. (l.) grassei	(32)	49.43 ± 12.62	(16)	14.97 ± 9.2
R. (l.) banyulensis	(15)	56.6 ±20.09	(10)	35.97 ± 27.36
R. (l.) lucifugus	(6)	68.97 ± 26.67	(18)	15.82 ± 7.72
R. flavipes	(25)	7.49 ± 8.09	(26)	3.57 ± 5.07
R. malletei	(11)	10.9 ± 15.08	(13)	19.5 ±10.4
R. virginicus	-	-	(10)	16.5 ± 7.0

Coptotermes formosanus. In *Reticulitermes*, the colonies of *R. santonensis* in France (Clément, 1978a), and *R. flavipes* in U.S. (Bagnères et al., 1990) are always open. Regardless of the season, workers do not show any aggression to individuals from different colonies. The aggression index is low in summer as well as winter. This is consistent with the low genetic distance between colonies. Experimental studies using individuals from different *Reticulitermes santonensis* colonies could be mixed with little or no reaction (Clément, 1986). Like those of *R. santonensis* and *R. flavipes,* colonies of *R. malletei,* the second American species to be studied, were found to be open (Clément et al., 1986).

Ethological data in *Reticulitermes (lucifugus) banyulensis* demonstrate that the situation can be more complex, since colonies can be open or closed, but intercolonial aggression is usually high. Fifty to 60% of colonies are monogynous and genetically distant from one another. Clément (1986) showed that the index of aggression was inversely related to the rate of inbreeding. Nevertheless it is noteworthy that although the aggression index in *R. (l.) banyulensis* colonies is high, there is a great deal of variability as shown by the high Standard Deviation values. This indicates great heterogeneity in the behavior of one colony with another (Bagnères, 1989).

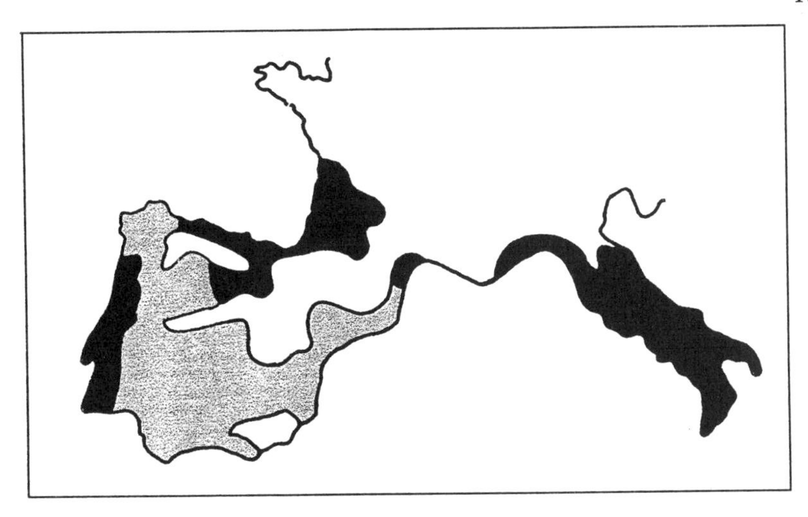

FIGURE 6.1. Internest agonism in *Reticulitermes* species of the *lucifugus* complex in Europe: black areas indicate regions in which index of agressivity between nests and populations is low (Ag $\leq$ 25). Grey dotted areas are inhabited by closed societies (Ag > 25). White areas are not inhabited by termites.

Another example of the complexity of the colonial recognition has been observed in *R. (lucifugus) lucifugus* in Italy, and *R. (lucifugus) grassei* from southwest France and from the Iberian Peninsula (north and west regions). It is possible to bring into contact in summer, when all nymphs exhibit short pterotheca, workers from any colonies in these different geographic areas of Italy, France, Spain and Portugal, without observing any aggressive behavior (Figure 6.1), These areas can be compared to 3 panmictic populations or to 3 supercolonies (Clément, 1981). During the winter, colonies of *R. (lucifugus) grassei* and *R. (lucifugus) lucifugus* are closed. In the dryer zones, on the edges of the distribution area, because of the more severe climatic conditions, aggressivity is rather like in *R. (l.) banyulensis*. A particular feature of *R. (l.) grassei* is that colonies appear to have a defined character i.e. a dominant colony remains dominant even though aggressivity may vary seasonally depending on the swarming date (Bagnères, 1989).

Recognition Signals

Evidence of Recognition Signals. Because termites are living in lightless galeries, and are blind most of their life, they communicate mainly by mechanical (acoustical) and chemical signals. Howse (1964) proved that sounds produced by termites serve as alarm signals and not for colonial recognition. However a recent paper by Kirchner et al. (1994) on vibrational alarm communication in the damp-wood termite *Zootermopsis nevadensis*, provoded no clear answer to the question of the functional significance and perception by the nestmates of this humming. The authors simply show that in the laboratory, the animals respond preferentially to temporal patterns similar to those of the natural signals: temporal rather spectral cues seem to be used for signal discrimination by workers and soldiers.

Different ethological studies indicate that all castes contribute to maintaining a closed colony. Workers exhibit intraspecific agonistic behavior in closed societies (Nel, 1968; Pickens, 1934; Clément, 1981a; Thorne, 1982b; Haverty and Thorne, 1989).

A number of authors have speculated that recognition is based on a "chemical identification card". Tactile contact has been observed before aggressive behavior (Stuart, 1970, 1975, 1976). Clément (1978a, 1981a) showed that termites perform a basic movement of contact when they pass each other in a gallery (Figure 6.2). A filmed study (Clément and Devez, 1981) indicated that aggressivity was always preceded by one or more sequences of this ritual-like behavior with application of the antennae or buccal palps all over the body of the insects.

Observations Supporting a Chemical Signature. Behavioral studies showed that termites exchange a constant flow of chemical signals through their antenna and palps (Verron, 1960, 1963; Abushama,

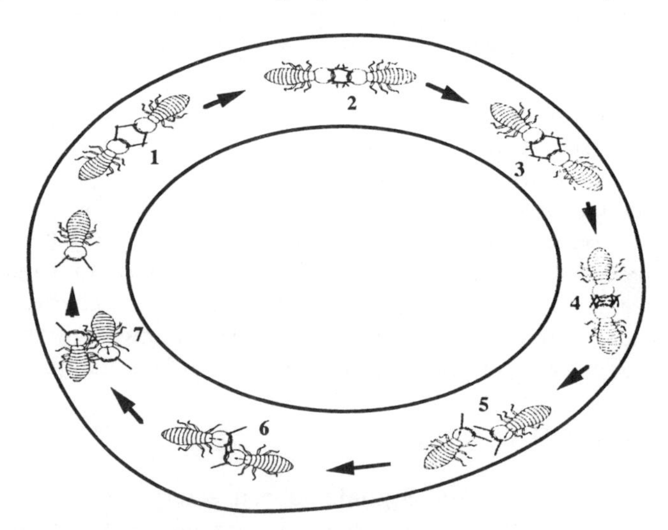

FIGURE 6.2. Basic contact sequence (movements 1 to 7) observed when *Reticulitermes* workers pass each other in a gallery.

1964a,b; Quennedey, 1978). In *Reticulitermes* it has been demonstrated that chemical substances in the cuticle are detected by olfactory organs (Clément, 1980, 1982c). These observations were confirmed by Binder (1988), and Haverty and Thorne (1989).

In agreement with Howse (1975), Clément (1982c, 1986) suggested that variations in the composition of cuticular products provided the basis for recognition inducing aggression.

The only experiments proving the existence of a chemical signature involve interspecific recognition behavior in *Reticulitermes*. The first ethological experiment was performed in *R. virginicus* by Howard et al. (1982), and suggested by analysis of the chemical mimicry used by termitophiles (Howard, 1979, 1993, Howard et al., 1982). Bagnères et al. (1991) have shown that a lure covered with organic extract from one species triggered aggression in another. The apolar fraction of the cuticular extract containing hydrocarbons has the greatest effect, but the polar fraction accounted for the chemical signature to a lesser degree. These experiments implicating cuticular hydrocarbons in interspecific recognition were confirmed by Takahashi and Gassa (1995). Thus it seems that the cuticle of termites has a chemical signature which is read by contact and triggers aggression.

A strong body of evidence confirmed the involvement of chemical cues in nestmate recognition.

Source and Description of Chemical Signals

Chemical studies showed that, like other social insects, termites have exocrine glands that release the chemical mediators involved in inter- and intraspecific recognition (Howse, 1975; Quennedy, 1975; Stuart, 1969, 1975, 1976). Termite glands are classically divided into 2 categories: digestive tract glands and tegumentary glands. Both categories of glands can release chemical substances which can be spread on the cuticle.The cuticle acting as a lipophilic layer can retain and homogenize exocrine secretions which can play in this way a role in colonial recognition (Blum, 1987). Although, even if it is not considered as a gland, the cuticle itself can be considered as a major source of signals involved in chemical communication processes.

Digestive Glands

Mandibular Glands. The mandibular glands are usually pluricellular structures with an intracellular reserve and a central duct ending at the base of each mandible. They contain characteristics secretory cells, with crystals probably of protein origin (Grassé, 1982). Unlike those in other insect orders such as Lepidoptera, Hymenoptera, or Blattoidea, the function of mandibular glands in termites is unclear. Their role in the production of saliva appears unlikely. They are inactive in workers and active in soldiers and functional sexuals. Grassé (1982) speculated that the glands could produce one or more pheromones. This hypothesis was based on Lebrun's studies (1972) concerning the inhibition of formation of sexuals, and on the fact that in many insects, e.g. coachroaches, these sort of glands produce gregarious pheromones (Brossut and Sreng, 1985).

Salivary (or Labial) Glands. Salivary gland generally occur in pairs, each with a large reservoir. The glands and reservoirs are connected to the hypopharynx by long ducts. Billen et al. (1989) in *Macrotermes bellicosus* and Kaib (1990) in *Schedorhinotermes lamanianus* recently described 3 types of secretory cells in the salivary gland of these 2 species. They mainly produce saliva which is used as social food. In some species they can produce a sort of glue or sticky saliva which can be used to kill opponents by immobilizing them, as in *Microtermes*. No data is available about production of this gland. According to Kaib (1990), molecules from these glands (or from their reservoir) are more likely to be hydrosoluble, especially in *Schedorhinotermes lamanianus*, and could act as a gregarious pheromone and appetite stimulator interactively with the sternal glands which would play the opposite role. We can speculate too that they could carry substances like hydrocarbons since termites show intense licking activity. The colonial odor could come from passive transfer of pheromonal substances by contact.

Tegumentary Glands

Unicellular Glands. Unicellular glands are found throughout the tegument but especially the head (Quennedey, 1975). Their exact function is unclear especially when they are found in glandular fields at the level of the cuticle. In other social insects some glandular structures apparently have an "active reservoir" function, e.g. postpharyngeal glands in ants (Bagnères and Morgan, 1991) or Dufour glands in wasps (Dani et al. 1994), and are implicated in nestmate recognition. Equivalent structures have not been described in termites, and it can be speculated that these unicellular glands could act as reservoirs. According to Quennedey (1978) in *Cubitermes fungifaber*, several of these cells can be seen with pores opening to the outside, near the fontanelle.

Pluricellular Glands. *Labral glands:* The labral glands were first discovered by Holmgren (1909) and further studied by Feytaud (1912). Their function is unclear but they are observed in all castes. *Abdominal pleural glands:* The abdominal pleural glands were discovered in *Cubitermes fungibaber* (Ampion, 1980). They are composed of a glandular structure with chemoreceptor hairs. No equivalent has been observed in any other species. *Tarsian glands:* In *Rhinotermitidae,* tarsian glands can be found at various levels of the tibias, in pits on the cuticle (Bacchus, 1979). Given the presence of the emptying canals, and their strategic position, these pluricellular glands can be ruled out as sources of recognition signals.*Tergal glands:* Tergal glands are found only under the abdominal tergites of imagos. They are absent in other castes. Lebrun (1971) reported the results of an extensive study of the tergal glands in *Kalotermes flavicollis*. In *Kalotermes flavicollis,* tergal glands regress in old sexual reproductives and in neotenics. They are not found in several families including that containing the genus *Reticulitermes* (*Heterotermitinae*). A comparative and evolutionary study on different families was conducted by Ampion (1980) who concluded that tergal glands could be implicated in reproductive activity. In coachroaches, which are the insects most closely related to termites, these glands release aphrodisiacs (for a review see Brossut and Sreng, 1985). Ampion (1980) also stated that these glands are present in *Schedorhinotermes lamanianus* but Kaib (1990) did not mention them in a recent review of chemical signals in this termite. No description of a pheromone has ever been published. *Sternal gland:* The morphology of the sternal gland was studied by Noirot and Noirot-Timothée (1965), Quennedey (1978), and also Lebrun (1971, 1972). This gland is found in all termites and forms a glandular pad on the medial side of the abdominal sternites, in a very primitive termite such as *Mastotermes* as well as in *Rhinotermitidae* or in *Termopsidae*. Size varies according to caste but it is always small. The nerve supply of the sternal gland is much more complex than that of the tergal gland. Laduguie (1993) recently compared the structure and activity of the sternal gland in three phylogenetically distant species, i.e. *Reticulitermes santonensis* (*Rhinotermitidae*: *Heterotermitinae*), *Pseudocanthotermes springer* (*Termitidae*: *Macrotermitinae*), and *Nasutitermes lujae* (*Termitidae*: *Nasutitermitinae*) at different developmental stages. This author concluded that the sternal gland was one of the most implicated glands in chemical communication.

In earlier works, Matsumura et al. (1968) and Howard et al. (1976) described substances found in the sternal gland of *Reticulitermes virginicus* as trail-pheromones. This finding was confirmed by Yamaoka et al. (1987) in *Reticulitermes speratus*. The most widely described substance is 3Z,6Z,8E-dodecatrien-1-ol. This compound was found also in *Pseudocanthotermes spiniger* by Bordereau et al. (1991).

Observations made during behavioral studies indicate that the gland contains several sorts of compounds of high and low molecular weight (Ritter et al., 1969; Traniello, 1982; Kaib et al., 1982; Runcie, 1987). The functional versatility of this gland (orientation, recruitment, food attraction, etc.) has been documented in numerous species (Kaib et al., 1982; Oloo and Leuthold, 1979; Kaib, 1990) including *Reticulitermes santonensis* in which Ladugie et al. (1994) stated that dodecatrienol serves as both a trail-following pheromone for workers and a sexual pheromone for imagos. Clément (1982a) showed that specific isolation of 3 species of *Reticulitermes* was at least partly due to sex-specific pheromones released at long distances. In *Reticulitermes flavipes*, Clément et al. (1989) showed that the sex pheromone involved in attraction was the *n*-tetradecyl propionate in the sternal gland. This specific isolation is also due to another family of compounds, i.e. contact pheromones on the cuticle. Grace et al. (1995) suggested that different colonies of *Reticulitermes hesperus* may produce different quantities of trail pheromone, but the data provided no evidence of colony recognition compounds. *Frontal gland:* Holmgren (1909) used the frontal gland as a basis for the currently used classification of termites. It is specific for Isoptera regrouping all genera of *Rhinotermitidae*. The Protermitidae group, now broken down into 4 families (*Mastotermitidae, Kalotermitidae, Hodotermitidae,* and *Termopsidae*), does not have a frontal gland. When present, this gland is found in some imagos, absent in workers and achieves fullest development in soldiers (Feytaud, 1946; Grassé, 1982). In genera presenting several classes of soldiers (minor and major), e.g. *Schedorhinotermes lamanianus*, the gland is present in the two forms (Kaib, 1990). In *Nasutitermes* the gland occupies must of the skull and has a characteristic pear shape (nasute).

Following the work of Moore (1964, 1968), the secretions of the frontal gland have been studied in a great number of termite species. Most of these secretions are terpenes, either simple monoterpenes in *Nasutitermes* including α-pinene (Vrkoc et al., 1978), often in combination with diterpenes, which probably serves as a solvent function (Braekman et al., 1983; Braekman et al., 1984; Gush et al., 1985), or sesquiterpene mixtures (Goh et al., 1990; Everaerts et al., 1993). Aldehydes, acids, and alkaloids have also been found in several species of Brazilian termites (Blum et al., 1982), and vinylketones as well as ethyl-ketones and aliphatic dienones in *Schedorhinotermes* (Prestwich et al., 1975). Many saturated and unsaturated hydrocarbons have also been identified and can dissolve mucopolysaccharides

(Moore, 1968; Prestwich et al., 1977). The great majority of secretions even in European and American subterranean termites are complex mixtures of terpenes which can be used to separate close species and subspecies (Bagnères et al., 1990). Great differences have been observed between soldiers from different colonies of European *Reticulitermes* (Parton et al., 1981). The major role of these products is defense since only the soldier caste, ensuring defense outside of the nest, exhibit the gland and its secretions. Terpene products have also been observed in imaginal sexuals of *Reticulitermes* in a much less concentrated form (Bagnères and Clément, unpublished results).

These chemicals can be included in the common nest chemical signature and serve both as alarm pheromones and defensive substances. Most of these molecules are volatile and the relative proportions vary from one colony to another. They can be found throughout the colony.

Cuticle

The cuticle is usually a thin tegument that covers most of the body of all castes of termites. The cuticle of insects has been described by a number of authors (Delachambre, 1973; Locke, 1974; Hadley, 1981; Wigglesworth, 1990) and in termites in particular by Bordereau (1979). In this chapiter we will be referring to the cuticle only in so far as it bears signals mainly on the epicuticle, i.e. the topmost layer. The epicuticule is crossed by small ducts that may bring to the surface products synthesized under the epidermis (and/or vice versa). Larvae and white soldiers renew their chemical apparatus at each molt. In addition to trail following and sexual attraction, interindividual interspecific and intraspecific communication is achieved by contact with the cuticle using the antennae as described above (Clément, 1982c, 1986).

The chemical composition of the epicuticle, at least of the apolar fraction, is now well-documented (for general review, see Lockey, 1988) although there are still questions about biosynthetic pathways, transport modes, and regulation processes. Oenocytes (epidermic cells modified to allow lipid metabolism) seem to be the most likely sites of biosynthesis, and haemolymph lipophorins serve for transportation of cuticular hydrocarbons (Katase and Chino, 1982, 1984). The hydrocarbons are synthesized from classical fatty acids, the precursors including acetate, propionate, succinate, malonate (Blomquist et al., 1980; Blomquist et al., 1982). The findings of Blomquist and coworkers have also contributed greatly to understanding of these processes in different insects (see general review in Blomquist et al., 1987, and de Renobales et al., 1991). In termites, *Reticulitermes flavipes* (Howard et al., 1978) and *Zootermopsis angusticollis* (Blomquist et al., 1979b) were the first species in which cuticular lipids were studied. The main role attributed to these lipids is that of a barrier to prevent desiccation and keep toxins out. The first evidence for the role of lipids in the recognition processes, which was suspected by Clément since 1978(a), was obtained in termites and confirmed in 1982 by Howard et al. (1982) and by Bagnères et al. (1991). Cuticular lipids are now currently considered as the main factors in chemical communication and as contact pheromones, and particularly in termites (Clément et al., 1988; Howard, 1993).

Cuticular hydrocarbons in termites are saturated (alkanes) and/or unsaturated (alkenes) long-chain hydrocarbons (generally 20 to 40 carbon atoms). They can be branched, the main ramifications being mono- and dimethylated in the lateral or central position. Alkenes are monounsaturated (monoenes), diunsaturated (dienes) or triunsaturated (trienes). The presence of conjugated dienes is uncommon in insects but was demonstrated for the first time in termites, i.e. *R. flavipes* (Howard et al., 1978) and *R. santonensis* (Bagnères et al., 1990) in the form of 7,9-pentacosadiene. Likewise trienes of cuticular origin, which are also uncommon in insects, have been described in *Reticulitermes malletei* (Bagnères, 1989).

Intraspecific and Intercolonial Variation in Chemical Signals

Assuming that chemical signals are responsible for colonial identity, as indicated by the studies of Howard et al. (1982) and Bagnères et al. (1991) concerning specific identity, it is necessary to measure qualitative and quantitative changes in molecular extracts from different castes and colonies of termites. To play a role in the colonial signature these molecules would have to be shared by all castes in the same pattern, despite the fact that the cuticle, because of its lipid nature, can be contaminated by molecules from other castes. Molecules must show the same variations in all individuals in a given colony.

Variations in Secretions from the Frontal Gland. Data reported by Parton et al. (1981) and Lemaire et al. (1990), in experiments on four species of *Reticulitermes*, were among the first to show that the nature and relative proportions of terpenes (monoterpenes: α and β-pinene, limonene, sesquiterpenes and a diterpene alcohol: geranyl linalool) and short-chain hydrocarbons secreted by soldiers, vary from one species, population, and colony to another. Variations between colonies of the same population are significant and are characteristic of each colony and population. In some populations of *R. (l.) banyulensis*, some molecules (sesquiterpenes) are absent in some colonies, whereas their neighbours have large amounts. These variations, associated with data obtained by studying enzyme polymorphism, can be used to calculate an index of similarity between populations.

Another example of variation in terpenes involves soldier phenotypes in *R. flavipes* (Bagnères et al., 1990). Multivariate analysis, using cuticular hydrocarbons as variables, separated the different phenotypes defined in function of the relative proportion of defensive substances (monoterpenes: α-pinene and β-pinene, and limonene; sesquiterpenes, α-cadinene, and γ-cadinene, and geranyl linalool).

In six species of *Nasutitermes*, mainly from New Guinea, Everaerts et al. (1988a, b) and Everaerts (1989) used correlation indexes to demonstrate that the monoterpene fraction is fairly constant in the colonies of some species, while it is highly variable in others. In most of the species studied by Everaerts there appeared to be a high degree of heterogeneity in secretion depending on the geographic distribution of the colonies. Other intraspecific chemical variability was described in *Longipeditermes longipes* (Goh et al.,1984).

It is interesting to note that the cuticle of workers in *Reticulitermes*, and probably in most other species, seems to be kept free from assumed contamination by terpenes and hydrocarbons secreted by the frontal gland of soldiers. This observation implies that there is a rapid biochemical cleansing process to limit this type of chemical contamination of the cuticle.

Although frontal gland secretions do not seem to be directly involved in nestmate recognition in most cases, nevertheless Chuah et al. (1983) suggested that the monoterpene/diterpene ratio might play a role in determining the soldier-to-worker ratio and/or the soldier-to-worker ratio could influence behavior and therefore the secretion of defensive substances by *Hospilitermes umbricus* soldiers.

Variations in Cuticular Products. Cuticular hydrocarbons composition changes during ontogenesis in insects (Trabalon et al., 1988). Early works by Howard et al. (1978, 1982) and Blomquist et al. (1979) and more recent ones by Haverty et al. (1990), Watson et al. (1989), Bagnères (1989), and Bagnères et al. (1988, 1990) show that different castes have different proportions of specific cuticular products. Chemical variation between castes is great considering overall intraspecific variations. This observation seems reasonable in social insects in which the different castes of each colony play a specific role. In addition to frontal gland secretions, another distinguishing factor for the soldier caste in *Reticulitermes* is cuticular hydrocarbons (Bagnères et al., 1988; Bagnères, 1989). In each caste an intracolonial homogeneity is present, but variations appear between nests.

We attempted to determine the influence of the environment in these differences by principal component analysis of different colonies of five species of *Reticulitermes* using cuticular hydrocarbons as variables then adding extrinsic data such as geographic location, the type of wood from which the insects were collected, the date of extraction, the date of analysis, etc. None of these variables was able to explain the intrinsic variations in each colony (Bagnères, 1989). Thus each colony seems to have its own odor even though, as in other insects, it is not possible to determine the relative role of each. The qualitative differences observed in terms of proportions and quantities (in the nanogram to microgram range) allow identification of one colony from another, product by product. In the following example, the quantity of 5-methylheptacosane has been calculated in 18 different colonies of *R. (l.) banyulensis* (Figure 6.3).

We can conclude that of the three gland systems, secreting chemical signals which have been thoroughly studied in termites, i.e. the sternal gland, frontal gland, and cuticle, only two families of compounds show significant intercolonial variations. They are defensive secretions and cuticular hydrocarbons.

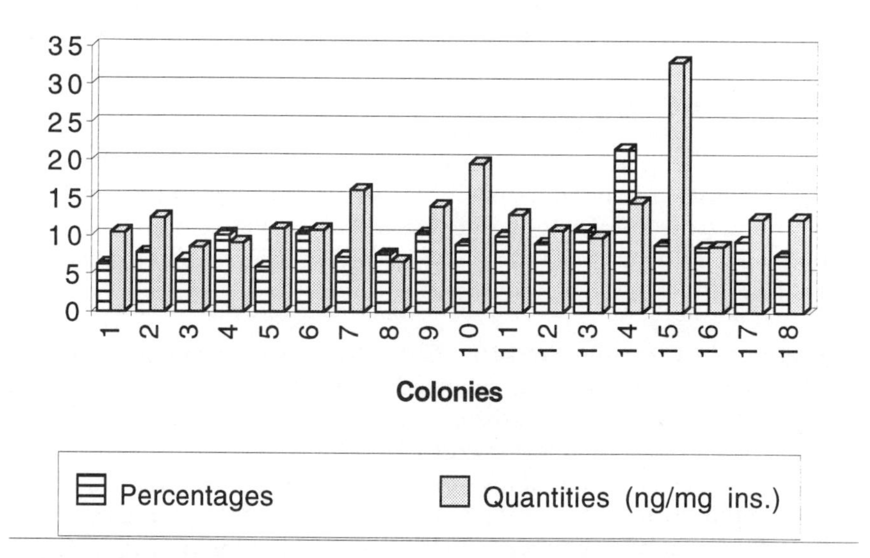

FIGURE 6.3. Quantities and relative proportions of the 5-MeC27 measured in workers from 18 nests of *Reticulitermes (lucifugus) banyulensis* from France (Roussillon).

Qualitative and quantitative variations have been noted between the defensive secretions of soldiers from different colonies but no trace of these molecules has been found in the cuticle of workers, even though the latter have a colonial chemical signature. It can thus be concluded that the signal involved in colonial recognition is cuticular hydrocarbons.

Thus, it can be hypothesized that, as in ants (Bonavita-Cougourdan et al., 1987, Nowbahari et al., 1990, Morel et al., 1988, Meskali et al., 1995) and wasps (Espelie et al., 1990; Lorenzi et al., 1996), cuticular hydrocarbons serve as chemical signals in the colonial signature.

Variations in the relative proportion of some cuticular products characterize only the nest. However given the great difference in genetic structure of colonies, these signals must be strongly regulated.

Regulation of the Colonial Chemical Signature. Molecular genetic studies in *Reticulitermes* (Clément, 1981) showed that all genetic structures ranging from the family structure to the panmictic population are possible. Although uncommon even more or less consanguineous tribal structures can be observed. Ethologic studies confirmed these data by measurement of the aggressive behavior between nests. When the society is closed, under ecological conditions prohibiting budding, the genetic distance is great between colonies which remain monogamous. In such cases the variations of chemical cuticular signals between colonies is great (Bagnères et al., 1988). If secretion of cuticular hydrocarbons depended only on the genome, there would be great individual differences in accordance with Mendelian laws. In fact the differences between cases were purely phenotypic. There is no difference between individuals belonging to the same caste. Similarly, the level of aggression is high for individuals from other colonies which should be chemically similar as a result of the gene allele process that codes the molecular proportions of molecules in the pheromonal blend. These finding strongly suggest that there must be an internal regulation mechanism to specifically standardize the blend of chemical signals that serves as the colonial chemical signature in all individuals in the colony. This has been measured in ants of the genus *Leptothorax* by Provost et al. (1993).

In open colonies with a population type genetic structure, there are numerous, genetically different functional reproductives. It can be speculated that exposure to a wide range of chemical signatures during larval olfactory training enables individuals in these colonies to accept different patterns. However, this hypothesis must be ruled out since there is little chemical difference between individuals in colonial populations. Thus it must be assumed that as in colonies with a family structure there is a regulation process that acts only on the cuticular hydrocarbon blends that characterize each small population. Thus the colonial chemical signature is standardized within the "open" zone.

It can be assumed that this mechanism is present in all species and that the openness of a colony is due to a combination of factors: genetic similarity, budding, and population density due to the availability of food. This mechanism reduces the difference between colonial chemical signatures and limits agonistic behavior, thus resulting in an open colony. *Reticulitermes (lucifugus) grassei* which lives in the north of its distribution area provides a good illustration of this process. Molecular genetic studies showed that despite their genetic closeness (mean DI = 0.018), 82% are family structures because it is not possible to occupy the whole area due to climatic conditions. Yet when workers from different colonies are confronted, there is no aggression. For all 100% closed family colonies in the Mediterranean area, the genetic distance is high (mean DI = 0.59) and the level of aggression is great (100%). Thus it appears there is a genetically determined signal that is standardized within the colony to obtain a colonial signature.

Various experiments and observations seem to indicate that contacts between members of each society result in a constant transfer of cuticular hydrocarbons. Some cues may be obtained by transfer after contact with other members of the colony (Vauchot, 1995; Vauchot et al., 1996). These molecules are rapidly broken down and resynthesized in proportions depending on the caste and above all the blend of mixtures characteristic of each colony. The mechanism can be explained as follows: the specificity of the synthetic enzymes of the hydrocarbons is low (Ferveur, 1991; Ferveur and Jallon, 1993) and the products of the first degradation of exogenous or endogenous molecules become substrates for the final stages of synthesis for only those which contribute to the colonial signature, thus modulating the quantities synthesized on the cuticle and leading to a homogeneous pattern in all members of a given colony.

It can be speculated that this degradation takes place directly on the cuticle given the presence of extracellular lipases. The products can be easily internalized or, alternatively, hydrocarbons may be constantly internalized for degradation and then reincorporated in new cuticular compounds. The absence of molecules from the defensive glands of soldiers on the worker cuticle suggests that defensive compounds are broken down as has been observed in ants with a non specific compound (Meskali et al., 1995).

Conclusion

Termite colonies can present many genetic structures (royal couple, small group of primary reproductives or neotenics, panmictic population society). They correspond to open or closed societies. As a general rule it can be stated that the family societies are closed and that population societies are open, but there are many subtleties. The genetic distance between colonies, when they have been determined, confirms this rule. In open colonies, the genetic distance between individuals is low without inbreeding, but the colonial chemical signatures are similar. In closed colonies, nestmate recognition is due to chemical signals constituting a genuine chemical signature. Various findings indicate that this chemical signature is based on blends of cuticular hydrocarbons. These chemical signatures may be standardized by an as yet unclear process that masks genetic differences in the colony.

At the present time, there is no ethological and chemical proof for the use in nestmate recognition of these hydrocarbons in termites as in ants and wasps, but just correlations.

Many questions remain to be answered. What molecules are actually involved in the colonial chemical signature? What is the regulation mechanism? What are the different steps in the synthesis of the molecules forming the colony odor? How do termites recognize these molecules using their antennae and palps? Are their specific receptors for these molecules? Are they taken into account by pheromone binding proteins? How is regulation achieved within a given colony?

References

Abushama, F.T. 1964a. The olfactory receptors on the antenna of the dampwood termite, *Zootermopsis angusticollis*. *Ent. Monthly M. Pagaz* 100:145-157.

Abushama, F.T. 1964b. Electrophysiological investigations on the antennal olfactory receptors of the moist wood termite, *Zootermopsis angusticollis*. *The Entomologist* 97:148-150.

Adams, E.S. 1991. Nestmate recognition based on heritable odors in the termite *Microcerotermes arboreus*. *Proc. Natl. Acad. Sci. USA,* 88:2031-2034.

Ampion, M. 1980. Les glandes tergales des imagos de termites: Etude comparative et signification évolutive. Thèse de 3ème cycle de l'Université de Dijon, France, 107p.

Andrews, E.A. 1911. Observations on termites in Jamaica. *J. Anim. Behav.* 1:193-228.

Bacchus, S. 1979. New exocrine gland on the legs of some *Rhinotermitidae* (Isoptera) *Intern. J. Morphol. Embry.* 8:135-142.

Bagnères, A.-G. 1989. Les hydrocarbures cuticulaires des insectes sociaux. Détermination et rôle dans la reconnaissance spécifique coloniale et individuelle. Thèse de Doctorate de l'Université Pierre et Marie Curie, Paris, France, 167p.

Bagnères, A.-G., and Morgan, E.D. 1991. The postpharyngeal glands and the cuticle of Formicidae contain the same characteristic hydrocarbons. *Experientia* 47:106-109.

Bagnères, A.-G., Clément, J.-L., Lange, C., and Joulie, C. 1988. Les hydrocarbures cuticulaires des *Reticulitermes* français: Variations spécifiques et coloniales. *Act. Coll. Ins. Soc.* 4:34-42.

Bagnères, A.-G., Clément, J.-L., Blum, M.S., Severson, R.F., Joulie, C., and Lange, C. 1990. Cuticular hydrocarbons and defensive compounds of *Reticulitermes flavipes* (Kollar) and *R. santonensis* (Feytaud): Polymorphism and chemotaxonomy. *J. Chem. Ecol.* 16:3213-3244.

Bagnères, A.-G., Killian, A., Clément, J.-L., and Lange, C. 1991. Interspecific recognition among termites of the genus *Reticulitermes*: Evidence for a role for the cuticular hydrocarbons. *J. Chem. Ecol.* 17:2397-2420.

Bequaert, J. 1921. Insects as food. How they have augmented the food supply of man kind in early and recent times. *Nat. Hist.* 21:191-200.

Billen, J., Joye, L. and Leuthold, R.H., 1989. Fine structure of the labial gland in *Macrotermes bellicosus*. *Acta Zool.* 70:37-45

Binder, B.F. 1988. Intercolonial aggression in the subterranean termite *Heterotermes aureus*. *Psyche* 95:123-137.

Blomquist, G.J., Howard, R.W., and McDaniel, C.A. 1979a. Structure of the cuticular hydrocarbons of the termite *Zootermopsis angusticollis* (Hagen). *Insect Biochem.* 9:365-370.

Blomquist, G.J., Howard, R.W., and McDaniel, C.A. 1979b. Biosynthesis of the cuticular hydrocarbons of the termite *Zootermopsis angusticollis* (Hagen). Incorporation of propionate into dimethylalkanes. *Insect Biochem.* 9:371-374.

Blomquist, G.J., Nelson, D.R, and de Renobales, M. 1987. Chemistry, biochemistry, and physiology of insect cuticular lipids. *Arch. Insect Bioch. Physiol.* 6:227-265.

Blomquist, G.J., Chu, A.J., Nelson, J.H., and Pomonis, J.G. 1980. Incorporation of succinate into methyl-branched alkane in a termite. *Arch. Biochem. Biophys.* 204:648-650.

Blomquist, G.J., Dwyer, L.A., Chu, A.J., Ryan, R.O., and de Renobales, M. 1982. Biosynthesis of linoleic acid in a termite, cockroach and cricket. *Insect Biochem.* 12:349-353.

Blum, M.S. 1987. Specificity of pheromonal signals: A search for its recognitive bases in terms of a unified chemisociality, pp. 401-405, *in* J. Eder and H. Rembold (eds). Chemistry and Biology of Social Insects, Verlag J. Peperny, München.

Blum, M.S., Jones, T.H., Howard, D.F., and Overal, W.L. 1982. Biochemistry of termite defenses: *Coptotermes*, *Rhinotermes* and *Cornitermes* species. *Comp. Biochem. Physiol.* 71B:731-733.

Bonavita-Cougourdan, A., Clément, J.-L., and Lange C. 1987. Nestmate recognition: The role of cuticular hydrocarbons in the ant *Camponotus vagus* (Scop). *J. Entomol. Sci.* 22:1-10.

Bordereau, C. 1979. La cuticule de la reine de termite. Ultrastructure, composition chimique, étude de sa croissance en l'absence de mue. Thèse Université de Dijon, France, 252p.

Bordereau, C., Robert, A., Bonnard, O., and Le Quéré, J.-L. 1991. (3Z,6Z,8E)-3,6,8-Dodecatrien-1-ol: Sex pheromone in a higher fungus-growing termite, *Pseudacanthotermes spiniger* (Isoptera: Macrotermitinae). *J. Chem. Ecol.* 17:2177-2191.

Braekman, J.C., Daloze, D., Dupont, A., Pasteels, J.-M., and Josens, G. 1984. Diterpene composition of defense secretion of four West African *Trinervitermes* soldiers. *J. Chem. Ecol.* 10:1363-1370.

Braekman, J.C., Daloze, D., Dupont, A., Pasteels, J.-M., Lefeuve, P., Bordereau, C., Declercq, J.P., and Van Meerssche, M. 1983. Chemical composition of the frontal gland secretion from soldiers of *Nasutitermes lujae* (Termitidae: Nasutitermitinae). *Tetrahedron* 39:4237-4241.

Brossut, R., and Sreng, L. 1985. L'Univers chimique des Blattes. *Bull. Soc. Entomol. Fr.* 90:1266-1280.

Broughton, R.E. and Kistner, D 1991. A DNA hybridization study of the termite genus *Zootermopsis*. *Sociobiology* 19:15-40.

Brown, W.V., Watson, J.A.L., Carter, F.L., Lacey, M.J., Bareti, R.A., and McDaniel, C.A. 1990. Preliminary examination of cuticular hydrocarbons of worker termites as chemotaxonomic characters for some Australian species of *Coptotermes*. *Sociobiology* 16:305-328.

Buchli, H.H.R. 1950. Recherche sur la fondation et le développement des nouvelles colonies chez le termite lucifuge. *Phys. Comp. Oec.* 2:145-160.

Buchli, H.H.R. 1958. L'origine des castes et les potentialités ontogénétiques des Termites européens du genre *Reticulitermes*. *Ann. Sc. Nat. Zool. Biol. Anim.* 20:263-429.

Chu, A.J., and Blomquist, G.J. 1980. Decarboxylation of tetracosanoic acid to *n*-tricosane in the termite *Zootermopsis angusticollis*. *Comp. Bioch. Physiol.* 66B:313-317.

Chuah, C.H., Goh, S.H., Prestwich, G.D., and Tho, Y.P. 1983. Soldier defense secretions of the Malaysian termite *Hospitalitermes umbrinus* (Isoptera: Nasutitermitinae). *J. Chem. Ecol.* 9:347-356.

Clément, J.-L. 1977a. Caryotype des *Reticulitermes* français. *C. R. Acad. Sc. Paris.* 284:2355-2357.

Clément, J.-L. 1977b. Ecologie des *Reticulitermes* français: position systématique des populations. *Bull. Soc. Zool. Fr.* 102:169-185.

Clément, J.-L. 1978a. L'aggression inter- et intraspécifique des espèces françaises du genre *Reticulitermes*. *C. R. Acad. Sc. Paris.* 286:351-354.

Clément, J.-L. 1978b. Nouveaux critères taxonomiques dans le genre *Reticulitermes*, description de nouveaux taxons français. *Ann. Soc. Entom. Fr.* 14:131-141.

Clément, J.-L. 1979a. Etude biométrique des populations de *Reticulitermes* (Isoptères) français (*R. (l.) grassei, R. santonensis et R. (l.) banyulensis*) et de populations américaines de *R. flavipes. Arch. Zool. Exp. Gen.* 120:65-87.
Clément, J.-L. 1979b. Hybridation expérimentale entre *Reticulitermes santonensis* Feytaud et *Reticulitermes lucifugus grassei. Ann. Sc. Nat. Zool.* 1:251-260.
Clément, J.-L. 1980. Agression intra- et interspécifique dans le genre *Reticulitermes*. Séquences comportementales de reconnaissance coloniale. *Biol. Ecol. Méd.* 3:157-158.
Clément, J.-L. 1981a. Comportement de reconnaissance individuelle dans le genre *Reticulitermes. C. R. Acad. Sc. Paris* 292:931-933.
Clément, J.-L. 1981b. Autoécologie comparée et distances génétiques des populations du complexe *Reticulitermes lucifugus. Vie et Milieu* 31:261-270.
Clément, J.L., 1981c. Enzymatic polymorphism in the European populations of various *Reticulitermes* species, pp. 49-62 *in* Howse P.E. and Clément J.L. (eds). Biosystematics of Social Insects. Academic Press. London, New-York.
Clément, J.-L. 1982a. Phéromones d'attraction sexuelle des termites européens du genre *Reticulitermes.* : Mécanismes comportementaux et isolement spécifique. *Biol. Behav.* 7:55-68.
Clément, J.-L. 1982b. Variabilité biométrique inter- et intraspécifique des Termites européens du genre *Reticulitermes* (Isoptères). *Arch. Zool. Exp. et Gén.* 122:397-410.
Clément, J.-L. 1982c. Signaux responsables de l'agression interspécifique des termites du genre *Reticulitermes* (Isoptères). *C. R. Acad. Sc. Paris.* 294:635-638.
Clément, J.-L. 1982d. Spéciation des *Reticulitermes* européens, espèces et mécanismes d'isolements. *C. R. Soc. Biogéographie* 58:145-158.
Clément, J.-L. 1984. Diagnostic alleles of species in *Reticulitermes. Experientia* 40:283-285.
Clément, J.-L. 1986. Open and closed societies in termites of the genus *Reticulitermes*, geographic variations and seasonality. *Sociobiology* 11:311-323.
Clément, J.-L., and Devez, 1981. Agression et génétique. Film Video 16mm. Société Française du Film de la Recherche Scientifique.
Clément, J.-L., Howard, R.W., Blum, M.S., and Lloyd, H. 1986. L'isolement spécifique des termites du genre *Reticulitermes* du sud-est des Etats-Unis. Mise en évidence grâce à la chimie et au comportement d'une espèce jumelle de *R. virginicus - R. malletei sp* Nov. et d'une semi-species de *R. flavipes. C.R. Acad. Sc.* 302:67-70.
Clément, J.-L., Lange, C., and Blum, M.S. 1985. Chimosystematique du genre *Reticulitermes* (Isoptères) aux U.S.A. et en Europe. *Act. Coll. Ins. Soc.* 2:123-31.
Clément, J.L., Lemaire, M., Nagnan, P., Escoubas, P., Bagnères, A.-G., and Joulie, C. 1988. Chemical ecology of European termites of the genus *Reticulitermes.* Allomones, pheromones and kairomones. *Sociobiology* 14:165-172.
Clément J.L., Lloyd H., Nagnan P., and Blum M.S. 1989. *n*-Tetradecyl propionate: Identification as a sex pheromone of the Eastern Subterranean Termite *Reticulitermes flavipes. Sociobiology* 15:19-24.

Coaton, W.G.M. 1949. Queen removal in termite control. *Farming in South Africa* 24:335-338

Dani, F.R., Oldham, N.J., Turillazzi, S., and Morgan, E.D. 1994. A preliminary investigation of the Dufour's gland secretion of *Polistes dominulus* and its social parasite *Polistes sulcifer* (Hymenoptera: Vespidae), p. 407, *in* A. Lenoir, G. Arnold and M. Lepage (eds). Les Insectes Sociaux. Publications Université Paris Nord, France.

Delachambre, J. 1973. La formation de la cuticule adulte chez *Tenebrio molitor* L. (Insecte, Coleoptère): Ultrastructure et histochimie, étude expérimentale de la sclérification. Thèse de Doctorat ès-sciences. Université de Dijon, France, 303p.

De Renobales, M., Nelson, D.R., and Blomquist, G.J. 1991. Cuticular lipids. *in* K. Binnington, A. Retnakaran (eds.). Physiology of Insect Epidermis, CSIRO, Australia.

Dietz H.F., and Snyder, T.E. 1923. Biological notes on the termites of the canal zone and adjoining parts of the Republic of Panama. *U.S.D.A. J. Agric. Res.* 26:279-302.

Dudley, P.H., and Beaumont, J. 1889a. The termites or so called white ants of the Isthmus of Panama. *J. New York Microscop Soc.* 5:111-112.

Dudley, P.H., and Beaumont, J. 1889b. Observations on the termites or white ants of the Isthmus of Panama. *Trans New York Acad. Sci.* 1:85-114.

Dudley, P.H. and, Beaumont, J. 1890. Termites of the Isthmus of Panama II. *Trans New York Acad. Sci.* 9:157-180.

Espelie, K.E., Wenzel, J.W., and Chang, G. 1990. Surface lipids of social wasp *Polistes metricus* and its nest and nest pedicel and their relation to nestmate recognition. *J. Chem. Ecol.* 16:2229-2241.

Everaerts, C. 1989. La secrétion frontale des *Nasutitermes*. Composition chimique, fonction, variations intra- et interspécifiques. Thèse de Doctorat de l'Université de Dijon, France, 111p.

Everaerts, C., Pasteels, J.-M., Roisin, Y., and Bonnard, O. 1988. Variation intra- et interspecifique des secrétions défensives de divers *Nasutitermes* de Nouvelle-Guinée. *Act. Coll. Ins. Soc.* 4:43-50.

Everaerts, C., Pasteels, J.M., Roisin, Y., and Bonnard, O. 1988. The monoterpenoid fraction of the defensive secretion in *Nasutitermitinae* from Papua New-Guinea. *Bioch. Syst. Ecol.* 16:437-444.

Everaerts, C., Roisin, Y., Le Quéré, J.-L., Bonnard, O. and, Pasteels, J.-M. 1993. Sesquiterpenes in the frontal gland secretions of *Nasute* soldier termites from New Guinea. *J. Chem. Ecol.* 19:2865-2879.

Ferveur, J.-F. 1991. Genetic control of pheromones in *Drosophila simulans*. I Ngbo, a locus on the second chromosome. *Genetics* 128:293-301.

Ferveur, J.-F., and Jallon, J.-M 1993. Genetic control of pheromones in *Drosophila simulans*. II Kété, a locus on the X chromosome. *Genetics* 133:561-567.

Feytaud, J. de 1912. Contribution à l étude du Termite lucifuge (Anatomie). Fondation de colonies nouvelles. *Arch. Anat. Micros.* 13:481-607

Feytaud, J. de 1946. Le peuple des Termites - Que sais-je ? (ed.). 127 p.

Goh, S.H., Chuah, C.H., Tho, Y.P., and Prestwich, G.D. 1984. Extreme intraspecific chemical variability in soldier defense secretions of allopatric and sympatric colonies of *Longipeditermes longipes*. *J. Chem. Ecol.* 10:929-944.

Goh, S.H., Chuah, C.H., Vadivello, J., and Tho, Y.P. 1990 Soldier defense secretions of Malaysian free-ranging termite of the genus *Lacesititermes* (Isoptera: Nasutitermitinae). *J. Chem. Ecol.* 16:619-630.

Grace, J.K., Wood, D.L., and Kim, M. 1995. Behavioural and chemical investigation of trail pheromone from the termite *Reticulitermes hesperus* (Bank) (Isoptera: Rhinotermitidae). *J. Appl. Ent.* 119:501-505.

Grassé, P.-P. 1982. Termitologia, Tome I: Anatomie-Physiologie-Reproduction des termites. Chap. IV, V, VII. Masson (ed.), Paris.

Grassé, P.-P. 1984. Termitologia, Tome II: Fondation des sociétés. Construction. Chap. I, Masson (ed.), Paris.

Grassé, P.-P. 1986. Termitologia, Tome III: Comportement-Socialité- Ecologie-Evolution-Systématique. Chap. V , Masson (ed.), Paris.

Gush, T.J., Bentley, B.L., Prestwich, G.D., and Thorne, B.L. 1985. Chemical variation in defensive secretions of four species of *Nasutitermes*. *Biochem. Syst. Ecol.* 13:329-336.

Hadley, N. 1981. Cuticular lipids of terrestrial plants and arthropods: a comparison of their structure, composition and waterproofing function. *Biol. Rev.* 56:23-47..

Harms 1927. Koloniegründung bei *Macrotermes gilvus*. *Zool. Anz.* 74:221-236.

Haverty, M.I., and Thorne, B.L. 1989. Agonistic behavior correlated with hydrocarbon phenotypes in dampwood termites, *Zootermopsis* (Isoptera: Termopsidae). *J. Insect Behav.* 2:523-543.

Haverty, M.I., Nelson, L.J. and, Page, M., 1990. Cuticular hydrocarbons of four populations of *Coptotermes formosanus* (Shiraki) in the US. Similarities and Origins of Introduction. *J. Chem. Ecol.* 16:1635-1647.

Haverty, M.I., Nelson, L.J., and Page, M. 1991. Preliminary investigations of the cuticular hydrocarbons from North American *Reticulitermes* and tropical and subtropical *Coptotermes* (Isoptera: Rhinotermitidae) for chemotaxonomic studies. *Sociobiology* 19:51-76.

Haverty, M.I., Thorne, B.L., and Page, M. 1990. Surface hydrocarbon components of two species of *Nasutitermes* from Trinidad. *J. Chem. Ecol.* 16:2441-2450.

Haverty, M.I., Page, M., Nelson, L.J., and Blomquist, G.J. 1988. Cuticular hydrocarbons of dampwood termites *Zootermopsis* : Intra and intercolony variation and potential as taxonomic characters. *J. Chem. Ecol.* 14:1035-1058.

Holmgren, N. 1909. Termitenstudien. I Anatomische Untersuchungen. *Köng. sv. Vet. Akad. Handl.* 44:1-215.

Howard, R.W. 1979. Mating behavior of *Trichopsenius frosti*: Physogastric *Reticulitermes flavipes* queens serve as sexual aggregation centers. *Ann. Entomol. Soc. Am.* 72:127-129.

Howard, R.W. 1993. Cuticular hydrocarbons and chemical communication, pp. 177-226. *in* D.W. Stanley-Samuelson and D.R. Nelson (eds). Insect Lipids: Chemistry, Biochemistry and Biology. University of Nebraska Press, Lincoln.

Howard, R.W., Matsumura, F., and Coppel, H.C. 1976. Trail-following pheromones of the *Rhinotermitidae*: Approaches to their authentication and specificity. *J. Chem Ecol.* 2:147-166.

Howard, R.W., McDaniel, C.A., and Blomquist, G.J. 1978. Cuticular hydrocarbons of the eastern subterranean termites *Reticulitermes flavipes*. *J. Chem. Ecol.* 4:233-245.

Howard, R.W., McDaniel, C.A., and Blomquist, G.J. 1980. Chemical mimicry as an integrating mechanism: Cuticular hydrocarbons of a termitophile and its host *Science* 210:431-433.

Howard, R.W., McDaniel, C.A., and Blomquist, G.J. 1982. Chemical mimicry as an integrationg mechanism for three termitophiles associated with *Reticulitermes viginicus*. *Psyche* 89:157-167.

Howard, R.W., Thorne, B.L., Levings, S.C., and McDaniel, C.A. 1988. Cuticular hydrocarbons as chemotaxonomic characters for *Nasutitermes corniger* and *N. ephratae*. *Ann. Entom. Soc. Am.* 81:395-399.

Howard, R.W., McDaniel, C.E., Nelson, D.R., Blomquist, G.J., Gelbaum, L.T., and Zalkow, L.H. 1982. Cuticular hydrocarbons of *Reticulitermes virginicus* (Banks) and their role as potential species and caste recognition cues. *J. Chem. Ecol.* 8:1227-1239.

Howick, C.D., and Creffield, J.W. 1980. Intraspecific antagonism in *Coptotermes acinaciformis* (Froggatt). *Bull. Ent. Res.* 70:17-23.

Howse, P.E. 1964. The significance of the sound produced by the termite *Zootermopsis angusticollis*. *Anim. Behav.* 12:284-300.

Howse, P. 1975. Chemical defences of ants, termites and other insects: Some oustanding question, p. 23-40, *in* Noirot, C., Howse, P.E., and Le Masne, G. (eds.). Pheromones and defensive secretions in social insects, Int. Symp. of IUSSI - Dijon Université Press.

Kaib, M. 1990. Intra- and interspecific chemical signals in the termite *Schedorhinotermes*. Production sites, chemistry, and behaviour, p. 26-32, *in* Crisbakin, F.G., Wiese, K., and Popov, A.V. (eds) Sensory Systeme and Communication in Arthropods. Advances in Life Sciences, Birkhäuser Verlag, Basel.

Kaib, M., Bruisma, O., and Leuthold, R.H. 1982. Trail-following in termites: evidence for a multicomponent system. *J. Chem. Ecol.* 8:1193-1205.

Kaiser, P. 1956. Die Homonalorgane der Termiten mit der Enstehung ihrer Kasten. *Mitt- Hamburgischen Zool. Mus. Inst.* 54:129-178.

Katase H., and Chino, H. 1982. Transport of hydrocarbons by the lipophorin of insect haemolymph. *Bioch. Biophys. Acta* 710:341-348.

Katase H., and Chino, H. 1984. Transport of hydrocarbons by the haemolymph lipophorin in *Locusta migratoria*. *Insect Biochem.* 14:1-6.

Kettler, R., and Leuthold, R.H. 1995. Inter- and intraspecific alarm response in the termite *Macrotermes sybhyalinus* (Rambur). *Ins. Soc.* 42:145-156.

Kirchner, W.H., Broecker, I., and Tautz, J. 1994. Vibrational alarm communication in the damp-wood termite *Zootermopsis nevadensis*. *Physiol. Entomol.* 19:187.

Korman, A.K., and Pashley, D.P. 1991 Genetic comparisons among U.S. populations of Formosan subterranean termites. *Sociobiology* 19:41-50.

Korman, A.K., Pashley, D.P., Haverty, M.I., and LaFage, J.P. 1991. Allozymic relationships among cuticular hydrocarbon phenotypes of *Zootermopsis* species. *Ann. Entomol. Soc. Am.* 84:1-9.

Laduguie, N. 1993. Phéromone de piste et phéromone sexuelle chez les Termites. Etude chez *Reticulitermes santonensis* (Feytaud), *Pseudacanthotermes spiniger* (Sjöstedt) et *Nasutitermes lujae* (Wasmann). Thèse de Doctorat de l'Université de Dijon, France, 130p.

Laduguie, N., Robert, A., Bonnard, O., Vieau, F., Le Quéré, J.-L., Semon, E., and Bordereau, C. 1994. Isolation and identification of (3Z,6Z,8E)-dodecatrien-1-ol in *Reticulitermes santonensis* Feytaud (Isoptera: Rhinotermitidae): Roles in worker trail-following and in alate sex-attraction behavior. *J. Insect Physiol.* 40:781-787.

Lebrun, D. 1971. Glandes tergales et surfaces cuticulaires correspondantes chez le termite à cou jaune. *Calotermes flavicollis* (Fabr.). *C. R. Acad. Sc. Paris* 272:3162-3164.

Lebrun, D. 1971. Différenciation cuticulaire au niveau de la glande sternale de *Calotermes flavicollis* (Fabr.). *C. R. Acad. Sc. Paris* 273:959-961.

Lebrun, D. 1972. Organisation cuticulaire propre aux régions antérieures et postérieures de la glande sternale des termites du genre *Reticulitermes* Holmgren. *C. R. Acad. Sc. Paris* 275:779-781.

Levings, S.C., and Adams, E.S. 1984. Intra- and interspecific territoriality in *Nasutitermes* in a Panamanian mangrove forest. *J. Anim. Ecol.* 53:705-714.

Lemaire, M., Nagnan, P., Clément, J.-L., Lange, C., Peru, L., and Basselier, J.-J. 1990. Geranyllinalool (dipterpene alcohol): An insecticidal component of pine wood and termites (Isoptera: Rhinotermitidae) in four European ecosystems. *J. Chem. Ecol.* 16:2067-2079.

Locke, 1974. The structure and formation of the integument in insects, pp. 123-213, *in* Rockstein M. (ed.). The physiology of insects, vol VI, Academic Press, Lond.

Lockey, K.H. 1988. Lipids of the insect cuticle: origin, composition and function. *Comp. Bioch. Physiol.* 89B:595-645.

Lorenzi, M.-C., Bagnères, A.-G., and Clément, J.-L. 1996. The role of cuticular hydrocarbons in social insects: is it the same in paper-wasps?, pp. 178 -189, *in* S. Turillazzi and M.J. West-Eberhard (eds.). Natural History and Evolution of Paper-Wasps. Oxford Univ. Press.

Matsumura, F., Tai, A., and Coppel, H.C. 1969. Termite trail-following substance, isolation and purification from *Reticulitermes virginicus* and fungus-infected wood. *J. Econ. Entomol.* 62:599-603.

Mayr, E. 1963. Animal species and evolution. Harvard University Press, Cambridge, 797 p.

Meskali, M., Bonavita-Cougourdan, A., Provost, E., Bagnères, A.-G., Dusticier G., and Clément J.-L. 1995. Mechanism underlying cuticular hydrocarbon homogeneity in the ant *Camponotus vagus* (Scop.) (Hymenoptera: Formicidae): Role of postpharyngeal glands. *J. Chem. Ecol.* 21:1127-1148.

Miller, E.M. 1969. Caste differentiation in the lower termites, p. 289-310, *in* Krishna, K. and Weesner, F.M. (eds.). Biology of Termites I. Academic Press, NY.

Moore, B.P. 1964. Volatile terpenes from *Nasutitermes* soldiers (Isoptera: Termitidae). *J. Insect Physiol.* 10:371-375.

Moore, B.P. 1968. Studies on the chemical composition and function of the cephalic gland secretion in Australian termites. *J. Insect Physiol.* 14:33-39.

Morel, L., VanderMeer, R.K., and Lavine, B.K. 1988. Ontogeny of nestmate recognition cues in the red carpenter ant *Camponotus floridanus*. *Behav. Ecol. Sociobiol.* 22:175-183.

Nel, J.J.C. 1968. Aggressive behavior of the harvester termites *Hodotermes mossambicus* and *Trinervitermes trinervoides*. *Ins. Soc.* 2:145-156.

Noirot, C. 1956. Les sexués de remplacement chez les termites supérieurs (Isoptera: Termitidae). *Ins. Soc.* 3:145-158.

Noirot, C. 1969. Formation of castes in higher termites, pp. 311-350, *in* Krishna K. and Weesner F.M. (eds.). Biology of Termites I. Academic Press, N. Y.

Noirot C. and Noirot-Timothée C. 1965. The sternal glands of termites: Segmental pattern, phylogenetic implications. *Ins. Soc.* 42:321-323.

Nowbahari, E., Lenoir, A., Clément, J.-L., Lange, C., Bagnères, A.-G., and Joulie, C. 1990. Individual, geographical and experimental variation of cuticular hydrocarbons of the ant *Cataglyphis cursor.* Their use in nest and subspecies recognition. *Biochem. Syst. Ecol.* 18:63-73.

Nutting, W.L. 1969. Flight and colony foundation, pp. 233-282, *in* Krishna, K. and Weesner, F.M. (eds). Biology of Termites I. Academic Press, N.Y.

Nutting, W.L. 1970. Composition and size of some termite colonies in Arizona and Mexico. *Ann. Ent. Soc. Am.* 63:1105-1110.

Oloo, G.W., and Leuthold, R.H. 1979. The influence of food on trail-laying and recruitment behaviour in *Trinervitermes Bettonianus* (Termitidae: Nasutitermitinae). *Ent. Exp. Appl.* 26:267-278.

Parton, A.H., Howse, P.E., Baker, R., and Clément, J.-L. 1981. Variation in the chemistry of the frontal gland secretion of European *Reticulitermes* species, pp. 193-209, *in* P.E. Howse and J.-L. Clément (eds). Biosystematics of Social Insects. Academic Press, London.

Pickens, A.L. 1934. The biology and economic significance of the western subterranean termite *Reticulitermes hesperus,* pp. 157-183, *in* Kofoid C.A. (ed). Termites and Termites Control. Univ. Calif. Press., Berkeley Calif.

Prestwich, G.D., Bierl, B.A., Devilbiss, E.D., and Chaudhury, M.F.B. 1977. Soldier frontal glands of the termite *Macrotermes subhyalinus*: morphology, chemical composition, and use in defense. *J. Chem. Ecol.* 3:579-590.

Prestwich, G.D., Kaib, M., Wood, W.F., and Meinwald, J. 1975. 1,13-tetradecadi-en-3-one and homologs, new natural products isolated from *Schedorhinotermes* soldiers. *Tet. Lett.* 52:4701-4707.

Provost, E., Rivière, G., Roux, M., Morgan, E.D., and Bagnères, A.-G. 1993. Change in the chemical signature of the ant *Leptothorax lichtensteini* with time. *Insect Biochem. Molec. Biol.* 23:945-957.

Quennedey, A. 1975. Morphology of exocrine glands producing pheromones and defensive substances in subsocial and social insects, pp. 1-21, *in* Noirot, C., Howse, P.E., and Le Masne, G. (eds.). Pheromones and defensive secretions in social insects. Int. Symp. of IUSSI. Dijon Univ. Press.

Quennedey, A. 1978. Les glandes exocrines des termites. Ultrastructure comparée des glandes sternales et frontales. Thèse de Doctorat ès-Sciences, Univ. Dijon, 254p.

Reilly, L.M. 1987. Measurements of inbreeding and average relatedness in a termite population. *Am. Naturalist* 130:339-349.

Ritter, F.J., and Coenen-Saraber, C.M.A. 1969. Food attractants and a pheromone as trail-following substances for the saintonge termite: Multiplicity of the trail following substances in *L. trabea*-infected wood. *Ent. exp. appl.* 12:611-622.

Roonwal, M.L., and Chotani, O.B. 1962. Termite *Odontotermes obesus*: royal chamber with four queens and two kings. *J. Bombay Nat. Hist. Soc.* 59:975-976.

Roonwal, M. L., and Gupta, S.D. 1952 An unusual royal chamber with two kings and two queens in the Indian mound-building termites, *Odontotermes obesus*. *J. Bombay Nat. Hist. Soc.* 51:293-294.

Roy-Noel, J. 1974. Contribution à la connaissance des reproducteurs de remplacement chez les Termites supérieurs. Observations sur *Bellicositermes natalensis*. *C. R. Acad. Sci. Paris* 278:481-482.

Runcie, C.D. 1987. Behavioral evidence for multicomponent trail pheromone in the termite *Reticulitermes flavipes* (Kollar) (Isoptera: Rhinotermitidae). *J. Chem. Ecol.* 13:1967-1978.

Skaife, S.H. 1954a. The black mound termite of the Cape *Amitermes atlanticus*. *Trans. Roy. Soc. South. Africa* 34:251-271.

Skaife, S.H. 1954b. Caste differentiation among termites. *Trans. Roy. Soc. South Africa* 34:345-353.

Skaife, S.H. 1955. Dwellers in Darkness Longmans, Green and Co, Ltd London 134p.

Strong, K.L., and Grace, J.K. 1993. Low allozyme variation in Formosan subterranean termite colonies in Hawaii. *Pan Pacific Entomologist* 69:51-56.

Stuart, A.M. 1969. Social behavior and communication, pp. 193-232, *in* Krishna, K. and Weesner, F.M. (eds.). Biology of termites I. Academic Press, New York and London.

Stuart, A.M. 1970. The role of chemicals in termite communication, pp. 79-106, *in* Johnson, J.W., Moulton DG. Turk A. (eds.). Advances in chemoreception Vol. I. Communication by chemical signals. Appleton-Century-Crofst. N.Y.

Stuart, A.M. 1975. Chemoreception, behaviour and polyethism in termites, pp. 343-348, *in* Olfaction and Taste V, Academic Press.

Stuart, A.M. 1976. Some aspects of communication in termites, pp. 400-405, *in* Proceeding XV Int. Cong. Entom.

Su, N.Y., and Haverty, M.I. 1991. Agonistic behavior among colonies of the Formosan subterranean termite, *Coptotermes formosanus* from Florida and Hawaii: Lack of correlation with cuticular hydrocarbon composition. *J. Insect Behav.* 4:115-128.

Su, N.Y., and Scheffrahn, R.H. 1988. Intra- and interspecific competition of the Formosan and eastern subterranean termite: evidence from field observation. *Sociobiology* 14:157-164.

Takamashi, S and, Gassa, A. 1995. Roles of cuticular hydrocarbons in intra- and interspecifique recognition behavior of two *Rhinotermitidae* species. *J. Chem. Ecol.* 21:1837-1845.

Thorne, B.L. 1982a. Polygyny in Termites: Multiple primary queens in colonies of *Nasutitermes corniger*. *Ins. Soc.* 2:102-117.

Thorne, B.L. 1982b. Termite-termite interactions: Workers as an agonistic caste. *Psyche* 89:133-150.

Thorne, B.L., and, Haverty, M. 1991. A review of intracolony, intraspecific and interspecific agonism in Termites. *Sociobiology* 19:115-145.

Trabalon, M., Campan, M., Clément, J.-L., Lefebvre, J., Lange, C., and Thon, B. 1988 Changes in cuticular hydrocarbon composition in the female *Calliphora vomitoria* (Diptera) during ontogenesis. *Behav. Proc.* 17:107-115.

Traniello, J.F.A. 1982. Recruitment and orientation components in a termite trail pheromone. *Naturwissenschaften* 69:343-345.

Traniello, J.F.A., and Beshers, S.N. 1985. Species specific alarm/recruitment responses in a Neotropical termite. *Naturwissenschaften* 72:491-492.

Vauchot, B. 1995. Transfert et régulation de la signature chimique chez deux espèces de termites, *Reticulitermes santonensis* et *Reticulitermes (lucifugus) grassei* vivant en colonie mixte expérimentale. Thèse de Doctorat de l'Université de Provence, Marseille, France, 73 pages

Vauchot, B., Provost, E., Bagnères, A.-G., and Clément J.-L. 1996. Regulation of the chemical signatures of two termite species, *Reticulitermes santonensis* and *Reticulitermes (lucifugus) grassei* living in mixed colonies. *J. Insect Physiol.* 42: 309-321.

Verron, H. 1960. La perception des odeurs chez *Calotermes flavicollis*. *C. R. Acad. Sc. Paris* 250:2931-2932.

Verron, M. 1963. Rôle des stimulis chimiques dans l'attraction sociale chez *Calotermes flavicollis*. *Ins. Soc.* 10:167-335.

Vrkoc, J., Krecek, J., and Hrdy, I. 1978. Monoterpenic alarm pheromones in two *Nasutitermes* species. *Acta Entomol. Bohem.* 75:1-8.

Wallis, D.I. 1964. Aggression in social insects, pp. 15-22, *in* Carthy, J.D. and Ebling, F.J. (eds). The Natural History of Aggression. Academic Press, London.

Watson, J.A.L., Brown, W.V., Miller, F.L., Carter, F.L., and Lacey, M.J. 1989. Taxonomy of *Heterotermes* (Isoptera: Rhinotermitidae) in south-eastern Australia: Cuticular hydrocarbons of workers, and soldier and alate morphology. *Syst. Entomol.* 14:299-325.

Weyer, F., 1930a. Uber Ersatzgeschlechtstiere bei Termiten. *Z. Morph. Okol. Tiere* 19:364-380.

Weyer, F., 1930b. Brobachtung über die Enstehung neuer Kolonien bei tropischen Termiten. *Zool. Jahrb.* 60:327-380.

Wigglesworth, V.B. 1990. The distribution, function and nature of "cuticulin" in the insect cuticle. *J. Insect Physiol.* 36:307-313.

PART THREE

Social Insect Releaser Pheromones

7

Pheromone Directed Behavior in Ants

Robert K. Vander Meer and Leeanne E. Alonso

Introduction

All ants are eusocial, which means there are overlapping generations in which adult workers (normally sterile) assist their mother in rearing sisters and brothers, and there are reproductive and non-reproductive castes (division of labor). There are about 11,000 known ant species that represent a large percentage of the earth's animal biomass. Ants occupy virtually every ecological niche on earth and are among the leading predators and scavengers of insects and small mammals. Concomitant with such variety comes tremendous diversity in colony structure and social organization. A large component of ant social organization is mediated by pheromonal communication.

The term pheromone was originally proposed by Karlson and Luscher (1959) and is defined as a substance secreted by an organism outside its body that causes a specific reaction in a receiving organism of the same species. Pheromones are classified into two broad categories: releaser types that result in an immediate behavioral response and primer types, where the initial perception results in the initiation of a complex physiological response. This chapter deals with releaser pheromones (see Vargo, this book, for ant primer pheromones).

Pheromone research has two basic elements. The first is observational, where a particular behavior is studied and experimental evidence suggests that pheromones are involved. The second part is the chemistry of the pheromones. A bioassay developed from behavioral observations is used to guide the isolation of active compounds. There are many papers that investigate the chemistry of glandular products without associated behavior (Cavill and Houghton

1974; Meinwald et al. 1983; Blum et al. 1987; Billen et al. 1988; do Nascimento et al. 1993a). This research is important, providing information on inter- and intraspecific chemical diversity and variation, and occasionally yielding novel chemistry. Similarly, there are many observational papers that report the probable existence of pheromones, but have not addressed the pheromone chemistry (Breed et al. 1987; Yamauchi and Kawase 1992; Passera and Aron 1993; Hölldobler et al. 1994a). These papers, too, have great value. They define the complexity and diversity of ant behavior and set the stage for the ultimate interaction between biologists and chemists.

Papers on ant behavior or pheromones account for over 18% of the 20,000 references currently in the comprehensive ant bibliography, FORMIS (Porter 1995). The number of pheromone specific publications (3.4%) have increased dramatically, since the term was coined in 1959, but now have leveled off to about 30/year. This large body of literature cannot be dealt with in this chapter, so we selected focus topics, and limit our discussion of other areas.

Types of Pheromone Interactions

Many pheromone induced behaviors associated with ants effectively maintain colony social structure, cohesiveness, and productivity. Worker/worker interactions focus on recruitment, colony immigration and alarm. Worker/queen associations involve worker attraction to queen produced pheromones and associated queen grooming, feeding, and protection. The queen can also influence the competitive functionality of sexuals, potential sexuals or other queens in the colony directly or through the workers (see primer pheromone Chapter by Vargo). Brood produced pheromones are thought to be produced because of the preferential treatment of brood by workers. Lastly, there are many potential pheromone mediated behaviors between male and female sexuals associated with mating activities. Morgan and Billen (this Book) deal in depth with the many exocrine glands available to the ant for the biosynthesis and distribution of pheromones.

Mating Flights and Male/Female Sexual Interactions

There are numerous points during mating flights and the interaction of male and female sexuals that have a high probability of being mediated by pheromones or more generally semiochemicals. These include (1) initiation of mating flight activity; (2) interaction of workers and alates during mating flight activity; (3) the actual flight of male and female alates (often synchronized); (4) cohesion of a male lek, if formed; (5) female location of the male lek; (6) and/or male location of females; (7) female mate selection (if this occurs); (8) inseminated female alate's choice of landing site; (9) newly mated queen's choice of nuptual chamber site; and (10) there are other possibilities that can match the many mating strategies of ants. As illustrated below, there is little actually known about the chemistry and behavior of these processes, perhaps because observations (and therefore bioassay development) are difficult, e.g. the fire ant, *Solenopsis invicta*, mates 300 meters in the air. However, we anticipate excellent future progress in this very important area of pheromone communication.

Sex Pheromones

The attraction of males to females (usually) for mating is especially well studied and understood in lepidopteran species; however, in ants there are only a few reports dealing with this phenomenon. Like most ant species,*Xenomyrmex floridanus*, male and female sexuals leave the nest on nuptual flights. The details of these flights are a puzzle. However, males are strongly attracted to the poison gland secretion of female sexuals (Hölldobler 1971a) and the secretion elicits male copulatory activity. Worker poison glands also attract males. It is unknown how this secretion and the behaviors noted are used in mating flights. Sexuals of the ant social parasite, *Formicoxenus nitidulus,* mate within the nest and are more amenable to observation (Buschinger 1976). As in the previous study, female poison gland products attract and release copulatory behavior in males. Here the alate female raises her gaster, and releases a droplet of venom through her extended sting. Males (wingless) are attracted to female alates over a short distance and attempt to copulate. The poison sac contents of other ant social parasite females, *Harpagoxenus canadensis, H. sublaeviv*, and *H. americanus*, act similarly to call and stimulate male copulation (Buschinger and Alloway 1979). The preceding examples appear to be ready and waiting for the isolation and identification of the sex pheromone components. Wingless female sexuals, ergatoids, of the primitive ant, *Rhytidoponera metalica,* gather at their nest entrance prior to male mating flights. They arch their gaster exposing the tergal gland. Compounds released from this gland attract males flying from nearby nests. Workers and reproductives have this gland, thus either the behaviors released are context specific or the chemistry produced by each caste is different (Hölldobler and Haskins 1977).

The sex pheromone of *Formica lugubris* has been isolated and identified (Walter et al. 1993). The chemistry is not novel, but the

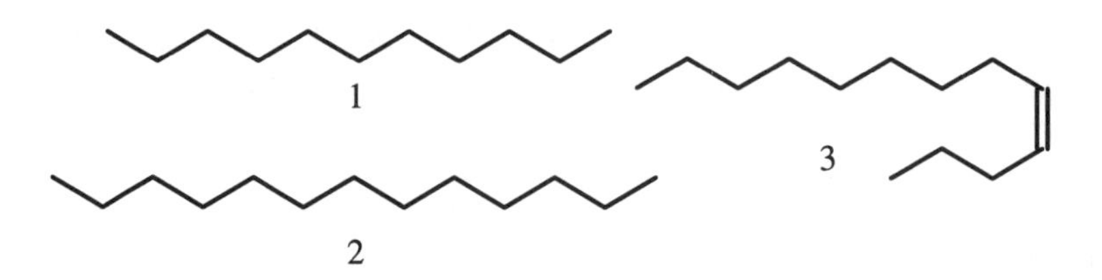

chemistry/behavior is exciting. Male and female alates take nuptial flights and mate in open places on the forest floor. The females call males by releasing the following products from their Dufour's gland, undecane (**1**); tridecane (**2**); and Z-4-tridecene (**3**) (100:5.32:4.25, respectively). All are significantly attractive, although the attractant power of undecane by itself is not statistically different from the three compounds together. Undecane and the other two compounds are also found in the Dufour's glands of workers. Again, the context under which exocrine gland products are released affects the behavior elicited.

Mating Flight Activities

Heightened worker excitement and aggression at the onset of mating flights has been observed in several ant species. Male mandibular gland products of *Acromyrmex* and *Atta* species elicit worker excitement during nuptual flights (Fowler, 1982). This chemistry may initiate mating flight activity, protect sexuals during the flight, and through heightened worker aggressive behavior toward landing newly mated queens, prevent colony foundation near established colonies. Obin and Vander Meer (1994) induced *S. invicta* flights in the laboratory and showed that chemical cues from both male and female alates, but not from workers, attracted workers, induced alarm-recruitment behaviors in the workers, and promoted alate retrieval by workers. They proposed that volatile substances produced by the alates were responsible for eliciting the worker reactions. Alonso and Vander Meer (In press) determined that the glandular source of these excitant pheromones were the mandibular glands. The chemistry of this secretion is under investigation.

The role of "alarm" pheromones in mating flight activity is supported by related findings that the chemical concentration and/or composition of alarm pheromones often differ between workers, female alates, and males (e.g. Law et al. 1965, Brand et al. 1973, Lloyd et al. 1975). In particular, Pasteels et al. (1980) showed that heads of *Tetramorium caespitum* male and female alates contain 4-methyl-3-hexanone, a compound not found in workers. The ketone attracts workers and a role in mating flight activities was suggested. Alarm pheromones identified from the large mandibular glands of *Atta sexdens rubropilosa* (Blum et al. 1968) vary in concentration, but not composition, between female and male alates (do Nascimento et al. 1993b). Workers and alates differ in the concentration of alarm pheromones from the mandibular glands of *Camponotus abdominalis* (Blum *et al*. 1988), but not *C. schaefferi* (Duffield and Blum 1975). See also Alarm Pheromones in this Chapter.

Queen / Worker Interactions

Queens of most species are surrounded by workers, probably for protection and tending (Hölldobler and Bartz 1985). This process does not necessarily require an attractant pheromone, since workers could randomly contact the queen and be stimulated to aggregate around her. Unfortunately, many studies have used bioassays that cannot distinguish between aggregation and attraction.

Researchers investigating queen/worker interactions have come up with an extensive array of behaviors and descriptors, e.g. attraction, aggregation, arrestant, worker retrieval, recognition, and queen tending. Stumper (1956) showed that workers from *Lasius alienus* and *Pheidole pallidula* were "attracted" to their respective queens. A solvent extract of queens was also attractive to workers of the same species. Pheromone specificity was demonstrated by the application of *Pheidole* queen extract onto cadavers of *Lasius* queens. *Pheidole* workers adopted the *Lasius* queens for about one day. This was the first demonstration, by chemical extraction, of queen produced pheromones acting on workers. The bioassay placed workers and the treated surrogate queen in a common arena, then worker behaviors were observed. This kind of bioassay can detect attraction and/or aggregation, but it is difficult to distinguish one from the other.

Similarly, the "attraction" of workers to queens of several species of army ants was investigated (Watkins and Cole 1966). Queens were left in specific locations on filter paper, removed, and then workers were introduced. Again this type of bioassay confounds attraction and aggregation responses. Workers from all species tested showed some cross-specificity. It was interesting that workers generally preferred their own queen over a conspecific queen from another colony. Similar intra-colony queen preferences were observed for the fire ant, *Solenopsis invicta* (Jouvenaz et al. 1974). Polygyne species become more complicated, for example, *Myrmica rubra*, (Evesham 1984) workers are faced with multiple queens producing an attractant pheromone. The attractant pheromone is not different for each queen, therefore, workers are attracted to all queens in a colony. Other recognition factors come into play that allow workers to identify and interact differently with the first queen they encounter. Evesham (1984) states that workers may be attracted to and visit many queens, but are "loyal" to one.

Fowler and Roberts (1982) appropriately described an "entourage" pheromone from the carpenter ant, *Camponotus pennsylvanicus*. This pheromone induces workers to cluster around their queen. In their assay they exposed blotter paper to queens and observed workers forming aggregations where the queens had been previously. No chemical isolation was attempted.

Oecophylla longinoda and *O. Smaragdina* queens are usually covered by a retinue of major workers. The queen produces a variety of releaser and primer pheromones from its head and abdominal intersegmental glands (Hölldobler and Wilson 1983). The pheromones are said to be "attractive" (aggregation?), induce worker regurgitation, and they induce workers to present trophic eggs to the queen (laid by major workers). See Vargo, this book, for more on ant primer pheromones.

Iridomyrmex humilis queens produce a worker "attractant" (Keller and Passera 1989). Their assay consisted of an arena with nestmate and non-nestmate queens confined by wire mesh, such that workers could contact the queen but the queen could not get out. Aggregation versus attraction is again an issue. Interestingly, the source of the queen's action on workers was determined to be the thorax, either from epidermal gland cells or from thoracic glands (Cariouetienne et al. 1992). Later, these researchers recognized that they could not distinguish between attraction and aggregation effects and defined "queen power" as the combined result of these two behaviors (Cariouetienne and Passera 1993). They determined that "queen power" was related to the egg-laying process and not insemination and dealation. More on this later.

The Chemistry of Queen/Worker Interactions

Although the bioassays in the above studies should be adequate to commence the isolation and identification of the pheromones, the chemistry involved in the interactions is mostly unknown. *Solenopsis invicta* is an outstanding exception. As already mentioned Jouvenaz et al. (1974) first noted that *S. invicta* workers aggregated at spots where their queens had been placed. Bioassays were developed to measure the behavioral effects (Glancey 1980, Lofgren et al. 1983), which were useful in determining the pheromone source (Vander Meer et al. 1980). The bioassays were not behavior specific. A surrogate queen bioassay could not differentiate between attraction or aggregation. An "olfactometer" bioassay allowed worker contact with a cotton tipped swab through which volatiles from test samples were being passed, but since the cotton tipped swab could trap these volatiles aggregation was again a possibility. However, these assays were used to determine that the elicited behaviors were derived from compounds produced from the queen's poison sac (Vander Meer et al. 1980). The pyrone, (E)-6-(1-pentenyl)-2H-pyran-2-one (**4**), invictolide (**5**), and dihydroactinidiolide (**6**) were subsequently isolated and identified from whole queen extracts (Rocca et al. 1983a, Rocca et al. 1983b). A "disrupted colony" field bioassay was developed where a colony was dug into, and the soil, workers, and brood were placed in an observation tray. Surrogate queen (rubber septum) test treatments were placed in the tray and attraction, clustering, movement of brood around the surrogate queen, development of a "queen trail", and movement of the surrogate queen back into the nest were observed. Components **4** and **5** were clearly required for the behavioral response; however, the role of component **6** remains unclear (Glancey et al. 1984).

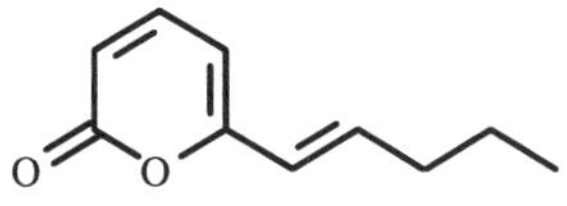

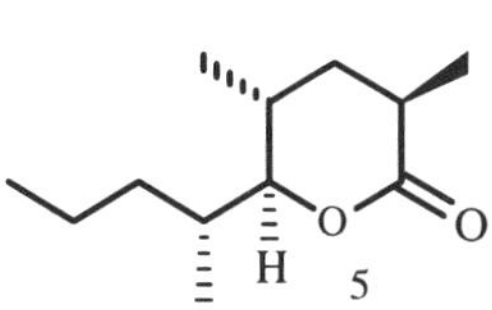

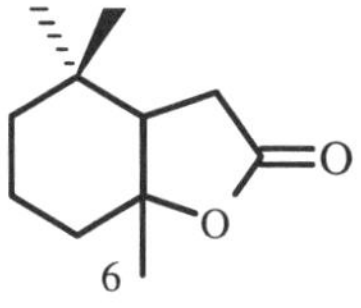

The other example comes from the pharaoh's ant, *Monomorium pharaonis*. Fertile queens produce (E,E,E)-1-isopropenyl-4,8,12-trimethylcyclotetradeca-3,7,11-triene (neocembrene) (Edwards and Chambers 1984). This compound is queen specific, produced in the Dufour's gland, and must be released through the sting. Neocembrene is attractive to Pharaoh's ant workers and *may* serve as a "queen recognition" pheromone.

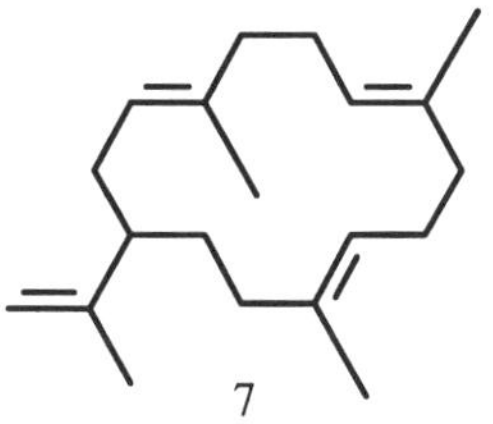

Fire ant queens deposit poison sac contents on their eggs as they are laid (Vander Meer and Morel, 1995). This marks the eggs with the queen/worker pheromone and antimicrobial alkaloids. The deposition process involves extension of the sting, movement of the egg to the base of the sting, where the sting is then drawn across the egg. Release of pheromone is related to the queen's egg-laying rate, thus the workers have a mechanism for assessing their queen's fecundity based on the amount of queen pheromone released. A relationship between the queen's fecundity and the magnitude of dealation inhibition has been demonstrated (Fletcher and Blum 1983). Fire ant queens treated with the insect growth regulator, fenoxycarb, do not lay eggs (Glancey and Banks 1988), nor do they inhibit dealation or attract workers (Obin et al. 1988). However, queen attractant was present in the poison sacs of fenoxycarb treated queens, but they were not releasing the pheromone because the egg laying process was stopped.

This egg laying behavior was also observed in *Monomorium pharaonis* (Vander Meer and Morel, 1995). Pharaoh's ant queens produce a worker attractant (see above) that is released through the sting (Edwards and Chambers 1984). Worker ant assessment of their queen's fecundity results in caste regulation by the workers. If the queen lays many eggs only workers are produced, but if the queen's egg production is low or the queen dies, sexuals are produced (Edwards 1987). Thus, the egg-laying process may be involved in queen / worker communication. *Iridomyrmex humilis* queens exhibit a significant increase in "queen power" when egg-laying commences (Cariouetienne and Passera 1993). A *Myrmica rubra* queen's "aggregative (queen recognition) power" was correlated with the sum of the diameters of eggs occupying the terminal zone of each ovariole, thus probably related to the egg production rate. There are many other examples (see Keller and Nonacs 1993; Hölldobler and Bartz 1985; Fletcher and Ross 1985) of a worker's ability to determine its queen's reproductive capability.

Alarm Pheromones

Alarm pheromones constitute a major evolutionary development for eusocial species that permits the collective resources of the colony to be rapidly exploited in response to stresses or perturbations to the colony (Blum 1985). Therefore, the function of alarm signals is not to ward off intruders or benefit the emitter, but to alert other members of the colony. Goetsch (1957) astutely pointed out that alarm signals themselves have no descriptive content, but merely transmit a state of excitement. What the alarmed ants do after receiving the alarm signal depends on other stimuli. It may be difficult to dissociate the other stimuli, thus alarm is often associated with attraction, recruitment, and defense (Hölldobler and Wilson 1990). These behaviors may be elicited by the alarm pheromones themselves, or may result from other chemicals released at the same time (Blum 1969). Gabba and Pavan (1970) suggest that alarm behavior is the initial stage of a complex of coordinated defensive acts. However, we believe that recognition of an object (animate or inanimate) as a non-nestmate represents the initiating process (see Vander Meer and Morel, this book).

Behavioral Responses to Alarm

Wilson and Regnier (1971) divided alarm behavior into two basic behavioral categories: "panic alarm," characterized by rapid non-directional movement and excited bursts of running, and "aggressive alarm," in which workers not only run excitedly but also move toward the source of the alarm substance, assume a defensive posture, and often attack.

In general, alarm behaviors differ between species (Hölldobler and Wilson 1990). In addition, it is often difficult to characterize alarm behaviors of even a single species because reactions vary with a multitude of factors (Parry and Morgan 1979). Alarm behaviors in some species consist only of attraction to the pheromone source or simply of raised heads and outstretched antennae (Table 7.1). Other species react with increased speed and orientation, frenzied erratic movement, or with aggressive behaviors such as spread mandibles, biting, stinging, and spraying of poison gland products (Table 7.1).

Reactions to alarm pheromones depend on pheromone concentration, the length of time workers are exposed to the alarm substance, as well as context (Gabba and Pavan 1970). Wilson (1958a) distinguished different behaviors released in *Pogonomyrmex badius* workers by low and high concentrations of its alarm pheromone. At low concentrations (10^{10} molecules/cm^3), ants moved toward the source of the pheromone while at higher concentrations ($>10^{11}$ molecules/cm^3) the ants reacted in an aggressive frenzy. Wilson and Bossert (1963) proposed a model of alarm pheromone diffusion in which receiving ants need to contact the inner, active space of the chemicals released to give a full reaction. If ants contact the chemicals at the periphery of the chemical plume, where the pheromone is less concentrated, their reaction may be lessened. This has been observed for several ant species, particularly for species in which several chemicals are involved in eliciting the alarm reaction. Each chemical has a different dispersion rate from the source (Regnier and Wilson 1968, Bradshaw et al. 1979a).

TABLE 7.1 Observed Alarm Behaviors of Workers and Related References

Behavior / References
Alertness: Wave Antennae/ Raise Head
Wilson 1958a; Duffield et al. 1977; Blum et al. 1988.
Colony Disperses/ Runs From Nest
Leuthold and Schlunegger 1973; Janssen et al. 1995.
Attraction To Source Of Pheromone
Wilson 1962; Cammaerts et al. 1978, 1981, 1985; Cammaerts and Mori 1987; Kugler 1979; Blum 1985; Blum et al. 1988; Hölldobler et al. 1990.
Increased Speed Of Movement
Wilson 1958a; Cammaerts et al. 1978, 1981, 1983, 1985; Cammaerts and Mori 1987; Kugler 1979; Blum et al. 1988; Scheffrahn et al. 1984; Hölldobler et al. 1990.
Frenzied Running/Nondirectional Running/Increased Sinuosity
Goetsch 1957; Wilson 1958a; Bergstrom and Lofqvist 1970; Leuthold and Schlunegger 1973; Duffield et al. 1977; Scheffrahn et al. 1984; Tomalski et al. 1987.
Arrested Motion
Olubajo et al. 1980
Aggressive Posture (Mandible Gaping, Defensive Stance)
Wilson 1958a; Blum et al. 1988; Tomalski et al. 1987; Kugler 1979.
Aggression: Bite, Sting, Or Attack Alien Object
Duffield and Blum 1973; Leuthold and Schlunegger 1973; Kugler 1979; Blum 1985; Tomalski et al. 1987; Blum et al. 1988; Hölldobler et al. 1990.
Mark Intruder With Chemical
Blum 1985

The Context of Alarm

The context in which alarm is released adds another dimension to the definition of alarm behavior. Disturbance to the colony results in alarm, whether the disturbance is mechanical, chemical, or from an enemy or predator. The response of workers varies greatly in relation to the severity of the disturbance, the concentration of the chemicals released, as well as with the nature of the ant species. Alarm can also be released by ants that are in trouble, e.g. ants trapped under soil after a cave-in, to call for help from nestmates.

The age of the ant colony, the state of activity or inactivity of the workers, and many other factors can also affect the type and intensity of the alarm reaction (Gabba and Pavan 1970). This proves a problem for defining alarm with any precision. Instead of a tight definition, as can be used in recruitment reactions, a range of behaviors are generally included in the definition of "alarm behavior" (Table 7.1). For example, near the nest individuals usually are attracted to the odor source and display aggressive behaviors (Shorey 1973). Further away from the nest, the same species may behave differently, perhaps fleeing from the disturbance instead of confronting it (Shorey 1973).

Another problem defining alarm based behavior is that similar behaviors may be observed in contexts that do not involve danger or disturbance. For example, alarm behaviors documented in association with mating flights (See Mating Flight Activity section, this Chapter).

Alarm and Aggression

The release of alarm pheromones from mandibular and poison glands, both associated with defensive structures (mandibles and sting) led to the hypothesis that alarm pheromones may have evolved from defensive compounds (Wilson and Regnier 1971, Hölldobler and Wilson 1990). Species with large, densely concentrated colonies may not be able to disperse readily when disturbed. These species appear to have evolved to meet danger head on and attack the source of disturbance (Regnier and Wilson 1968). Ant species with small, mobile colonies often disperse (Regnier and Wilson 1968) or freeze (Olubajo et al 1980) in response to disturbance or danger. Execution of these behaviors is facilitated by the ability of ants and other social insects to alert nestmates through alarm pheromones and then recruit others to the source, something that solitary animals cannot do.

In most cases aggression is triggered by an interaction of the pheromone stimulus with other appropriate stimuli coming from the intruding enemy (Shorey 1973). However, some authors report that the alarm pheromone alone can cause aggressive behavior (Moser et al. 1968, Regnier and Wilson 1968, Fales et al. 1972). When ants exhibit an aggressive alarm reaction they usually move towards the source of a disturbance where they attack alien objects (Wilson and Regnier 1971). Most bioassays that have shown aggressive reactions to alarm pheromones were conducted with test substances applied to alien material that could be contacted by the workers (e.g. Tomalski et al. 1987, Kugler 1979, Hölldobler et al. 1990). Bioassays that prohibited contact between the ants and a physical object usually resulted in rapid panic alarm or attraction to the source, but no defensive posture or attacking (e.g. Wilson 1958a, Duffield et al. 1977, but see Blum et al. 1968). This suggests that in addition to the alarm pheromone, physical contact may be required for aggressive behavior to occur. To explain why aggressive behavior is not directed toward nestmates, Tricot et al.

(1972) proposed that the alarm pheromones of *Myrmica rubra* may be comprised of inhibitory and stimulative components that direct the aggressiveness of alarmed workers toward the cause of the disturbance instead of toward the worker emitting the alarm pheromone.

Bioassays Used to Detect "Alarm"

A wide range of behavioral bioassays have been used in the laboratory to show alarm behaviors in different ant species and to determine the source and chemical compounds responsible for eliciting alarm reactions. The bioassays are usually conducted on ant colonies housed in the laboratory with tests occurring in a foraging arena attached to the colony's nest tray. Few assays are conducted on colonies in their natural setting (but see Adams 1994 and Janssen et al. 1995). Development of a quick, quantifiable, and reproducible bioassay is essential for pheromone isolation and identification. There have been few studies that start with investigation of natural alarm behaviors, and then identify the chemical releaser. However, there have been several studies that first identify a chemical compound from an ant gland and then try to determine what behaviors the compound(s) elicits. The two approaches are not the same.

Most assays test substances presented to ants on a substrate such as a small filter paper disc that the ants are allowed to contact (Morgan et al. 1978; Cammaerts et al. 1978, 1981, 1983, 1985, 1988; Bradshaw et al. 1975, 1979a, 1979b; Kugler 1979; Blum et al. 1981; Scheffrahn et al 1984; Tomalski et al. 1987; Blum et al. 1988; Scheffrahn and Rust 1989; Hölldobler et al. 1990). In another type of assay, test materials are presented on a substrate that is held or pinned above the ants so that the ants do not come in direct contact with the substrate or test material (Wilson 1958a, Leuthold and Schlunegger 1973, Duffield et al. 1977, Duffield et al. 1980, Olubajo et al. 1980, Pasteels et al. 1989). Other bioassays involve holding a crushed body part directly over a group or trail of workers (Duffield et al. 1980) or applying a test solution onto paper that covers the floor of the foraging arena of an ant colony (Adams 1994). Blum et al. (1968) used a bioassay set-up in which air was drawn at a constant rate over the test material before it reaches ants enclosed in a small arena. Janssen et al. (1995) presented test chemicals on wooden applicator sticks that were inserted into the soil at nest openings of a laboratory contained colony. Alonso and Vander Meer (et al.) used a syringe to draw air from vials containing test samples. They then applied this air to small groups of workers.

Most alarm pheromone studies have not quantified the alarm response. In fact, many studies do not report how they analyzed the alarm reaction but simply state that alarm behavior was observed (Bergstrom and Lofqvist 1970; Duffield et al. 1977, 1980; Blum et al. 1968; Blum et al. 1981; Blum et al. 1988; Pasteels et al. 1989). However, assays have been developed to quantify the alarm reaction of workers. The work of Cammaerts and colleagues on ants in the genera *Myrmica* and *Manica*, may be the most precise. They measured the increase in linear and angular speed of workers exposed to alarm substances and measured the orientation of worker movement by calculating the angles between the workers' course and a direct line to the stimulus (Cammaerts et al. 1978, 1981, 1983, 1985, 1988; Morgan et al. 1978; Cammaerts and Mori 1987). Other methods of quantifying alarm have included counting the number of ants that aggregate at test substance vs. the number at the control (Tomalski et al. 1987, Scheffrahn and Rust 1989, Hölldobler et al. 1990, Adams 1994), time that workers stay at the substance (Tomalski et al. 1987), number of workers that move in response to the test substance (Olubajo et al. 1980, Salzemann et al. 1992), time to and duration of movement (Wilson 1958a), number of ants that are attracted or repelled by the substance or that attack the substance (Bradshaw et al. 1975, 1979a, 1979b; Kugler 1979), the number of ants leaving their nest (Janssen et al. 1995), or evaluation of responses on an alarm scale (from no reaction to full alarm reaction; Leuthold and Schlunegger 1973, Alonso and Vander Meer (In press).

Glandular Sources of Alarm Pheromones

The glandular sources of alarm pheromones have been identified from seven of the 11 extant subfamilies of ants, with some similarity between subfamilies in the sources and chemistry of the alarm pheromones. Ant species in the Dolichoderinae release alarm substances only from their pygidial glands (=supra-anal or anal glands?) (Parry and Morgan 1979). Alarm substances have so far been found only from the mandibular glands of ants in the Ecitoninae (Brown 1960, Torgerson and Akre 1970) and Pseudomyrmecinae (Blum 1969). In contrast, alarm substances are released from the mandibular gland, Dufour's gland, and poison gland in ant species of the Formicinae and Myrmecinae (Parry and Morgan 1979). Ants in the Myrmeciinae release alarm compounds from their mandibular glands, Dufour's glands, and rectal glands (Blum 1969, Parry and Morgan 1979). In these subfamilies, full alarm behavior sometimes is released only by a combination of chemicals from multiple glandular sources (Blum 1969).

Chemistry and Concentration of Alarm Pheromones

As predicted by Bossert and Wilson (1963), chemicals that elicit alarm reactions typically are in the C5-C10 range and have a molecular weight of 100-200. These chemicals volatilize rapidly and most have a boiling point between 151-263°C (Blum 1969). Hölldobler and Wilson (1990; Table 7.4) provide an excellent tabular review of ant alarm pheromones. Alarm signals usually require an immediate and quick response to a perceived danger, which should not last long. Thus, alarm pheromones are likely released as a quick puff, with a short fade time (Bossert and Wilson 1963). Wilson and Bossert (1963) hypothesized that alarm pheromones have an "active space" in which the concentration of the volatile compounds are at or above the threshold to release alarm behaviors. As pheromone diffuses through space the concentration diminishes, finally falling below the response threshold (Regnier and Wilson 1968). They also developed models that could predict the threshold concentration of alarm substances in the air. Using these models they estimated that alarm pheromones in the ant *Acanthomyops claviger* had a threshold concentration of 10^{10}-10^{12} molecules/cm^3 and that of *Pogonomyrmex badius* to be about 10^{13} molecules/cm^3. Using a different method, Moser et al. (1968) estimated the detection threshold of *Atta texana* to be approximately 10^6 molecules/cm^3 and the alarm threshold to be 10^8 molecules/cm^3.

Alarm pheromones represent a wide range of chemical structures including terpenoids, alcohols, aldehydes, ketones, esters, nitrogen heterocycles, and even sulfur containing compounds. Examples are shown below (see also Figure 1.19, Chapter 1, this book). Pyrazine nitrogen heterocycles and sulfur compounds have been found in the Ponerinae, e.g. 2-6-dimethyl-3-propylpyrazine (**8**) from *Odontomachus brunneus* (Wheeler and Blum 1973) and dimethyl disulfide (**9**) from *Paltothyreus tarsatus* (Casnati et al. 1967). Cyclic ketones have been found in Dolichoderinae, e.g. 2-methyl cyclopentanone (**10**) from *Azteca* spp. (Wheeler et al. 1975). Like the previous types of compounds, terpenoids appear to be more specific to subfamily, in this case, Formicinae. Geraniol (**11**) is found in *Cataglyphis* spp. (Hefetz 1985) and *Oecophylla longinoda* (Bradshaw et al. 1975). Other classes of compounds may be found in several subfamilies. For example, aldehydes (hexanal, **12**; *Oecophylla longinoda*; Bradshaw et al. 1975), aliphatic ketones (3-decanone, **13**; *Manica bradleyi*; Fales et al. 1972, and 6-methyl-5-hepten-2-one, **14**; *Conomyrma pyramica*; McGurk et al. 1968; *Tapinoma melanocephalum;* Tomalski et al. 1987; and others). Esters (e.g. methyl anthranilate, **15**; *Aphaenogaster fulva* and *Xenomyrmex floridanus*; Duffield et al. 1980), and alcohols (e.g. 4-

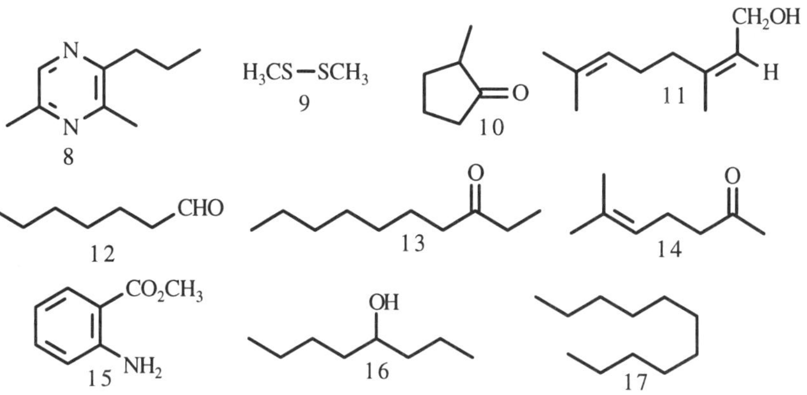

CH_2OH N $H_3CS-SCH_3$ 9 O 11 H N 10 8 O O CHO 12 13 14 CO_2CH_3 OH 15 NH_2 16 17

heptanol, **16**; *Zacryptocerus varians*; Olubajo et al. 1980), each occur in several different subfamilies. Formic acid is the only carboxylic acid that induces an alarm response and is found in several species, including *Polyrhachis* and *Cataglyphis* spp. (Hefetz and Orion 1982). The hydrocarbon n-undecane (**17**) is also found in several species, e.g. *Cataglyphis* spp. (Hefetz and Orion 1982).

Specificity of Alarm Pheromones

Alarm pheromones are generally considered to be the least species-specific of all pheromone classes (Blum 1969, Parry and Morgan 1979). Many of the same alarm compounds are found in a number of ant species, even multiple subfamilies (Hölldobler and Wilson 1990; Table 7.4) and a compound can often cause an alarm reaction in species other than the one in which it was found. As many researchers have pointed out, it would seem to be a defensive advantage to an ant colony to be able to detect the alarm pheromone of other ant species, as well its own (Blum 1969, Wilson 1971, Parry and Morgan 1979). However, the advantage would only be realized in species that were sympatric and competed for territory and resources. On the other hand, an equally justified case can be made for species-specific alarm pheromones. In fact, it has been shown that some ant species react with full alarm only to their natural alarm pheromone and not to closely related chemical compounds (Blum 1969, Parry and Morgan 1979). When *Iridomyrmex pruinosis* was exposed to its natural alarm substance as well as other compounds with similar physical properties, full alarm behavior was released by the natural substance, a lesser response was observed to related substances and no response to unrelated compounds.

Cammaerts et al. (1978, 1981, 1983, 1985) isolated chemicals from the mandibular glands of *Myrmica* species that function as alarm pheromones. They found that the mandibular glands of six *Myrmica* species contained 3-octanone (**18**) and 3-octanol (**19**), but in different proportions between species (Crewe and Blum 1970). Furthermore, the alarm response of each species depended on different ratios and mixtures of the two chemicals. 3-Decanol (**20**) was required in *M. lobicronis* (Cammaerts et al. 1983) and *M. scabrinodis* (Morgan et al. 1978) to produce a full alarm reaction that included orientation toward the source. The six *Myrmica* species were differentiated based on their mandibular gland chemistry and alarm responses. In *M. rubra, M. ruginodis, M. lobicornnis,* and *M. sulcinodis*, 3-octanone (**18**) attracted and increased speed in workers, with 3-octanol (**19**) acting as a synergist (Cammaerts et al. 1978, 1981). In contrast, 3-octanol (**19**) caused

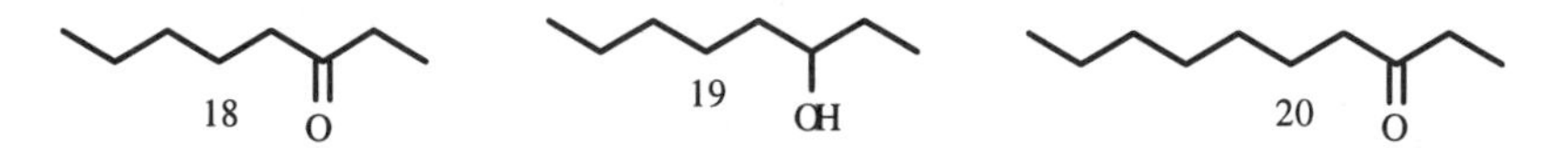

attraction and increased movement in *M. scabrinodis* and *M. sabuleti* (Cammaerts et al. 1978, 1981, 1983). Chemicals from multiple glands in one species were found to contain alarm compounds, with chemicals from the mandibular gland, Dufour's gland, and venom gland necessary to cause a full alarm reaction in *Myrmica rubra* (Tricot et al. 1972, Cammaerts-Tricot 1974). Earlier, the ant, *Myrmica rubida,* was identified by its mandibular gland chemistry to differ significantly from other *Myrmica* species and was subsequently placed in the genus *Manica* (Maschwitz 1966). Further studies are needed to determine the relative frequency of specific versus general alarm pheromones in ants.

We anticipate that future research in ant alarm pheromones will focus on a critical definition of the behavior to be measured, e.g. distinguish between attraction, repellency, aggressive alarm, and passive alarm. Emphasis will be on developing close links between behavior and chemistry -- with a behavioral assay guiding the chemical isolation. Also, it will be interesting to see whether or not the instances where alarm pheromones lack specificity, impart advantages or disadvantages to the ants in real world situations.

Brood Pheromones

By definition a brood pheromone is a chemical or mixture of chemicals released by immature stages that elicits a behavioral response in conspecific adults. There are several reported behavioral responses associated with brood pheromones: (1) Brood recognition -- worker ants recognize the various immature stages (Le Masne 1953). This implies that each immature stage is recognized by workers through differences in its surface chemistry (pheromones); (2) Attraction -- brood release a volatile pheromone that attracts adults (Glancey and Dickens 1988); (3) Aggregation -- Worker ants cluster around a brood pheromone that has a settling affect on the workers (Watkins and Cole 1966); and (4) Brood-tending -- a general term for feeding, grooming, and when a nest is disturbed, brood retrieval.

Brood pheromone literature prior to 1988 was critically reviewed by Vander Meer and Morel (1988; see also Morel and Vander Meer 1988). They identified several problems in demonstrating the existence of brood pheromones, such as bioassay inadequacy due to confounding food responses and aggregation with brood retrieval and attraction, respectively. Other problems revolved around the reliability of controls, lack of quantitation, and/or inappropriate chemical techniques. The interspecific adoption of brood is widely reported (Jaisson 1971). This creates a basic problem, since by definition a pheromone is a substance that causes a specific behavioral or physiological reaction in a receiving organism of the same species (Nordlund and Lewis 1976). Brood from one species may receive similar differential treatment from workers of several species. Are there alternative non-pheromonal explanations?

Vander Meer and Morel (1988) offer an explanation for the differential treatment of brood that does not involve pheromones. The hypothesis is based on (1) reports that the basic immature stages (egg, larvae, pupae) are often morphologically distinguishable (Robinson and Cherrett 1974; Brian 1975; Petralia and Vinson 1979); (2) unlike workers, brood are immobile and lack the ability for agonistic display, thus it is predicted that con- or hetero-specific worker aggression toward brood should be significantly less then toward other workers; (3) the cuticular chemistry comprising nestmate recognition cues, changes continuously; (4) the developmental stages can offer a reward to workers through regurgitation or secretions. Thus, the lack of brood/worker aggression allows newly introduced interspecific brood time to acquire their new colony's recognition profile. Associative learning based on the above characters (morphological, behavioral and chemical) and a reward, provide the basis of the non-pheromone differential treatment of brood by workers. For details see Vander Meer and Morel 1988).

What Is New in Brood Pheromones

Carlin (1988) states in his review of brood discrimination in ants that there are strong generalized brood stimuli, these stimuli are similar enough to facilitate acceptance across kin, colony and species categories; and there is sufficient variation in brood stimuli so that non-exclusive discrimination of kin, nestmates and species can occur.

The purpose of the Vander Meer and Morel (1988) review was to stimulate interest and research on brood pheromones; however, the past seven years have not brought us closer to the identification of a brood pheromone. *Rhytidoponera confusa* workers could not distinguish between their own larvae and conspecific or allospecific larvae (Crosland 1988). Similarly, slave-making ant, *Formica sanguinea,* workers cared for homo- and heterospecific (slave *Formica cunicularia*) cocoons without preference, regardless of the early social experience of *F. sanguinea* (Mori et al. 1992).

The pre-imaginal experience of workers in *Camponotus floridanus* influenced nestmate brood recognition for only a short time, after which, no preference in conspecific brood was observed (Carlin and Schwartz 1989). Similarly, teneral *Ectatomma tuberculatum* worker experience has a temporary effect on their brood discrimination (Fénéron and Jaisson 1995). This illustrates the transient nature of pre-imaginal and early adult experience. Brood tending *Ectatomma tuberculatum* workers (2-10 weeks old) preferentially tend nestmate rather than non-nestmate brood, however, older workers (>10 weeks) showed no preference (Fénéron and Jaisson 1992; 1995).

Workers from queenright *Monomorium pharaonis* colonies accept conspecific worker larvae, but cannibalize nestmate and introduced conspecific sexual larvae (Edwards and Abraham 1991). If the queens are absent, sexual brood are allowed to develop and conspecific sex brood are accepted. Worker larvae are covered with bifurcated hairs, whereas sexual larvae are hairless. The author suggests that this morphological difference may allow workers to differentiate between sexual and worker larvae.

The bottom line is we are still awaiting definitive experimental evidence for brood pheromones or alternative explanations.

Recruitment Pheromones

Ants have evolved a wide variety of mechanisms for recruitment and orientation using pheromones. One of the least complicated is tandem running, exemplified by *Leptothorax* spp. (Möglich et al. 1974). After finding a food source and returning to its nest a scout worker raises its gaster and releases poison sac contents through its extended sting. As soon as an attracted worker contacts the scout "tandem running" begins. The recruited worker closely follows and maintains contact with the scout by touching the gaster and hind legs. In this case the poison gland is the source of the pheromone; however, in *Pachycondyla obscuricornis* (Traniello and Hölldobler 1984) tandem running is mediated by pheromones released by the pygidial gland. In these examples, orientation pheromones are not needed. Going up the complexity ladder is *Camponotus socius*, whose scouts lay a chemical trail from the hindgut, then they perform a mechanical waggle that can induce tens of other workers to follow the initiating scout back to the food source (Hölldobler 1971b). In other species, e.g. *Formica fusca*, scouts lay an orientation trail back to their nest and use a mechanical waggle to excite and induce nestmates to follow that trail unaided by the originating scout (Möglich and Hölldobler 1975). Analogous to the latter system is that of *Solenopsis invicta*, except that pheromones released by the Dufour's gland are responsible for orientation (Vander Meer et al. 1981), attraction (Vander Meer et al. 1988), and orientation induction (Vander Meer et al. 1990). The recruitment system of the weaver ant, *Oecophylla longinoda* (Hölldobler and Wilson 1978), is extremely complicated and uses at least five different systems. There is long range recruitment to intruders, invoked by rectal gland odor trails and by antennation and jerking physical cues; short range recruitment to territorial intruders where the pheromones are produced by the sternal gland; recruitment to new terrain elicited by pheromones produced from the rectal gland and the tactile stimulus of antennation; recruitment to new food sources, that is mediated by trail pheromones produced by the rectal gland and modulated by tactile stimuli; and finally emigration to new sites (Hölldobler and Wilson 1978).

Glandular Sources

Recruitment pheromones are diverse in orgin, although almost all are derived from abdominal exocrine glands. The glands thus far identified as sources of recruitment pheromones are the (1) poison, (2) Dufour's, (3) cloacal, (4) hindgut, (5) pygidial, (6) rectal, (7) sternal, and (8) tibial glands (see Chapter 1, this book for illustrations, and Hölldobler and Wilson [1990] for specific citations). In most species the recruitment pheromones are from a single glandular source; however, multiple glands may be involved. For example, *Megaponera foetens* uses products from the pygidial gland to recruit workers, and poison gland secretions for a long lasting orientation effect (Hölldobler et al. 1994b). The variety of sources leads to a complex variety of structures.

The Chemistry of Recruitment Pheromones

There has been significant activity in the isolation and identification of recruitment pheromones in the last five to six years. Thus, this section will only deal with recruitment pheromones reported since Hölldobler and Wilson (1990) and the reader is referred to the latter reference and to Chapter 1, this book, for additional information.

Scout ants of *Myrmicaria eumenoides* release a recruitment pheromone from their poison gland via the sting (Kaib and Dittebrand 1990). In this species when the scout worker finds prey or an enemy, it recruits other workers to the site of encounter with an attractant pheromone. The major volatile compound from the poison sac was determined to be (+)-limonene (**21**). This compound by itself induced the alarm recruitment observed with the whole gland extract. There is also an alkaloid dominant fraction of lower volatility that may serve to regulate the release rate of the volatile limonene. This species applies venom to prey or enemy, instead of injecting the venom. So for prey suppression or defense the limonene helps the venom components penetrate the cuticle of prey or enemy.

The trail pheromone of *Tetramorium meridionale* consists of four pyrazines: methylpyrazine (**22**), 2,5-dimethylpyrazine (**23**), trimethylpyrazine (**24**), and 3-ethyl-2,5-dimethylpyrazine (**25**) and indole (**26**). As in the previous example the source of this pheromone blend is the poison gland. The pyrazines (1-10 pg each per individual) and indole (ca 1 ng per individual) work together synergistically (Jackson et al. 1990). Indole (**26**) and trimethylpyrazine (**24**) each release some trail following behavior, but at a much weaker level then the mixture of pyrazines and indole.

In contrast to the above, the source of *Tapinoma simrothi*'s trail pheromone is the pygidial gland (Simon and Hefetz 1991). Interestingly, the pygidial gland also has been reported to be the source of alarm pheromones for this species (Hefetz and Lloyd 1983). Iridomyrmecin (**27**) and iridodials (**28**) were found to elicit both alarm and trail following behavior, depending on the concentration and whether the compounds were released from a point source or streaked as a trail. The higher concentration released alarm, while lower concentrations released trail following (Simon and Hefetz 1991). This is a good example of context (and concentration) dependent behaviors derived from the same glandular source and products.

After finding appropriate prey, *Leptogenys diminuta* scouts lay a chemical trail back to their nest using poison and pygidial gland secretions. The scout enters the nest and recruits many workers, then leads them to the prey along the trail (Wilson 1958b). The (3R,4S) 4-methyl-3-heptanol (**29**) isomer elicits trail following. The other three stereoisomers were not present in poison gland extracts, but neither did they interfere with the activity of the natural isomer when a synthetic mixture of all four possible isomers was presented to the ants (Attygale et al. 1988; Steghaus-Kovâc et al. 1992). The recruitment part of the system is associated with the pygidial gland, and *cis*-isogeraniol (**30**) was identified as the active agent (Attygale et al. 1988; 1991). Electrophysiology studies (Kern and Bestmann 1993) clearly showed the selective activity of trail pheromone component (**29**) and that only recruitment pheromone (**30**) was detected by the ant's antennae among several isomers presented to the ants. The electroantennagram results were much more definitive than the behavioral assays and is illustrative of how useful this technique can be in cutting through chemical complexity. This is a complex recruitment system, utilizing products from two exocrine glands.

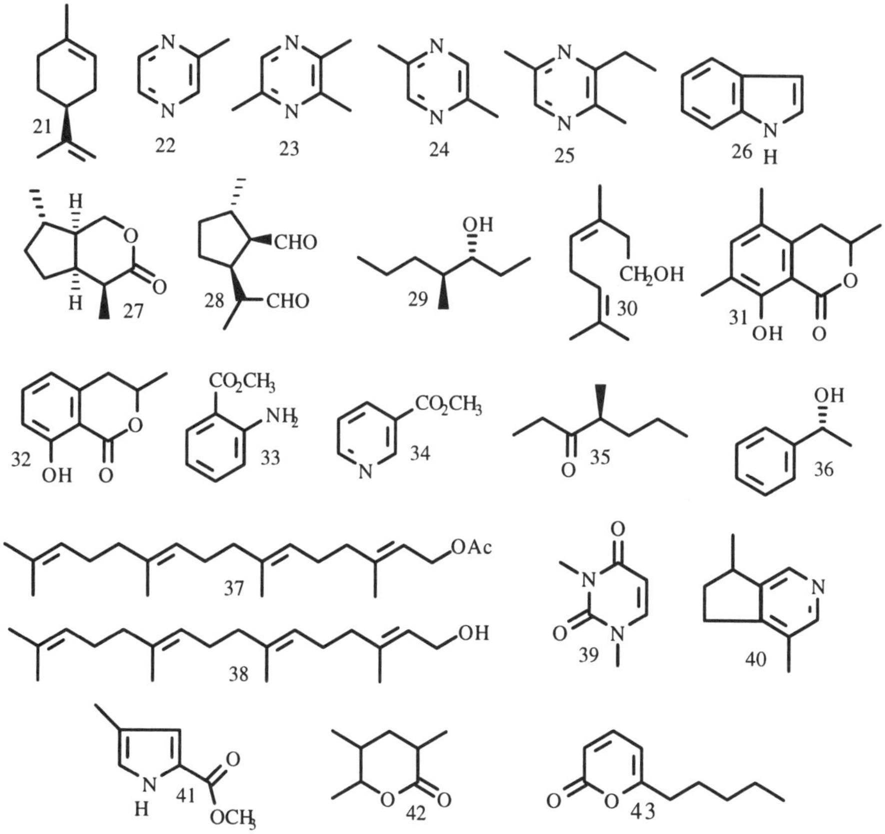

A new class of trail pheromone was reported from *Lasius niger* (Bestmann et al. 1992). The pheromone is produced in the rectal sac and was determined, at least in part, to be (R) 3,4-dihydro-8-hydroxy-3,5,7-trimethylisocoumarin (**31**). The related isocoumarin, mellein (**32**), was isolated from the rectal sacs of *Formica rufa*. This compound elicited trail following in *F. rufa* workers (Bestmann et al. 1992).

Army ants are blind, yet carry out sophisticated foraging raids, which have been assumed to be chemically mediated. Oldham et al. (1994) reported the first chemical dissection of this process using an *Aenictus* species. The source of the pheromone is the postpygidial gland, which contains at least two active components. Interestingly this system is analogous to the orientation, attraction, orientation primer situation in *Solenopsis invicta* (Vander Meer et al. 1981; 1988; 1990; Alvarez et al. 1987), where the sub-categories of recruitment are released by different pheromones from the Dufour's gland for *S. invicta*. For this *Aenictus* species, methyl anthranilate (**33**) released trail orientation behavior, but only if the workers ants had already been following a natural trail. Further chemical analysis and behavioral bioassays revealed that methyl nicotinate (**34**) acted in concert with component **33** to prime or induce workers to follow the trail. By itself (**34**) had no behavioral effect. This example and that of *S. invicta* illustrate the potential chemical and behavioral complexity of the recruitment process.

Hölldobler et al. (1995) investigated the recruitment pheromones from *Aphaenogaster albisetosus* and *A. cockerelli*. The pheromone is derived from the poison gland and is composed of (S) and (R)-4-methyl-3-heptanone (**35**, S structure shown) in a ratio of 8:2 for *A. albisetosus*. However, *A. cockerelli*'s trail pheromone is (R)-1-phenylethanol (**36**); however, (S)-4-methyl-3-heptanone (**35**) is also present. These results explain the asymmetric interspecific trail following results previously reported (Hölldobler et al. 1978). *A. cockerelli* only follows its own trail, since *A. albisetosus* does not produce compound **36**, whereas *A. cockerelli* follows the *A. albisetosus* trail due to the presence of compound **35**. It should be noted that these classes of compounds are not usually produced by poison glands. This is an interesting case where vibrational signals are used by scouts in the nest to prime or modulate the recruitment process (Markl and Hölldobler 1978).

Bestmann and co-workers (1995a) isolated all-*trans* geranylgeranyl acetate (**37**) and the corresponding alcohol (**38**) from the Dufour's glands of *Ectatomma ruidum*. Both compounds generated electroantennogram responses, but the acetate response was more pronounced. Similarly, orientation bioassays of the two compounds at physiological concentrations showed that the acetate performed significantly better than the alcohol. The acetate then, is considered the main trail pheromone component.

Two interesting compounds, N,N-dimethyluracil (**39**) and actinidine (**40**) were isolated from the poison and pygidial glands, respectively, of *Megaponera foetens* (Jannsen et al. 1995). A combination of electroantennogram and behavioral bioassays was used to monitor biological activity. Electroantennogram results showed compounds **39** and **40** to be significantly active; however, behavioral assays could only establish a trail following role for **39**. Actinidine (**40**) caused worker excitement, but did not elicit the recruitment observed with whole pygidial gland extracts. This is another complex recruitment system involving compounds from at least two glandular sources.

The first trail pheromone was isolated and identified by Tumlinson et al. (1971), as methyl 4-methylpyrrole-2-carboxylate (**41**) from *Atta texana*. Subsequently, this compound was identified as the trail pheromone of other leaf-cutting ant species, as was 3-ethyl-2,5-dimethylpyrazine (**25**) (Cross et al. 1979). Do Nascimento et al. (1994) added to our knowledge of this economically important group by identifying compound **41** as the sole trail pheromone component from *Acromyrmex subterraneus subterraneus*. No pyrazines could be detected and when pyrazines were assayed with compound **41**, no enhanced activity was observed.

Many carpenter ant species are well known pests and have been studied extensively (see below). It was only recently that the recruitment pheromone of a carpenter ant species was identified from hindgut extracts (Bestmann et al. 1995b). The carpenter ant, *Camponotus herculeanus*, was found to use 2,4-dimethyl-5-hexanolide (**42**) as a trail orientation pheromone component. This compound has three asymmetric centers and of the four possible diastereomer pairs, two were found in the natural material.

Of note are two recent papers that have reinvestigated early investigations of the trail pheromones of *Lasius fuliginosus* (Huwyler et al. 1975) and *Pristomyrmex pungens* (Hayashi and Komae 1977). Both initial papers report the trail pheromones as a series of short chain fatty acids. Quinet and Pasteels (1995) investigated the trail following behavior of a myrmecophilous staphyinid beetle, *Homeusa acuminata*, whose host is *Lasius fuliginosus*. The authors demonstrated that the beetles and ants follow artificial trails of hindgut extracts at 0.03 hindgut equivalents/cm. However, when experiments were conducted with the reported six straight chain fatty acids (Huwyler et al. 1975; C_6-C_{10} + C_{12}), neither the ants or the beetles responded to physiological levels. The ants only followed trails containing very high fatty acid concentrations, ca. 3 x 10^3 hindgut equivalents/cm

(Quinet and Pasteels 1995). Thus either the fatty acids are not acting as trail pheromone components or other unidentified compounds are acting synergistically in concert with the carboxylic acids.

In a second paper Hayashi and Komae (1977) isolate fatty acids from whole *Pristomyrmex pungens* worker extracts. The carboxylic acid fraction was active in a trail bioassay and contained typical fatty acids found in animal tissue ($C_{14:0}$, $C_{16:0}$, $C_{16:1}$, $C_{18:0}$, $C_{18:1}$, $C_{18:2}$, $C_{18:3}$, $C_{20:5}$, and $C_{20:5}$). Janssen et al. (1997) reinvestigated the trail pheromone of *P. pungens* and found that the source of the trail pheromone is the poison gland. They further isolated an active compound, 6-*n*-amyl-2-pyrone (**43**), by bioassay, electroantennograms, and chemical analyses. Thus, in this case it is clear that the fatty acids originally reported (Hayashi and Komae 1977) are not involved as trail pheromone components.

As evidenced by the chemical structures and the above references, over the past five years there has been a tremendous amount of activity in elucidating the chemistry and behavior of ant recruitment pheromones.

Conclusion

Early work in ant pheromones centered primarily on alarm and other defensive semiochemicals, due to their greater abundance. However, in the past decade, with better instrumentation and new techniques, we have seen excellent progress in other aspects of ant pheromone chemistry, especially the recruitment pheromones. Yet what has been accomplished is not even the tip of the ant pheromone iceberg. We anticipate progress to continue at an accelerated rate in the years to come and look forward to new and exciting chemical ecology stories from the fascinating world of the ant.

References Cited

Adams, E.S., 1994. Territory defense by the ant *Azteca trigona*: maintenance of an arboreal ant mosaic. *Oecologia, 97*: 202-208.

Alonso, L. and R.K. Vander Meer, In press. Source of alate excitant pheromones in the red imported fire ant, *Solenopsis invicta* (Hymenoptera: Formicidae), *J. Insect Behavior.*

Alvarez, F. M., R.K. Vander Meer and C.S. Lofgren, 1987. Synthesis of homofarnesenes: Trail pheromone components of the fire ant, *Solenopsis invicta*. *Tetrahedron, 43:* 2897-2900.

Attygalle, A. B., O. Vostrowsky, H. J. Bestmann, S. Steghaus-Kovac and U. Maschwitz, 1988. (3R,4S)-4-Methyl-3-heptanol, the trail pheromone of the ant *Leptogenys dimminuta. Naturwissenschaften, 75*: 315-317.

Attygalle, A. B., S. Steghaus-Kovac, V. Ahmad, U. Maschwitz, O. Vostrowsky, and H. J. Bestmann, 1991. *cis*-Isogeraniol, a recruitment pheromone of the ant *Leptogenys dimminuta*. *Naturwissenschaften.* *78*: 90-2.

Bergstrom, G. and J. Lofqvist, 1970. Chemical basis for odour communication in four species of *Lasius* ants. *J. Insect. Physiol.* *16*: 2353-2375.

Bestmann, H. J., F. Kern, D. Schafer and M. C. Witschel, 1992. Pheromones 86. 3,4-dihydroisocoumarins, a new class of ant trail pheromones. *Angew. Chem. Int. Ed.* *31*: 795-796.

Bestmann, H.J., E. Janssen, F. Kern, B. Liepold and B. Hölldobler, 1995a. All *trans*-geranylgeranyl acetate and geranylgeraniol, recruitment pheromone components in the Dufour gland of the ponerine ant *Ectatomma ruidum*. *Naturwissenschaften*, *82*: 334-336.

Bestmann, H. J., U. Haak and F. Kern, 1995b. 2,4-Dimethyl-5-hexanolide, a trail pheromone component of the carpenter ant *Camponotus herculteanus*. *Naturwissenschaften* *82*: 142-144.

Billen, J.P. J., B.D. Jackson and E.D. Morgan, 1988. Secretion of the Dufour gland of the ant *Nothomyrmecia macrops* (Hymenoptera: Formicidae). *Experientia* *44*: 715-719.

Blum, M.S., 1969. Alarm pheromones. *Annu. Rev. Entomol.* *4*: 57-80.

Blum, M.S. , 1985. Alarm pheromones. In:, *Comprehensive Insect Physiology, Biochemistry, and Pharmacology* (Kerkut, G.A., and L.I. Gilbert, eds.) Vol. 9: Behaviour. Pergamon Press: New York. Pp. 193-224.

Blum, M.S., J.M. Brand and E. Amante, 1981. o-aminoacetophenone: identification in a primitive fungus-growing ant (*Mycocepurus goeldii*). *Experientia, 37*: 816-817.

Blum, M.S., F. Padovani, H.R. Hermann, and P.B. Kannowski, 1968. Chemical releasers of social behavior, XI: Terpenes in the mandibular glands of *Lasius umbratus*. *Ann. Entomol. Soc. Am., 61*: 1354-1359.

Blum, M. S., L. Morel, and H. M. Fales, 1987. Chemistry of the mandibular gland secretion of the ant *Camponotus vagus*. *Comp Biochem Physiol B Comp Biochem* *86*: 251-252.

Blum, M. S., R.R. Snelling, R.M. Duffield, H.R. Hermann, and H.A. Lloyd, 1988. Mandibular gland chemistry of *Camponotus (Myrmothrix) abdominalis*: Chemistry and chemosystematic implications (Hymenoptera: Formicidae). In: *Advances in myrmecology*, (J. C. Trager, ed.) E. J. Brill, Leiden, pp. 481-490.

Bossert, W.H. and E.O. Wilson, 1963. The analysis of olfactory communication among animals. *J. Theor. Biol.* *5*: 443-469.

Bradshaw, J.W.S., R. Baker and P.E. Howse, 1975. Multicomponent alarm pheromones of the weaver ant. *Nature, 258*: 230-231.

Bradshaw, J.W.S., R. Baker and P.E. Howse, 1979a. Multicomponent alarm pheromones in the mandibular glands of major workers of the African weaver ant, *Oecophylla longinoda*. *Physiol. Entomol.* *4*: 15-25.

Bradshaw, J.W.S., R. Baker, P.E. Howse, and M.D. Higgs, 1979b. Caste and colony variations in the chemical composition of the cephalic secretions of the African weaver ant, *Oecophylla longinoda*. *Physiol. Entomol.* *4*: 27-38.

Brand, J. M., R.M. Duffield, J.G. MacConnell, and M.S. Blum, 1973. Caste-specific compounds in male carpenter ants. *Science 179*: 388-389.

Breed, M.D., J.H. Fewell, A.J. Moore, and K.R. Williams. 1987. Graded recruitment in a ponerine ant. *Behav. Ecol. Sociobiol. 20*: 407-411.

Brian, M.V., 1975. Larval recognition by workers of the ant *Myrmica. Anim. Behav., 23*: 745-756.

Brown, W.L., 1960. The release of alarm and attack behavior in some New World army ants. *Psyche. 66*: 25-27.

Buschinger, A., 1976. Giftdrusensekret als Sexual-pheromon bei der Gastameise *Formicoxenus nitidulus* (Nyl.) (Hym., Form.). *Insectes Soc. 23*: 215-225.

Buschinger, A., and T.M. Alloway, 1979. Sexual behaviour in the slave-making ant, *Harpagoxenus canadensis* M. R. Smith, and sexual pheromone experiments with *Harpagoxenus canadensis, Harpagoxenus americanus* (Emery), and *Harpagoxenus sublaevis* (Nylander) (Hymenoptera; Formicidae). *Z. Tierpsychol. 49*: 113-119.

Cammaerts-Tricot, M.C., 1974. Recrutement d'ouvrieres, chez *Myrmica rubra,* par les pheromones de l'appareil a venin. *Behaviour, 50*: 111-122.

Cammaerts, M-C. and K. Mori, 1987. Behavioural activity of pure chiral 3-octanol for the ants *Myrmica scabrinodis* Nyl. and *Myrmica rubra* L. *Physiol. Entomol. 12*: 381-385.

Cammaerts, M.C., M.R. Inwood, E.D. Morgan, K. Parry and R.C. Tyler, 1978. Comparative study of the pheromones emitted by workers of the ants *Myrmica rubra* and *Myrmica scabrinodis. J. Insect Physiol. 24*: 207-214.

Cammaerts, M.C., R.P. Evershed and E.D. Morgan, 1981. Comparative study of the mandibular gland secretion of four species of *Myrmica* ants. *J. Insect Physiol. 72*: 225-231.

Cammaerts, M.C., R.P. Evershed and E.D. Morgan, 1983. The volatile components of the mandibular gland secretion of workers of the ants *Myrmica lobicornis* and *Myrmica sulcinodis. J. Insect Physiol. 29*: 659-664.

Cammaerts, M.C., A.B. Attygalle, R.P. Evershed, and E.D. Morgan, 1985. The pheromonal activity of chiral 3-octanol for *Myrmica* ants. *Physiol. Entomol. 10*: 33-36.

Cammaerts, M.C., A.B. Attygalle, O. Vostrowsky and H.J. Bestmann, 1988. Ethological studies of the mandibular gland secretion of the ant *Manica rubida* (Formicidae: Myrmicinae). *J. Insect Physiol. 34*: 347-350.

Cariouetienne, A., S. Aron and L. Passera, 1992. Queen attractivity in the Argentine ant *Iridomyrmex humilis* (Mayr.). *Behav. Process. 27*: 179-186.

Cariouetienne, A. and L. Passera, 1993. Queen power in relation to age and mating status in the argentine ant *Iridomyrmex humilis* (Mayr). *Insectes Sociaux 40*: 87-94.

Carlin, N.F. 1988. Species, kin and other forms of recognition in the brood discrimination behavior of ants. In:, *Advances in myrmecology*, 551 p. (J. C. Trager, Ed.) E.J. Brill, New York, pp. 267-295.

Carlin, N.F. And P.H. Schwartz, 1989. Pre-imaginal experience and nestmate brood recognition in the carpenter ant, *Camponotus floridanus*. *Anim. Behav. 38*: 89-95.

Casnati, G., A. Ricca and M. Pavan, 1967. Sulla secrezione difensiva delle glandole mandibolari di *Paltothyreus tarsatus* (Fabr.) (Hymenoptera: Formicidae). *Chimica e I'Industria (Milano) 49:* 57-58.

Cavill, G.W.K. and E. Houghton, 1974. Volatile constituents of the Argentine ant, *Iridomyrmex humilis*. *J. Insect Physiol. 20*: 2049-2059.

Crewe, R. M. And M.S. Blum, 1970. Alarm pheromones in the genus *Myrmica* (Hymenoptera: Formicidae). their composition and species specificity. *Z. Vgl. Physiol., 70*: 363-373.

Crosland, M. W. J., 1988. Inability to discriminate between related and unrelated larvae in the ant *Rhytidoponera confusa* (Hymenoptera: Formicidae). *Ann. Entomol. Soc. Am., 81*: 844-850.

Cross, J.H., R.C. Byler, U. Ravid, R.M. Silverstein, S.W. Robinson, P.M. Baker, J. Sabino de Oliveira,A.R. Jutsum, and M.J. Cherrett, 1979. The major component of the trail pheromone of the leaf-cutting ant, *Atta sexdens rubropilosa* Forel. 3-ethyl-2,5-dimethylpyrazine. *J. Chem. Ecol.5*: 187-203.

do Nascimento, R.R. , B.D. Jackson, E.D. Morgan, W.H. Clark and P.E. Blom, 1993a. Chemical secretions of two sympatric harvester ants, *Pogonomyrmex salinus* and *Messor lobognathus*. *J. Chem Ecol.19*: 1993-2005.

do Nascimento, R.R., E.D. Morgan, J. Billen, E. Schoeters, T.M.C. Della Lucia and J.M.S. Bento, 1993b. Variation with caste of the mandibular gland secretion in the leaf-cutting ant *Atta sexdens rubropilosa*. *J. Chem. Ecol. 19*: 907-918.

do Nascimento, R.R., Morgan, D.D.O. Moreira and T.M.C. Della Lucia, 1994. Trail pheromone of the leaf-cutting ant Acromyrmex subterraneus subterraneus (Forel). *J. Chem. Ecol. 20*: 1719-1724.

Duffield, R.M. and M.S. Blum, 1973. 4-Methyl-3-Heptanone: Identification and function in *Neoponera villosa* (Hymenoptera: Formicidae). *Ann. Entomol. Soc. Amer. 66*: 1357.

Duffield, R.M. and M.S. Blum, 1975. Identification, role, and systematic significance of 3-octanone in the carpenter ant, *Camponotus schaefferi* Whr. *Comp. Biochem. Physiol. 51B*: 281-282.

Duffield, R.M., J.M. Brand and M.S. Blum, 1977. 6-methyl-5-hepten-2-one in *Formica* species: identification and function as an alarm pheromone (Hymenoptera: Formicidae). *Ann. Entomol. Soc. Am. 70*: 309-310.

Duffield, R.M., J.W. Wheeler and M.S. Blum, 1980. Methyl anthranilate: identification and possible function in *Aphaenogaster fulva* and *Xenomyrmex floridanus*. *Florida Entomol. 63*: 203-206.

Edwards, J. P. 1987. Caste regulation in the pharaoh's ant *Monomorium pharaonis*: the influence of queens on the production of new sexual forms. *Physiol. Entomol. 12*: 31-39.

Edwards, J.P. and J. Chambers, 1984. Identification and source of a queen-specific chemical in the Pharaoh's ant, *Monomorium pharaonis* (L.). *J. Chem. Ecol. 10*: 1731-47.

Edwards, J.P. and L. Abraham, 1991. Caste regulation in the pharaoh's ant *Monomorium pharaonis*: recognition and cannibalism of sexual brood by workers. *Physiol. Entomol. 16*: 263-271.

Evesham, E.J.M. 1984. The attractiveness of workers towards individual queens of the polygynous ant *Myrmica rubra* L. *Biol. Behav. 9*: 144-156.

Fales, H.M., M.S. Blum, R.M. Crewe and J.M. Brand, 1972. Alarm pheromones in the genus *Manica* derived from the mandibular gland. *J. Insect Physiol. 18*: 1077-1088.

Fénéron, R. and P. Jaisson, 1992. Nestmate brood recognition among workers of different social status in *Ectatomma tuberculatum* Olivier (Formicidae, Ponerinae). *Behav. Process. 27*: 45-52.

Fénéron, R. and P. Jaisson, 1995. Ontogeny of nestmate brood recognition in a primitive ant, *Ectatomma tuberculatum* olivier (Ponerinae). *Anim. Behav. 50*: 9-14.

Fletcher, D.J.C. and M.S. Blum, 1983. The inhibitory pheromone of queen fire ants: effects of disinhibition on dealation and oviposition by virgin queens. *J. Comp. Physiol. A 153*: 467-475.

Fletcher, D.J.C. and K.G. Ross, 1985. Regulation of reproduction in eusocial Hymenoptera. *Annu. Rev. Entomol. 30*: 319-343.

Fowler, H.G. 1982. Male induction and function of workers' excitability during swarming in leaf-cutting ants (*Atta* and *Acromyrmex*) (Hymenoptera, Formicidae) pheromones. *Int J Invertebr Reprod, 4*: 333-335.

Fowler, H.G. and R.B. Roberts, 1982. Entourage pheromone in carpenter ant (*Camponotus pennsylvanicus*) (Hymenoptera: Formicidae) queens. *J. Kans. Entomol. Soc. 55*: 568-570.

Gabba, A. and M. Pavan, 1970. Researches on Trail and Alarm Substances in Ants (161-195). In *Advances in Chemoreception*. Vol. 1 *Communication by chemical signals*. (Johnston, J.W., Jr., Moulton, D.G. and Turk, A. Eds.) Appleton-Century Crofts, New York, 412 pp.

Glancey, B.M. 1980. Biological studies of the queen pheromone of the red imported fire ant. *Proc. Tall Timbers Conf. Ecol. Anim. Control Habitat Manage. 7*: 149-154.

Glancey, B.M. and W.A. Banks, 1988. Effect of the insect growth regulator Fenoxycarb on the ovaries of queens of the red imported fire ant (Hymenoptera: Formicidae). *Ann. Entomol. Soc. Am. 81*: 642-648.

Glancey, B.M., J. Rocca, C.S. Lofgren and J. Tumlinson, 1984. Field tests with synthetic components of the queen recognition pheromone of the red imported fire ant, *Solenopsis invicta*. *Sociobiology 9*: 19-30.

Glancey, B.M. and J.C. Dickens, 1988. Behavioral and electrophysiological studies with live larvae and larval rinses of the red imported fire ant, *Solenopsis invicta* Buren (Hymenoptera: Formicidae). *J. Chem. Ecol. 14*: 463-473.

Goetsch, W., 1957. *The Ants* (Manheim, R., Trans.) University of Michigan Press, Ann Arbor

Hayashi, N. And H. Komae, 1977. The trail and alarm pheromones of the ant *Pristomymex pungens* Mayr. *Experientia. 33*: 424-425.

Hefetz, A. 1985. Mandibular gland secretions as alarm pheromones in two species of the desert ant *Cataglyphis*. *Z. Naturforschung ser. C 40*: 665-666.

Hefetz, A. and T. Orion, 1982. Pheromones of ants of Israel: I. The Alarm-Defense system of some larger formicinae. *Israel J. Entomol. 16*: 87-97.

Hefetz, A. and H.A. Lloyd, 1983. Identification of new components from anal glands of *Tapinoma simrothi Pheonicium*. *J. Chem. Ecol. 9*: 607-613.

Hölldobler, B. 1971a. Sex pheromone in the ant *Xenomyrmex floridanus*. *J. Insect Physiol. 17*: 1497-1499.

Hölldobler, B. 1971b. Recruitment behavior in *Camponotus socius* (Hym. Formicidae). *Z. Vergl. Physiologie. 75*: 123-142.

Hölldobler, B. and C.P. Haskins, 1977. Sexual calling behavior in primitive ants. *Science 195*: 793-794.

Hölldobler, B. And E.O. Wilson, 1978. The multiple recruitment systems of the African weaver ant *Oecophylla longinoda* (Latreille) (Hymenoptera: Formicidae). *Behav. Ecol. Sociobiol. 3*: 19-60.

Hölldobler, B. and E.O. Wilson, 1983. Queen control in colonies of weaver ants (Hymenoptera: Formicidae). *Ann. Entomol: Soc. America* 76: 235-238.

Hölldobler, B. and S.H. Bartz, 1985. Sociobiology of reproduction in ants. *Fortschr. Zool. 31*: 237-257.

Hölldobler, B. and E.O. Wilson, 1990. *The Ants*. Cambridge, Mass.: Belknap Press.

Hölldobler, B., R.C. Stanton and H. Markl, 1978. Recruitment and food retrieving behaviour in *Novomessor* (Hymenoptera: Formicidae). I. Chemical signals. *Behav. Ecol. Sociobiol. 4*:163-181.

Hölldobler, B., J.M. Palmer and M.W. Moffett, 1990. Chemical communication in the dacetine ant *Daceton armigerum* (Hymenoptera: Formicidae). *J. Chem. Ecol., 16*: 1207-1219.

Hölldobler, B., C. Peeters and M. Obermayer, 1994a. Exocrine glands and the attractiveness of the ergatoid queen in the ponerine ant *Megaponera foetens*. *Insect. Soc. 41*: 63-72.

Hölldobler, B., U. Braun, W. Gronenberg, W.H. Kirchner and C. Peeters, 1994b. Trail communication in the ant *Megaponera foetens* (Fabr.) (Formicidae, Ponerinae) *J. Insect Physiol.40*: 585-593.

Hölldobler, B., N.J. Oldham, E.D. Morgan and W.A. König, 1995. Recruitment pheromones in the ants *Aphaenogaster albisetosus* and *A. Cockerelli* (Hymenoptera: Formicidae). *J. Insect Physiol. 41*: 739-744.

Huwyler, S.K. Grob and M. Viscontini, 1975. The trail pheromone of the ant *Lasius fuliginosus*: identification of six components. *J. Insect Physiol. 21*: 299-304.

Jackson, B.D., S.J. Keegans, E.D. Morgan, M.C. Cammaerts and R. Cammaerts, 1990. Trail pheromone of the ant *Tetramorium meridionale*. *Naturwissenschaften 77*: 294-296.

Jaisson, P. 1971. Experiences sur l'agressivite chez les fourmis. *Acad Sci Paris C R Ser D, 273*: 2320-2323.

Janssen, E., H.J. Bestmann, B. Hölldobler and F. Kern, 1995. N,N-dimethyluracil and actinidine, two pheromones of the ponerine ant *Megaponera foetens* (Fab.) (Hymenoptera: Formicidae). *J. Chem. Ecol. 21*: 1947-1955.

Janssen, E., B. Hölldobler, F. Kern, H.J. Bestmann and K. Tsuji, 1997. The trail pheromone of the myrmicine ant *Pristomyrmex pungens* (Hymenoptera: Formicidae). *J. Chem. Ecol.* 23: 1025-1034.

Jouvenaz, D.P., W.A. Banks and C.S. Lofgren, 1974. Fire ants: Attraction of workers to queen secretions. *Ann. Entomol. Soc. Am. 67*: 442-444.

Kaib, M. and H. Dittebrand, 1990. The poison gland of the ant *Myrmicaria eumenoides* and its role in recruitment communication. *Chemoecology 1*: 3-11.

Karlson, P. and M. Luscher, 1959. "Pheromones," a new term for a class of biologically active substances. *Nature (London) 183*: 155-176.

Keller, L. and L. Passera, 1989. Influence of the number of queens on nestmate recognition and attractiveness of queens to workers in the Argentine ant, *Iridomyrmex humilis* (Mayr). *Anim. Behav. 37*: 733-740.

Keller, L. and P. Nonacs. 1993, The role of queen pheromones in social insects: queen control or queen signal? *Anim. Behav.* 45: 787-794.

Kern, F. and H.J. Bestmann, 1993. Antennal elctrophysiological responsiveness of the ponerine ant *Leptogenus diminuta* to trail and recruitment pheromones and its structure analogs. *Naturwissenschaften 80*: 424-427.

Kugler, C. 1979. Alarm and defense: a function for the pygidial gland of the myrmicine ant, *Pheidole biconstricta* in coffee plantations in the foothills of the Sierra Nevada de Santa Marta in northern Colombia. *Ann. Entomol. Soc. Am., 72*: 532-536.

Law, J.H., E.O. Wilson and J.A. McCloskey, 1965. Biochemical polymorphism in ants. *Science. 149*: 544-546

Le Masne, G. 1953. Observations sur les relations entre le couvain et les adultes chez les fourmis. *Annales des Sciences Naturelles, ser. 1, 15*: 1-56.

Leuthold, R.H. And U. Schlunegger, 1973. The alarm behaviour from the mandibular gland secretion in the ant *Crematogaster scutellaris*. *Insectes Soc. 20*: 205-213.

Lloyd, H. A., M.S. Blum and R.M. Duffield, 1975. Chemistry of the male mandibular gland secretion of the ant, *Camponotus clarithorax*. *Insect Biochem. 5*: 489-494.

Lofgren, C.S., B.M. Glancey, A. Glover, J. Rocca and J. Tumlinson, 1983. Behavior of workers of *Solenopsis invicta* (Hymenoptera: Formicidae) to the queen recognition pheromone: Laboratory studies with an olfactometer and surrogate queens. *Ann. Entomol. Soc. Am. 76*: 44-50.

Markl, H. and B. Hölldobler, 1978. Recruitment and food retrieving behaviour in *Novomessor* (Hymenoptera: Formicidae). II. Vibrational signals. *Behav. Ecol. Sociobiol. 4*: 183-216.

Maschwitz, U. 1966. Alarm substances and alarm behavior in social insects. *Vitamins and Hormones, 24*: 267-290.

McGurk, D.J., J. Frost, G.R. Waller, E.J. Eisenbraun, K. Vick, W.A. Drew and J. Young, 1968. Iridodial isomer variation in dolichoderine ants. *J. Insect Physiol.14*: 841-845.

Meinwald, J., D.F. Wiemer and B. Hölldobler, 1983. Pygidial gland secretions of the ponerine ant *Rhytidoponera metallica*. *Naturwissenschaften 70*: 46-47.

Möglich, M. and B. Hölldobler, 1975. Communication and orientation during foraging and emigration in the ant *Formica fusca*. *J. Comp. Physiol. 101*: 275-288.

Möglich, M. U. Maschwitz and B. Hölldobler, 1974. Tandem calling: a new kind of signal in ant communication. *Science 86*: 1046-1047.

Morel, L. and R.K. Vander Meer, 1988. Do ant brood pheromones exist? *Ann. Entomol. Soc. Am. 81*: 705-710.

Morgan, E.D., M.R. Inwood and M.-C. Cammaerts, 1978. The mandibular gland secretion of the ant, Myrmica scabrinodis. *Physiol. Entomol. 3*: 107-114.

Mori, A., A. Zaccone and F. Le Moli, 1992. Experience-independent attraction to host-species ant cocoons in the slave-maker *Formica sanguinea* Latr. (Hymenoptera: Formicidae). *Ethol. Ecol. and Evol, Special Issue 2*: 85-89.

Moser, J.C., R.C. Brownlee, and R. Silverstein, 1968. Alarm pheromones of the ant Atta texana. *J. Insect Physiol. 14*: 529-535.

Nordlund, D.A. and W.J. Lewis, 1976. Terminology of chemical releasing stimuli in intraspecific and interspecific interactions. *J. Chem. Ecol.* 2: 211-220.

Obin, M.S. and R.K. Vander Meer, 1994. Alate semiochemicals release worker behavior during fire ant nuptial flights. *J. Entomol. Sci., 29*: 143-151.

Obin, M.S., B.M. Glancey, W.A. Banks and R.K. Vander Meer, 1988. Queen pheromone production and its physiological correlates in fire ant queens (Hymenoptera: Formicidae) treated with fenoxycarb. *Ann. Entomol. Soc. Amer. 81*: 808-815.

Oldham, N.J., E.D. Morgan, B. Gobin and J. Billen, 1994. First identification of a trail pheromone of an army ant (*Aenictus* species). *Experientia 50*: 763-765.

Olubajo, O., R.M. Duffield and J.W. Wheeler, 1980. 4-Heptanone in the mandibular gland secretion of the Nearctic ant, *Zacryptocerus varians* (Hymenoptera: Formicidae). *Ann. Entomol. Soc. Am. 73*: 93-94.

Parry, K. and E.D. Morgan, 1979. Pheromones of ants: a review. *Physiol. Entomol. 4*: 161-189.

Passera, L. and S. Aron, 1993. Factors controlling dealation and egg laying in virgin queens of the Argentine ant *Linepithema humile* (Mayr) (*=Iridomyrmex humilis). Psyche 100*: 51-63.

Pasteels, J.M., J.C. Verhaeghe, J.-C. Braekman, D. Daloze and B. Tursch, 1980. Caste-dependent pheromones in the head of the ant *Tetramorium caespitum*. *J. Chem. Ecol. 6*: 467-472.

Pasteels, J.M., D. Daloze and J.L. Boeve, 1989. Aldehydic contact poisons and alarm pheromone of the ant *Crematogaster scutellaris* (Hymenoptera: Myrmicinae): Enzyme-mediated production from acetate precursors. *J. Chem. Ecol. 15*: 1501-1511.

Petralia, R.S. and S.B. Vinson, 1979. Developmental morphology of larvae and eggs of the imported fire ant, *Solenopsis invicta*. *Ann. Entomol. Soc. Amer. 72*: 472-484.

Porter, S.D. 1995. *FORMIS: A Master Bibliography of Ant Literature.* Gainesville, FL. USA.

Quinet, Y. and J.M. Pasteels, 1995. Trail following and stowaway behaviour of the myrmecophilous staphylinid beetle *Homoeusa acuminata* during foraging trips of its host *Lasius fuliginosus* (Hymenoptera: Formicidae). *Insectes Sociaux 42*: 31-44.

Regnier, F.E. and E.O. Wilson, 1968. The alarm-defence system of the ant *Acanthomyops claviger*. *J. Insect Physiol. 14*: 955-970.

Robinson, S.W. and J.M. Cherrett, 1974. Laboratory investigations to evaluate the possible use of brood pheromones of the leaf-cutting ant *Atta cephalotes* (L.) (Formicidae, Attini) as a component in an attractive bait. *Bull. Entomol. Res. 63*: 519-529.

Rocca, J.R., J.H. Tumlinson, B.M. Glancey and C.S. Lofgren, 1983a. The queen recognition pheromone of *Solenopsis invicta*, preparation of (E)-6-(1-pentenyl)-2H-pyran-2-one. *Tetrahedron Lett. 24*: 1889-1892.

Rocca, J.R., J.H. Tumlinson, B.M. Glancey and C.S. Lofgren, 1983b. Synthesis and Stereochemistry of Tetrahydro-3,5-dimethyl-6-(1-methylbutyl)-2H-pyran-2-one, a component of the queen recognition pheromone of *Solenopsis invicta*. *Tetrahedron Lett. 24*: 1893-1896.

Salzemann, A., P. Nagnan, F. Tellier and K. Jaffe, 1992. Leaf-cutting ant *Atta laevigata* (Formicidae: Attini) marks its territory with colony-specific dufour gland secretion. *J. Chem Ecol. 18*: 183-196.

Scheffrahn, R.H., L.K. Gaston, J.J. Sims and M.K. Rust, 1984. Defensive ecology of *Forelius foetidus* and its chemosystematic relationship to *Forelius* (*=Iridomyrmex*) *pruinosus* (Hymenoptera: Formicidae: Dolichoderinae). *Environ. Entomol.*, *13*: 1502-1506.

Scheffrahn, R.H. and M.K. Rust, 1989. Attraction by semiochemical mediators and major exocrine products of the myrmicine ant, *Crematogaster californica* Emery. *Southwest. Entomol.*, *14*: 49-55.

Shorey, H.H. 1973. Behavioral responses to insect pheromones. *Annu. Rev. Entomol. 18*: 349-380.

Simon, T. and A. Hefetz, 1991. Trail-following responses of *Tapinoma simrothi* (Formicidae: Dolichoderinae) to pygidial gland extracts. *Insect. Soc. 38*: 17-25.

Steghaus-Kovac, S., U. Maschwitz, A.B. Attygalle, R.T.S. Frighetto, N. Frighetto, O. Vostrowsky and H.J. Bestmann, 1992. Trail-following responses of *Leptogenys diminuta* to stereoisomers of 4-methyl-3-heptanol. *Experientia 48*: 690-694.

Stumper, R. 1956. Sur les sécrétions attractives des fourmis femelles. *Compt. Rend. Acad. Sci. Paris 242*: 2487-2489.

Tomalski, M.D., M.S. Blum, T.H. Jones, H.M. Fales, D.F. Howard and L. Passera, 1987. Chemistry and functions of exocrine secretions of the ants *Tapinoma melanocephalum* and *T. erraticum*. *J. Chem. Ecol.*, *13*: 253-263.

Torgerson, R.L. and R.D. Akre, 1970. The persistence of army ant chemical trails and their significance in the ecitonine-ectophile association (Formicidae: Ecitonini). *Melanderia. 5*: 1-28.

Traniello, J.F.A. and B. Hölldobler, 1984. Chemical communication during tandem running in *Pachycondyla obscuricornis* (Hymenoptera: Formicidae). *J. Chem. Ecol. 10*: 783-794.

Tricot, M.-C., J.M Pasteels, and B. Tursch, 1972. Phéromones stimulant et inhibant l' agressivité chez *Myrmica rubra. J. Insect Physiol. 18*: 499-509.

Tumlinson, J.H., R.M. Silverstein, J.C. Moser, R.G. Brownlee, and J.M. Ruth, 1971. Identification of the trail pheromone of a leaf-cutting ant, *Atta texana*. *Nature, 234*: 348-349.

Vander Meer, R.K. and L. Morel,1988. Brood pheromones in ants. In *(Advances in myrmecology,* J. C. Trager, Ed.) E.J. Brill, New York pp. 491-513.

Vander Meer, R.K. and L. Morel, 1995. Ant queens deposit pheromones and antimicrobial agents on eggs. *Naturwissenschaften. 82*: 93-95.

Vander Meer, R.K., B.M. Glancey, C.S. Lofgren, A. Glover, J.H. Tumlinson and J. Rocca, 1980. The poison sac of red imported fire ant queens: Source of a pheromone attractant. *Ann. Entomol. Soc. Amer. 73*: 609-612.

Vander Meer, R.K., F.D. Williams and C.S. Lofgren, 1981. Hydrocarbon components of the trail pheromone of the red imported fire ant, *Solenopsis invicta*. *Tetrahedron Lett.* 22: 1651-1654.

Vander Meer, R.K., F. Alvarez and C.S. Lofgren, 1988. Isolation of the trail recruitment pheromone of *Solenopsis invicta*. *J. Chem. Ecol. 14*: 825-838.

Vander Meer, R.K., C.S. Lofgren and F.M. Alvarez, 1990. The orientation inducer pheromone of the fire ant *Solenopsis invicta*.. *Physiol. Entomol. 15*: 483-488.

Walter, F., D.J.C. Fletcher, D. Chautems, D. Cherix, L. Keller, W. Francke, W. Fortelius, R. Rosengren and E.L. Vargo, 1993. Identification of the sex pheromone of an ant, *Formica lugubris* (Hymenoptera, Formicidae). *Naturwissenschaften 80*: 30-34.

Watkins, J.F., II. and T.W. Cole, 1966. The attraction of army ant workers to secretions of their queens. *Tex. J. Sci. 18*: 254-265.

Wheeler, J.W. and M.S. Blum, 1973. Alkylpyrazine Alarm Pheromones in Ponerine Ants. *Science 182:* 501-503.

Wheeler, J.M., S.L. Evans, M.S. Blum, and R.L.Torgerson, 1975. Cyclopentyl ketones: Identification and function on *Azteca* ants. *Science 187*: 254-255.

Wilson, E.O. 1958a. A chemical releaser of alarm and digging behavior in the ant *Pogonomyrmex badius* (Latreille). *Psyche, 65*: 41-51.

Wilson, E.O. 1958b. The beginnings of nomadic and group-predatory behavior in the ponerine ants. *Evolution, 12*: 24-31.

Wilson, E.O. 1962. Chemical Communication Among Workers of the Fire Ant *Solenopsis saevissima* (Fr. Smith). *Animal Behaviour 10*: 134-164.

Wilson, E.O. 1971. *The insect societies.*, Harvard University Press, Cambridge, MA.

Wilson, E.O. and W.H. Bossert, 1963. Chemical communication among animals. *Recent Prog. Hormone Res. 19*: 673-716.

Wilson, E.O. and F.E. Regnier, Jr., 1971. The evolution of the alarm-defense system in the formicine ants. *Am. Nat. 105*: 279-289.

Yamauchi, K. and N. Kawase, 1992. Pheromonal manipulation of workers by a fighting male to kill his rival males in the ant *Cardiocondyla wroughtonii*. *Naturwissenschaften 79*: 274-276.

8

Releaser Pheromones in Termites

Jacques M. Pasteels and Christian Bordereau

Introduction

In termites, as in other social insects, pheromones are involved in probably all social activities from simple recognition between social partners to nest building (see Bruinsma 1979 for the role of pheromones in building behavior). However, only pheromones involved during foraging, sexual behavior and defense will be reviewed here, i.e. trail, sex and alarm pheromones. These pheromones have been studied in some detail recently, leading to significant advances both at the behavioral and chemical levels. Previous reviews were written at a very early stage of termite pheromone research (Stuart 1969, 1970, Moore 1974) and the most recent review is that of Prestwich (1983).

Trail Pheromones and Foraging

In termites which collect food at a distance from their nest, the discovery of food is not the achievement of scouts which singly explore the foraging field far from the nest. Exploration is above all a collective process: a collective trail or gallery grows from the nest or as an extension of an already preexisting network of trails or galleries. This collective trail seems to progress blindly until food is reached. When recruitment occurs, it is only at a very short distance from the ends of these networks. Food attractants diffusing in the soil could guide the trail towards the food source, but probably only over a very short distance. Collective exploration was described by Grassé and Noirot (1951) in a Macrotermitinae, *Odotontermes magdalenae,* and by Pasteels (1965) in a Nasutitermitinae, *Nasutitermes lujae*. It has since been reported in a diversity of species (Table 8.1). In this respect, termite foraging behavior is more similar to that of group hunting in ants than to any other foraging techniques reported in ants by Oster and Wilson (1978). The major differences between termite species concern the relative importance of underground and above ground foraging and, in the latter case, of foraging in the open air versus under the protection of tunnels built with earth or wood mixed with saliva and feces.

Another difference between termite species concerns which caste actually initiates trail building and/or recruitment. In *N. lujae* (Pasteels 1965), it was concluded that the oldest workers, which have the best developed sternal glands, initiate trail building. This was recently confirmed by Laduguie (1993). However, in *Nasutitermes costalis*, the soldiers initiate the trail and recruit (Traniello 1981). In *Schedorhinotermes lamanianus*, orientation trails are initiated by minor soldiers, but recruitment trails are laid by workers (Kaib 1990). In *Macrotermes bellicosus*, the small workers build the underground galleries, but trail recruitment is due to large workers which only recruit other large workers (Lys and Leuthold 1991).

Hodotermes mossambicus is unique among termites in that orientation during exploration and foraging during daylight hours is based more on visual (photo-menotactically) than on chemical cues (Leuthold et al. 1976). This termite lives in semiarid grassland, and food is discovered by scouts exploring the ground up to 3 m from the numerous exit holes of the underground network of galleries. A trail is progressively built by successful scouts. At night, orientation is purely chemical, the termites using trail pheromone gradients around holes and food sources before a more defined trail is eventually built (Heidecker and Leuthold 1983).

Although foraging along odorous trails is prominent in termites and was reported in the literature over 100 years ago (Dudley and Beaumont 1889), research on trail pheromones has progressed very slowly.

The sternal gland was first recognized as the source of trail pheromone in termites in the early sixties (Lüscher and Müller 1960, Stuart 1961), although this gland had been described already at the end of the last century (Grassi and Sandias 1893). Interestingly, the sternal gland is the only reported source of trail pheromone in termites, making termites far more homogenous in this respect than ants, in which at least eight different sources of trail pheromones are known (Hölldobler and Wilson 1990). Trail communication is certainly very ancient in termites and possibly evolved only once. Differences between species, however, are observed in the number and location of sternal glands, and in their structure (Ampion and Quennedey 1981). Sternal

TABLE 8.1 Collective exploration during foraging in termites.

Taxa	Food Collected	Structure Built	Site of Observation	Reference
Rhinotermitidae				
Rhinotermitinae				
Schedorhinotermes lamanianus	wood	open-air trails, then tunnels	laboratory	a
Termitidae				
Macrotermitinae				
Odontotermes magdalenae	grass	open-air trails	field	b
Macrotermes michaelseni	grass litter, wood, dung	open-air trails, surface and subterranean galleries	field	c
M. bellicosus	plant litter, wood, leaves	subterranean galleries	outdoor experimental design	d
Nasutitermitinae				
Nasutitermes lujae	wood	open-air trails, then tunnels	laboratory	e
Trinervitermes bettonianus	grass	open-air trails	laboratory	f
T. geminatus	grass	open-air trails	field	g

a = Kaib 1990; b = Grassé and Noirot 1951; c = Oloo 1984; d= Lys and Leuthold 1991; e = Pasteels 1965; f = Oloo and Leuthold 1979; g = Rickli and Leuthold 1987.

glands are located in front of sternites 3-5 in *Mastotermes darwiniensis*, in front of sternite 4 in the Hodotermitidae and Termopsidae, and in front of sternite 5 in all other termites (Noirot, 1995).

The first trail pheromone, (*Z*,*Z*,*E*)-3,6,8-dodecatrienol (**1**) was identified in *Reticulitermes flavipes* by Matsumura et al (1968), and today only one other compound, (*E*,*E*,*E*)-neocembrene (**2**), first identified in *Nasutitermes exitiosus* (Birch et al. 1972, Kodama et al. 1975), can be claimed with confidence to be a termite trail pheromone. These two compounds are structurally and biosynthetically unrelated.

OH

1

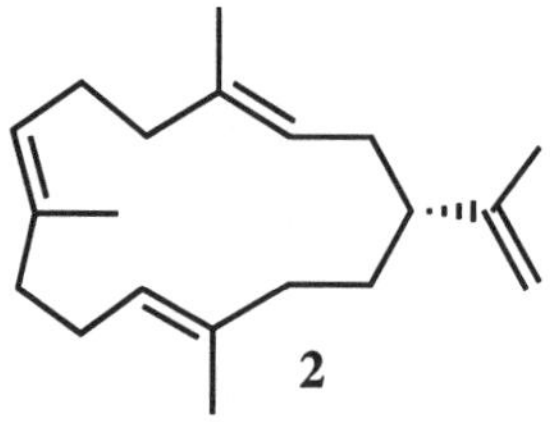

2

One major difficulty in termite trail pheromone research is that termites follow artificial trails produced using a wide diversity of compounds, albeit at concentrations that are physiologically unrealistic. The most caricatural example of such an unnatural response is the well-known report that some termites follow the ink of ball-point pens from which active glycol derivatives were isolated (Becker and Mannesmann 1968). In the same line, Hummel and Karlson (1968) observed that *Zootermopsis nevadensis* followed phthalates, contaminants found in termite extracts stored in plastic bottles.

Another difficulty lies in the fact that wood fed on by the termites may contain attractants which induce trail-following behavior (e.g. Ritter et al. 1973). These attractants may or not be identical to actual trail pheromones. Dodecatrienol (**1**) is present in wood infested by the fungus *Gloeophyllum trabeum*, sometimes in higher concentration than in termites (Matsumura et al. 1969), and neocembrene (**2**) or a closely related compound with similar pheromonal activity was isolated from the Indian incense cedar (Ritter et al. 1977).

Artifacts are thus frequent, and before it can be claimed with some confidence that a compound inducing trail-following behavior is actually a trail pheromone, two requirements must be met: it should be demonstrated that the compound is present in the sternal gland of the termite concerned and that it is active at physiologically realistic concentrations.

In Table 8.2 are reported the termite species for which dodecatrienol has been demonstrated to be a trail pheromone. It is active at concentrations as low as 10^{-1} or 10^{-2} pg/cm. This compound was observed in extracts of *Reticulitermes speratus* and *R. santonensis* fed with filter paper, and also in *Pseudacanthotermes spiniger* fed on food devoid of the compound. Moreover, it was actually observed in extracts of the sternal glands of workers of *R. speratus* and of alates of *P. spiniger* (see below; references in Table 8.2).

Interestingly, dodecatrienol was demonstrated to be the trail pheromone of several species within the Rhinotermitidae, belonging to very distinct subfamilies, and also in a member of the termitid subfamily Macrotermitinae. The Macrotermitinae were reported on other grounds to be more closely related to the Rhinotermitidae than are the other Termitidae (Deligne 1985).

Termite species producing neocembrene as trail pheromone are listed in Table 8.3. This compound was detected in sternal gland extracts of workers of *Trinervitermes bettonianus* as well as in extracts of natural trails laid by the termites. Although few quantitative data are available, neocembrene seems to be produced in much greater quantities (3 orders of magnitude higher) than dodecatrienol, and to require higher threshold concentrations for activity, at least in *T. bettonianus* (McDowell and Oloo 1984).

TABLE 8.2 Termite species secreting (*Z,Z,E*)-3,6,8-dodecatrienol (**1**) as trail pheromone.

Taxa	Threshold (pg/cm)	Amount (pg/worker)	Reference
Rhinotermitidae			
Heterotermitinae			
Reticulitermes flavipes	< 0.01	NA	a
R. virginicus	0.01	NA	b
R. speratus	0.02	3	c
R. santonensis	0.1	1-10	d
Coptotermitinae			
Coptotermes formosanus	0.02*	50	e
Termitidae			
Macrotermitinae			
Pseudacanthotermes spiniger	0.1	10*	f,g

NA = not available; * estimated. a=Matsumura et al. 1968; b=Tai et al. 1969; c=Tokoro et al. 1990, 1991; d=Laduguie et al. 1994b; e= Tokoro et al. 1989; f=Bordereau et al. 1993; g=Laduguie et al. 1994a.

TABLE 8.3. Termite species secreting neocembrene (**2**) as trail pheromone.

Taxa	Threshold (pg/cm)	Amount (ng/worker)	Reference
Nasutitermitinae			
Nasutitermes exitiosus	NA	NA	a
Trinervitermes bettonianus	1 or 500*	11	b

NA = not available; * = depending on the bioassay design. a=Birch et al. 1972; b=McDowell and Oloo 1984.

From Tables 8.2 and 8.3, it is tempting to conclude that a small number of trail pheromones evolved in termites and that these have remained evolutionary conservative, being observed in species belonging to quite distinct taxa (dodecatrienol) or in species from very different geographic origin (e.g. neocembrene in an Australian *Nasutitermes* and an African *Trinervitermes*). However, any speculation on the evolution of trail pheromones in termites is presently impossible, as illustrated by behavioral evidence. For example, Laduguie et al. (1994a) tested highly purified dodecatrienol on various termites (Table 8.4). As expected, dodecatrienol released trail-following behavior in all Rhinotermitidae tested, but only in one Macrotermitinae of 4 tested. Even more intriguing is the fact that two *Nasutitermes* species responded with higher sensitivity than the Rhinotermitidae and Macrotermitinae. Making the paradox even greater, Laduguie et al. (1994a) were unable to observe dodecatrienol in extracts of *N. lujae,* although their analytical procedure was sensitive enough to detect 0.1 pg of dodecatrianol. Neocembrene was not detected in extracts of whole *N. lujae* workers. It was neither observed in extracts of their sternal glands, nor in extracts of trails laid on filter paper. When neocembrene (purity >99%, ICI) was tested on this termite at three different concentrations (10, 50 and 500 ng/cm of artificial trail), only the concentration of 50 ng/cm induced trail-following behavior. This concentration is $5x10^3$ times higher than the threshold concentration of dodecatrienol (Lefeuve and Ladurie, pers. comm.). Preliminary data suggest that the pheromone of *N. lujae* could be closely related to dodecatrienol. *Kalotermes flavicollis* did not respond to dodecatrienol, nor did *Cryptotermes brevis* or *Zootermopsis angusticollis* when tested in other experiments (Matsumura et al. 1972), but several Termitinae and one Apicotermitinae did respond with various levels of sensitivity (Table 8.4).

TABLE 8.4. Trail-following response of various termites to artificial trails drawn with (*Z,Z,E*)-3,6,8-dodecatrienol (**1**) (from Laduguie et al. 1994a).

TAXA	THRESHOLD (pg/cm)
Kalotermitidae	
Kalotermes flavicollis	NR
Rhinotermitidae	
Heterotermitinae	
Heterotermes sp.	1
Rhinotermitinae	
Schedorhinotermes lamanianus	1
Termitidae	
Macrotermitinae	
Ancistrotermes cavithorax	1
Macrotermes bellicosus	NR
M. subhyalinus	NR
Microtermes sp.	NR
Nasutitermitinae	
Nasutitermes lujae	0.01
N. diabolus	0.01
Termitinae	
Amitermes evuncifer	10
Basidentitermes sp.	1
Cephalotermes rectangularis	1
Cubitermes fungifaber	1
C. severus	0.1
Microcerotermes sp.	0.1
Noditermes sp.	0.1
Ophiotermes sp.	0.1
Thoracotermes sp.	10
Apicotermitinae	
Astalotermes sp.	10

NR: no response in the range 10-0.01 pg/cm.

Behavioral evidence also suggests that termite trail pheromones could be multicomponent pheromones, as they are in many ant species. For example, the total activity of the sternal gland extracts of *R. santonensis* cannot be explained solely by the amount of dodecatrienol found in the extract. One to 10 pg of dodecatrienol were estimated per worker, yet the biological activity of the extract of a single gland is equalled by 100 pg of this substance. This suggests that additional compounds could act in synergy (Laduguie et al. 1994b).

Also, natural or artificial trails were reported to be species-specific even in species known to produce the same trail pheromone. Some specificity can be explained by quantitative differences in the amount produced and in the response thresholds (Bordereau et al. 1993, Laduguie et al. 1994a). However, Howard et al. (1976) demonstrated that extracts of several *Reticulitermes* and of *Coptotermes formosanus* are species-specific when tested just above threshold level, although they all secrete dodecatrienol, strongly suggesting that additional compounds could be responsible for this species-specificity (see also Moore 1974, Oloo 1981, Oloo and McDowell 1982, Kaib et al. 1982).

Soldiers and workers or minor and major workers play different roles during exploration, trail building and recruitment (see above). This could partly result from qualitative differences in trail pheromones between castes as suggested by Lys and Leuthold (1991) to explain why major workers only recruit major workers in *Macrotermes bellicosus*. Caste-specificity could be due to different blends of compounds, but strong evidence for this is lacking. On the contrary, Traniello and Busher (1985) and Kaib (1990) concluded from their experiments, with *N. costalis* and *S. lamanianus*, respectively, that differences in behavior between soldiers and workers during the early stages of foraging can be explained by quantitative differences in the secretion of and threshold response to the same pheromone.

More convincing is evidence for the production in the sternal glands of both an ephemeral recruiting trail component and a long-lasting component (stable during months or even years!) acting as orientation cue (Hall and Traniello 1985, Runcie 1987).

Although the trail pheromone from the sternal gland is undoubtedly of prime importance during recruitment, during the first stages of exploration and during the building up of the trail and gallery network, trails are later marked with feces, earth or wood mixed with saliva, or covered with tunnels made with these building materials. These galleries and tunnels offer the termites protection and a physical guide, but also the termites travel in a complex olfactory environment. It remains possible that compounds isolated from whole termites, from tunnels, or even from wood could have a biological significance in the orientation of termites and in the specificity of their trail network, even if they act as attractants or trail-following inducers only at much higher concentrations than the sternal gland pheromones, e.g., 3-hexenol in *Kalotermes flavicollis* (Verron and Barbier 1962), nonanol, decanol, undecanol and dodecanol also in the *K. flavicollis* (Klochov and Zhuzhikov 1990), or *n*-hexanoic acid in *Zootermopsis nevadensis* (Karlson et al. 1968).

Sex Pheromones and Postflight Behavior

The first detailed descriptions of postflight behavior of termites were given by Fuller (1915) who reported that females produce pheromones during their calling attitude and tandem running.

When soliciting mates, termites (females in most species, but males in some, see below), raise their abdomen exposing the sternal gland. However, it was not until the early seventies that it was experimentally demonstrated that the sternal glands release sex pheromones in *Kalotermes flavicollis* (Wall, 1971) and in *Zootermopsis nevadensis* (Pasteels 1972).

The first sex pheromone, neocembrene, was identified in 1984 by McDowell and Oloo in the sternal glands of dealates of *Trinervitermes bettonianus*. To date, only one other sex pheromone, (*Z*,*Z*,*E*)-3,6,8-dodecatrienol, has been unequivocally identified in the sternal glands of *Pseudacanthotermes spiniger* (Bordereau et al. 1991) and in *Reticulitermes santonensis* (Laduguie et al. 1994b) dealates.

These pheromones are identical to the trail pheromones produced in the sternal gland of workers, but are secreted in much higher quantities, especially in females whose sternal gland is larger (Table 8.5). These pheromones are used as trail pheromone during pairing and tandem running. In *T. bettonianus*, not only female dealates, but also males, lay powerful trails when walking, allowing unpaired individuals to encounter a partner (Leuthold and Lüscher 1974). In *P. spiniger*, however, females do not react to either the male or female extracts (Bordereau et al. 1991). In this species, workers responded much better to low pheromone concentrations than males in trail-following tests, but males continue to follow trails at higher concentrations at which workers no longer responded (Figure 8.1).

Dodecatrienol also attracts males at short range in *P. spiniger* (Bordereau et al. 1991). At high concentrations males display typical excitement behavior (Bordereau et al. 1993). A concentrated extract of 2000 workers corresponding to the extract of one female (Table 8.5) induced in males all behaviors induced by an extract of one female (Laduguie 1993). Although neocembrene was not tested as a male attractant in *T. bettonianus* by McDowell and Oloo (1984), this function is very likely. Leuthold (1975) reported that the female sternal gland secretes an airborne attractant, actives at shorter range (4 cm) than the attractant secreted by the female tergal glands (range: 10 cm).

TABLE 8.5. Volume of the sternal gland and amount of trail/sex pheromone in workers (w), males (m) and females (f) of three termite species.

Taxa (pheromone produced)	Volume of the sternal gland (10^6 μm^3)			Amount of trail/sex pheromone (ng)		
	w	m	f	w	m	f
Trinervitermes bettonianus (neocembrene)	0.47	4.4	31	11	680	12,000
Pseudacanthotermes spiniger (dodecatrienol)	0.74	12	150	0.01	1-3	15-20
Reticulitermes santonensis (dodecatrienol)	1.4	3.3	4.8	0.01	NA	0.1

NA : not available; combined from Leuthold and Lüscher (1974), Quennedey (1977), McDowell and Oloo (1984), Bordereau et al. (1991), Laduguie (1993), Laduguie et al. (1994b).

It would be, however, an oversimplification to consider that sex pheromones in female sternal glands are always identical to the trail pheromone produced by workers. In *Hodotermes mossambicus,* a calling attitude is taken by the males, which have an extraordinarily hypertrophied sternal gland. However, extract of male sternal gland, contrary to that from workers, does not induce trail-following behavior in workers. It was concluded that trail pheromone and sex pheromone must be different (Leuthold and Bruinsma 1977). Similarly, MacFarlane (1983) reported that in *Macrotermes michaelseni* extracts

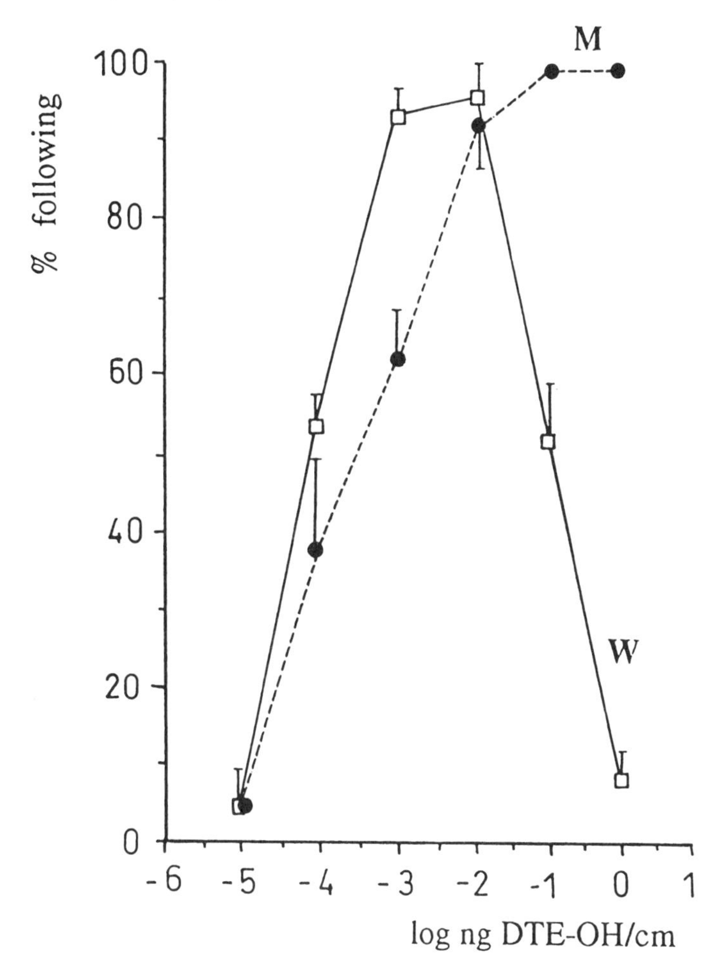

FIGURE 8.1 Responses of workers (W) and males (M) of *Pseudacanthotermes spiniger* to the trail pheromone, (*Z*,*Z*,*E*)-3,6,8-dodecatrienol (DTEOH). Modified from Bordereau et al. (1993).

of female sternal glands are less active in inducing trail-following behavior than those of workers, although the female glands are more developed than those of workers in all other Macrotermitinae studied so far. Caution is needed in the interpretation of such results. It could be that extracts of adult glands were too concentrated in pheromone to be able to induce trail-following response in workers. In other species of termites the response drops beyond an optimal concentration (Figure 8.1), as it does in ants (Pasteels et al 1985).

Several observations suggest that sex pheromones are multi-component systems. The total amount of dodecatrienol in the sternal gland of one *R. santonensis* female is estimated at 10-100 pg, but should contain much more (from 1-5 ng) to account for its total activity in bioassays (Laduguie et al. 1994b). Possibly other components act in synergy.

In European and North American *Reticulitermes* species, Clément and coworkers observed a species-specificity of the female sternal gland extracts, although probably all species or at least some of them produce dodecatrienol as trail pheromone (Clément 1982, Clément et al. 1986). Other components in female sternal glands could be responsible for this specificity. The compound n-tetradecylpropionate identified in male and female sternal gland extracts of *R. Flavipes* (10 times more in females) could be one of those synergists. It attracts males of *R. flavipes* (although only at concentrations much higher than those physiologically realistic), but not males of other species (Clément et al. 1989).

In some species, there are indications that females and males produce sex-specific pheromones. In *Zootermopsis angusticollis*, one of us (Pasteels 1972) observed, following the swarm of a few alates in a laboratory colony, one dealate male which adopted a typical calling attitude by raising its abdomen exposing its sternal gland. None of the other dealates were calling. However, Castle (1934) reported calling attitude only in females. At that time, calling females were thought to be the rule in termites and the report of Castle should be confirmed. Bioassays indicated that dealates were more attracted to sternal gland extracts of the other sex (Pasteels 1972). Similar results were reported by Stuart (1975) for *Zootermopsis angusticollis* and *Reticulitermes flavipes*. In Kenya, only males of *Hodotermes mossambicus* call, according to Leuthold and Bruinsma (1977). The sternal gland of males secretes an attractant for females, but apparently no trail pheromone, contrary to the sternal gland of females. However, Hewitt and Nel (1969) reported calling attitudes in both sexes of this species in South Africa.

Tergal glands also produce sex pheromone involved during postflight behavior. Their role is less well understood and could vary from species to species. First, their location varies: in tergites 3-10 or 6-10, 8-10, 9 and 10 or still restricted to tergite 10 according to the species. They are sometimes present in both sexes, sometimes present only in females, and in some species totally lacking (Ampion and Quennedey 1981).

Tergal gland secretion could help to maintain contact between sexes during tandem running, as suggested by their location and their frequent predominance in females. However, female tergal glands are also exposed during calling, and in *Trinervitermes bettonianus*, airborne attractants are produced by the tergal glands, whereas the female sternal gland is more important than the tergal glands in allowing tandem cohesion (Leuthold 1975).

Finally, Noirot (1969) and Ampion and Quennedey (1981) reported a sexual dimorphism in glands observed in adults and absent in neuters, e.g. posterior sternal glands in segments 6 and 7 in females, but in segments 6-9 in males of *Mastotermes darwiniensis*; posterior sternal glands in segments 8 and 9 in males, but not in females, of *Porotermes, Stolotermes, Prorhinotermes*; pleural glands in segments 3-5 in female *Cubitermes fungifaber*, but not in males or in other *Cubitermes* species. Nothing is known of the function of these glands, but the probability that they produce pheromones during postflight behavior seems high. Obviously, communication during postflight behavior is more diversified in termites than commonly believed.

Alarm Pheromones and Defense

In termites, defensive reactions are coordinated by various signals, either mechanical, such as vibratory movements and head banging (Stuart 1969), or chemical, i.e., trail and alarm pheromones. Defensive recruitment by trail pheromone was demonstrated by Stuart (1963) in *Zootermopsis nevadensis*, by Traniello (1981) in *Nasutitermes costalis* and recently postulated by Kettler and Leuthold (1995) for *Macrotermes subhyalinus*. It is probably a widespread feature in termites deserving further investigations. However, no defensive recruitment was observed in *Armitermes chagresi* (Traniello et al. 1984).

Occurrence of alarm pheromones in termites was first demonstrated by Ernst (1959) in a *Nasutitermes* species and has since been observed in several chemically defended termites (Moore 1968, 1969, Maschwitz and Mühlenberg 1972, Eisner et al. 1976, Kriston et al. 1977, Vrkoc et al. 1978, Stuart 1981, Kaib 1990, Roisin et al. 1990). Alarm pheromones are volatile constituents, often monoterpenes, of the soldier defensive secretion from the frontal gland. In *Schedorhinotermes lamanianus* minor, but not major, soldiers secrete alarm pheromones (Kaib 1990). In *Hodotermes mossambicus*, Wilson and Clark (1977) suggested that disturbed workers produce an alarm pheromone excreted in fecal material. This is the only report of an alarm pheromone in a termite species not chemically defended by soldier frontal gland secretion.

Some monoterpenes produced by soldiers are chiral compounds, sometimes produced in remarkably high enantiomeric purity (Everaerts et al. 1990, Lindström et al. 1990, Valterová et al. 1992, 1993). The level of enantiomeric purity varies considerably however between compounds, termite species, and for some species even between allopatric populations (Valterová et al. 1993).

α-Pinene is the alarm pheromone of *Nasutitermes princeps*. *N. princeps* secretes the (+)-enantiomer with great purity and this is more active in inducing alarm reaction than the (-)-enantiomer (Roisin et al. 1990). Carene is the most active monoterpene in inducing alarm response in *N. costalis* (Vrkoc et al. 1978) and only the (-) enantiomer is secreted (Valterová et al. 1992,1993). The alarm pheromone activity of the two enantiomers was not compared in this species.

Along the north coast of Papua New Guinea, *N. princeps* is widely distributed. Diterpene composition of soldier secretions varies to some extend between colonies, and this can be used to distinguish domes belonging to the same polydomous colony from domes belonging to neighboring foreign colonies (Roisin et al. 1987). The monoterpenic fraction used as alarm pheromone is far more similar from colony to colony (Pasteels et al 1988). Similarly, the monoterpenic fraction of *N. costalis*, which functions as alarm pheromone, is remarkably constant even in colonies from different populations (Valterová et al. 1993). In contrast, in species which do not use monoterpenes as alarm pheromone, considerable variation is observed between colonies from different populations, e.g. in *N. novarumhebridarum* (Pasteels et al. 1988) and in *N. ephratae* (Valterová et al. 1993). This suggests that the evolution of monoterpenes as alarm pheromone led to stabilizing selection on the monoterpenic fraction of the defensive secretions.

Soldiers and workers respond differently during defensive recruitment and have a complementary role in defense. Usually, soldiers are first locally recruited by alarm pheromones acting as short-range attractants (*Nasutitermes exitiosus*: Eisner et al. 1976; *N. corniger*: Stuart 1981; *N. costalis*: Traniello 1981; Traniello and Beshers 1985; *Schedorhinotermes lamanianus*: Kaib 1990). Workers are recruited later by trail pheromone in *N. corniger* (Stuart 1981) or by the soldier defensive secretion in *N. exitiosus* (Eisner et al. 1976). Workers participate in defense either by building partitions or covering enemies with fecal material, or actually by fighting, for example, biting enemies partially immobilized by soldier secretion (Stuart 1967, Eisner et al. 1976, Traniello and Beshers 1985, Kaib 1990).

The reaction of various castes to alarm pheromone was studied in some detail in *N. princeps* by Roisin et al. (1990). In this species, workers actually were more attracted to a point source of pheromone than soldiers, but only if they were tested without soldiers. Soldiers somewhat inhibited the reaction of the workers. Being much more active in exploring the environment, soldiers are usually the first to encounter the enemy. Workers finish the job begun by the soldiers which disable the enemy with defensive secretion and then patrol elsewhere. Only older, pigmented workers responded to the pheromone; workers of stage 1 were never recruited. In this species, as in many other termitids (Noirot 1955), workers molt and thus belong to different developmental stages. In mixed groups of aged major and minor workers, only the majors were attracted by the alarm pheromone.

Not all chemically defended termites secrete alarm pheromones. As stated above, no defensive recruitment was observed in *Armitermes chagresi*, although the soldiers have a well developed frontal gland secreting macrocyclic lactones (Traniello et al. 1984). This termite uses a "static warfare": no recruitment occurs, but the soldiers have a phragmotic head able to block the narrow galleries of the nest. *Nasutitermes* spp. demonstrate a "mobile warfare" implying recruitment of their numerous soldiers. The structure of *Nasutitermes* nests is correlated to this mobile warfare strategy (Deligne and Pasteels 1982). More surprising is the observation that in sympatric arboreal *Nasutitermes* spp., some secrete alarm pheromone but others do not. For example, in coconut plantations along the north coast of Papua New Guinea *N. princeps* secretes alarm pheromone (Roisin et al. 1990), but *N. novarumhebridarum* does not (Leponce and Roisin pers. comm.). Similarly, in Central and South American forests, *N. costalis* secretes alarm pheromone (Vrkoc et al. 1978), but *N. ephratae* does not (Maschwitz and Mülhenberg 1972). Both *N. costalis* and *N. princeps* are polydomous species occupying large territories (Roisin and Pasteels 1986, Roisin et al. 1987), in contrast to *N. Novarumhebridarum*, which is monodomous and exploits local resource offered by the dead coconut tree bearing its nest (Leponce et al. 1995). *Nasutitermes princeps* is able to form supercolonies of more than 100 nests and expand its territory by new buds, aggressively outcompeting other arboreal termites (e.g., *Microcerotermes biroi)* by invading their nests and killing their inhabitants (Leponce et al. 1994).

It would be premature and probably over simplistic to suggest that alarm pheromones only evolved in aggressive, expansionist, termite species and not in more timid species. Alarm pheromones are one component of the defensive and offensive strategies of the species. Correlates should be established in a number of species between the various components of these strategies for a better understanding of the evolution of alarm pheromones as a function of the behavioral ecology of termites.

Conclusion

Our present knowledge of releasing pheromones in termites remains far behind our understanding of releasing pheromones in other social insects. Recent developments, however, indicate lines of research that should be rewarding in the near future.

Trail pheromones seem far more accessible to analysis, now that it is recognized that trail pheromones are often produced in much higher quantities by females as sex pheromones. A strong emphasis should be given to the identification of minor compounds possibly acting as synergists in both trail and sex communication. Analytical chemical methods coupled with electro-physiological methods should be as rewarding in this context as they have proved to be in so many other areas of research on insect pheromones.

Detailed behavioral studies are also badly needed, especially in termite postflight behavior. Available data suggest that postflight behavior is quite diversified in termites. The exact role of the tergal glands remains poorly understood, and nothing is known of the function of other glands in alates, described by Ampion and Quennedey (1981). The diversity of these glands points to the diversity of termite behavior. It is amazing that it is not even known with certainty if both sexes call in *Zootermopsis*, the best studied termite genus in the United States. Undoubtedly, research on postflight behavior should be most rewarding, despite the fact that swarming is often inconspicuous in lower termites.

Research on termite alarm pheromones has rarely gone beyond the stage of crude bioassays. The role of these compounds should be analyzed in more realistic situations not only during defense against predators, but also during competition with other termites. Indeed, the behavioral ecology of termites remains understudied and presents great opportunities for important advances.

Acknowledgments

We thank M. Leponce, D. McKey, Ch. Noirot and Y. Roisin for useful suggestions and comments. J.M. Pasteels thanks the Belgian Fund for Joint Basic Research (grants No 2.4513.90 and 2.4558.92).

References Cited

Ampion, M. and A. Quennedey. 1981. The abdominal epidermal glands of termites and their phylogenetic significance. In: *Biosystematics of Social Insects* (P.E. House and J.-L. Clément, Eds.), Academic Press, London, pp.249-261.

Becker, G. 1966. Spurfolge-Reaktion von Termiten auf Glykol-Verbindung. *Zeit. Angew. Zool.* 53: 495-498.

Becker, G. and R. Mannesmann. 1968. Untersuchungen über das Verhalten von Termiten gegenüber einigen spurbildenden Stoffen. *Z. Angew. Entomol.* 62: 399-436.

Birch, A.J., W.V. Brown, J.E.T. Corrie and B.P. Moore. 1972. Neocembrene A, a termite trail pheromone. *J. Chem. Soc. Perkin Trans.* 1: 2653-2658.

Bordereau, C., A. Robert, O. Bonnard and J.L. Le Quéré. 1991. (3Z,6Z,8E)-3,6,8-dodecatrien-1-ol: sex pheromone in a higher fungus-growing termite, *Pseudacanthotermes spiniger* (Isoptera, Macrotermitinae). *J. Chem. Ecol.* 17: 2177-2191.

Bordereau, C., A. Robert, N. Laduguie, O. Bonnard, J.L. Le Quéré and R. Yamaoka. 1993. Détection du (Z,Z,E)-3,6,8-dodecatrien-1-ol par les ouvriers et les essaimants de deux espèces de termites champignonnistes: *Pseudacanthotermes spiniger* et *P. militaris* (Termitidae, Macrotermitinae). *Actes Coll. Ins. Soc.* 8: 145-149.

Bruinsma, O.H. 1979. An analysis of building behaviour of the termite *Macrotermes subhyalinus. Ph.D. Thesis, Wageningen,* 86 pp.

Castle, G.B. 1934. The damp-wood termites of western United States, genus *Zootermopsis* (formerly, *Termopsis*). In: *Termites and Termite Control,* 2nd. Ed. (C.A. Kofoid, Ed.), University of California Press, Berkeley, pp. 273-310.

Clément, J.L. 1982. Phéromones d'attraction sexuelle des termites européens du genre *Reticulitermes* (Rhinotermitidae). Mécanismes comportementaux et isolements spécifiques. *Biol. Behav.* 7: 55-68.

Clément, J.L., R. Howard, M.S. Blum and H. Lloyd. 1986. L'isolement spécifique des termites du genre *Reticulitermes* du Sud-Est des Etats-Unis. Mise en évidence grâce à la chimie et au comportement d'une espèce jumelle de *R. virginicus = R. malletei* sp. nov. et d'une semi-species de *R. flavipes*. *C. R. Acad. Sci. Paris* 302: 67-70.

Clément, J.L., H. Lloyd, P. Nagnan and M.S. Blum. 1989. n-Tetradecyl propionate: identification as a sex pheromone of the eastern subterranean termite *Reticulitermes flavipes. Sociobiology* 15: 19-24.

Deligne, J. 1985. Apport de la micromorphologie du labre à la compréhension de la phylogénèse des termites. *Actes Coll. Ins. Soc.* 2: 35-42.

Deligne, J. and J.M. Pasteels. 1982. Nest structure and soldier defense: an integrated strategy in termites. In: *The Biology of Social Insects* (M.D. Breed, C.D. Michener and H.E. Evans, Eds.), Westview Press, Boulder, pp. 288-289.

Dudley, P.H. and J. Beaumont. 1889. Observations on the termites or white ants of the isthmus of Panama. *Trans. N.Y. Acad. Sci.* 8: 85-114.

Eisner, T., I. Kriston and D.J. Aneshansley. 1976. Defensive behavior of a termite (*Nasutitermes exitiosus*). *Behav. Ecol. Sociobiol.* 1: 83-125.

Ernst, E. 1959. Beobachtungen beim Spritzakt der *Nasutitermes*-Soldaten. *Rev. Suisse Zool.* 66: 289-295.

Everaerts, C., O. Bonnard, J.M. Pasteels, Y. Roisin and W.A. König. 1990. (+)-α-Pinene in the defensive secretion of *Nasutitermes princeps* (Isoptera, Termitidae). *Experientia* 46: 227-230.

Fuller, C. 1915. Observations on some South African termites. *Ann. Natal Museum* 3: 329-505.

Grassé, P.-P. and C. Noirot. 1951. Orientation et routes chez les termites. Le "balisage des pistes". *Ann. Psychol.* 50: 273-280.

Grassi, B. and A. Sandias. 1893. Costituzione e sviluppo della società dei termitidi. *Atti Accad. Gioenia Sci. Nat. Catania* 6: 1-75.

Hall, P. and J.F.A. Traniello. 1985. Behavioral bioassays of termite trail pheromones. Recruitment and orientation effects of cembrene-A in *Nasutitermes costalis* (Isoptera: Termitidae) and discussion of factors affecting termite response in experimental contexts. *J. Chem. Ecol.*, 11: 1503-1513.

Heidecker, J.L. and R.H. Leuthold. 1983. The organisation of collective foraging in the harverster termite *Hodotermes mossambicus* (Isoptera). *Behav. Ecol. Sociobiol.* 14: 195-202.

Hewitt, P.H. and J.J. Nel. 1969. The influence of group size on the sarcosomal activity and the behaviour of *Hodotermes mossambicus* alate termites. *J. Insect Physiol.* 15: 2169-2177.

Hölldobler, B. and E.O. Wilson. 1990. *The Ants*, Springer-Verlag, Berlin, 732 pp.

Howard, R., F. Matsumura and H.C. Coppel. 1976. Trail-following pheromones of the Rhinotermitidae: approaches to their authentication and specificty. *J. Chem. Ecol.* 2: 147-166.

Hummel, H. and P. Karlson. 1968. Hexansäure als Bestandteil des Spurpheromons der Termite *Zootermopsis nevadensis* Hagen. *Hoppe-Seyler's Z. Physiol. Chem.* 349: 725-727.

Kaib, M. 1990. Intra- and interspecific chemical signals in the termite *Schedorhinotermes*-Production sites, chemistry, and behaviour. In: *Sensory Systems and Communication in Arthropods. Advances in Life Sciences*. (F.G. Gribahin, K. Wiese and A.V. Popov, Eds.), Birkhauser, Basel, pp. 26-31.

Kaib, M., O. Bruinsma and R.H. Leuthold. 1982. Trail-following in termites: evidence for a multicomponent system. *J. Chem. Ecol.* 8:1193-1205.

Karlson, P., M. Lüscher and H. Hummel. 1968. Extraktion und biologische Auswertung des Spurpheromones der Termite *Zootermopsis nevadensis*. *J. Insect Physiol.* 14: 1763-1771.

Kettler, R. and R.H. Leuthold. 1995. Inter-and intraspecific alarm response in the termite *Macrotermes subhyalinus* (Rambur). *Ins. Soc.* 42: 145-156.

Klochov, S.G. and D.D. Zhuzhikov. 1990. Termite trail pheromones: specificity and biosynthesis. In: *Sensory System and Communication in Arthropods. Advances in Life Sciences.* (F.G. Gribakin, K. Wiese and A.V. Popov, Eds.), Birkhauser, Basel, pp. 40-43.

Kodama, M., Y. Matsuki and S. Itô. 1975. Syntheses of macrocyclic terpenoids by intramolecular cyclization I. (±)-Cembrene-A, a termite trail pheromone, and (±)-nephthenol. *Tetrahedron Lett.* 35: 3065-3068.

Kriston, I., J.A.L. Watson and T. Eisner. 1977. Non-combative behaviour of large soldiers of *Nasutitermes exitiosus* (Hill): An analytical study. *Ins. Soc.* 24: 103-111.

Laduguie, N. 1993. Phéromones de piste et phéromones sexuelles chez les termites. *Thèse de doctorat, Université de Bourgogne*, 112 pp.

Laduguie, N., A. Robert, N. Salini and C. Bordereau. 1994a. Les termites et le dodecatrienol. *Actes Coll. Ins. Soc.* 9: 27-34.

Laduguie N., A. Robert, O. Bonnard, F. Vieau, J.L. Le Quéré, E. Semon and C. Bordereau. 1994b. Isolation and identification of (3*Z*,6*Z*,8*E*)-3,6,8-dodecatrien-1-ol in *Reticulitermes santonensis* Feytaud (Isoptera, Rhinotermitidae): Roles in worker trail-following and in alate sex-attraction behavior. *J. Insect Physiol.* 40: 781-787.

Leponce, M., Y. Roisin and J.M. Pasteels. 1994. Ecology of two arboreal-nesting termites in New Guinea coconut plantations. In: *Les Insectes Sociaux* (A. Lenoir, G. Arnold and M. Lepage, Eds.), Université de Paris Nord, Paris. p.307.

Leponce, M., Y. Roisin and J.M. Pasteels. 1995. Environmental influences on the arboreal nesting termite community in New Guinean coconut plantations. *Environ. Entomol.* 24: 1442-1452.

Leuthold, R.H. 1975. Orientation mediated by pheromones in social insects. In: *Pheromones and Defensive Secretions in Social Insects*. (C. Noirot, P.E. Howse and G. Le Masne, Eds.), Université de Dijon, Dijon, pp.197-211.

Leuthold, R.H. and O. Bruinsma. 1977. Pairing behavior in *Hodotermes mossambicus* (Isoptera). *Psyche* 84: 109-119.

Leuthold, R.H. and M. Lüscher. 1974. An unusual caste polymorphism of the sternal gland and its trail pheromone production in the termite *Trinervitermes bettonianus*. *Ins. Soc.* 21: 335-341.

Leuthold, R.H., O. Bruinsma and A. van Huis. 1976. Optical and pheromonal orientation and memory for homing distance in the harvester termite *Hodotermes mossambicus* (Hagen). *Behav. Ecol. Sociobiol.* 1: 127-139.

Lindström, M., T. Norin, I. Valterová and J. Vrkoc. 1990. Chirality of the monoterpene alarm pheromones of termites. *Naturwissenschaften* 77: 134-135.

Lüscher, M. and B. Müller. 1960. Ein spurbilendes Sekret bei Termiten. *Naturwissenschaften* 21: 503.

Lys, J.A. and R.H. Leuthold. 1991. Task-specific distribution of the two worker castes in extranidal activities in *Macrotermes bellicosus* (Smeathman): observation of behaviour during food acquisition. *Ins. Soc.* 38: 161-170.

MacFarlane, J. 1983. Observation on trail pheromone, trail-laying and longevity of natural trails in the termite *Macrotermes michaelseni*. *Insect Sci. Appl.* 4: 309-318.

Maschwitz, U. and M. Mühlenberg. 1972. Chemische Gefahrenalarmierung bei Termite. *Naturwissenschaften* 59: 516-517.

Matsumura, F., H.C. Coppel and A. Tai. 1968. Isolation and identification of termite trail-following pheromone. *Nature* 219: 963-964.

Matsumura, F., A. Tai, H.C. Coppel. 1969. Termite trail-following substance, isolation and purification from *Reticulitermes virginicus* and fungus infected wood. *J. Econ. Entomol.* 62: 599-603.

Matsumura, F., D.W. Jewett and H.C. Coppel. 1972. Interspecific response to synthetic trail-following substances. *J. Econ. Entomol.* 65: 600-602.

McDowell, P.G. and G.W. Oloo. 1984. Isolation, identification and biological activity of trail-following pheromone of termite *Trinervitermes bettonianus* (Sjöstedt) (Termitidae: Nasutitermitinae). *J. Chem. Ecol.* 10: 835-851.

Moore, B.P. 1968. Studies on chemical composition and function of the cephalic gland secretion in Australian termites. *J. Insect Physiol.* 14: 33-39.

Moore, B.P. 1969. Biochemical studies in termites. In: *Biology of Termites*, vol. 1 (K. Krishna and F.M. Weesner, Eds.), Academic Press, New York, pp: 407-432.

Moore, B.P. 1974. Pheromones in termite societies. In: *Pheromones* (M.C. Birch, Ed.), American Elsevier, New York, pp.250-266.

Noirot, C. 1955. Recherches sur le polymorphisme des termites supérieurs. *Ann. Sci. Nat. Zool., Paris, 11ème s.* 17: 399-595.

Noirot, C. 1969. Glands and Secretions. In: *Biology of Termites* (K. Krishna and F.M. Weesner, Eds.), Academic Press, New York, pp.89-123.

Noirot, C. 1995. The sternal glands of termites: segmental pattern, phylogenetic implications. *Ins. Soc.* 42: 321-323.

Oloo, G.W. 1981. Specificity of termite trails: Analysis of natural trails of *Trinervitermes, Macrotermes,* and *Odontotermes* from sympatric populations. *Entomol. Exp. Appl.* 29: 162-168.

Oloo, G.W. 1984. Some observations on the trail-laying behaviour of *Macrotermes michaelseni* (Sjöst) (Termitidae). *Insect Sci. Appl.* 5: 259-262.

Oloo, G.W. and R.H. Leuthold. 1979. Influence of food on trail-laying and recruitment behaviour in *Trinervitermes bettonianus* (Termitidae: Nasutitermitinae). *Entomol. Exp. Appl.* 26: 267-278.

Oloo, G.W. and P.G. Mc Dowell. 1982. Interspecific trail-following and evidence of similarity of trails of *Trinervitermes* species from different habitats. *Insect Sci. Appl.* 3: 157-161.

Oster, G.F. and E.O. Wilson. 1978. *Caste and Ecology in the Social Insects*, Princeton University Press, Princeton, 352 pp.

Pasteels, J.M. 1965. Polyéthisme chez les ouvriers de *Nasutitermes lujae* (Termitidae Isoptères). *Biologia Gabonica* 1: 191-205.

Pasteels, J.M. 1972. Sex-specific pheromones in a termite. *Experientia* 28: 105-106.

Pasteels J.M., J.L. Deneubourg, J.C. Verhaeghe, J.L. Boevé and Y. Quinet. 1985. Orientation along terrestrial trails by ants. In: *Mechanisms in Insect Olfaction Signals* (T. Payne and M. Birch, Eds.), Clarendon Press, Oxford, pp. 131-138.

Pasteels, J.M., Y. Roisin, C. Everaerts, O. Bonnard, J.C. Braekman and D. Daloze. 1988. Morphological and chemical criteria in the taxonomy of *Nasutitermes* from Papua New Guinea (Isoptera: Termitidae). *Sociobiology*, 14: 193-206.

Prestwich, G.D. 1983. Chemical systematics of termite exocrine secretions. *Ann. Rev. Ecol. Syst.* 14: 287-311.

Quennedey, A. 1977. An ultrastructural study of the polymorphic sternal gland in *Reticulitermes santonensis* (Isoptera, Rhinotermitidae), another way of looking at the true termite trail-pheromone. *Proc. VIIIth Int. Congr. IUSSI, Wageningen*, pp. 48-49.

Rickli, M. and R.H. Leuthold. 1987. Spacial organisation during exploration and foraging in the harvester termite, *Trinervitermes geminatus*. *Rev. Suisse Zoo.l.* 94: 545-551.

Ritter, F.J., A.M. van Oosten and C.J. Persoons. 1973. Attractants and trail-following compounds for the Saintonge termite: tricycloekasantalal and dihydroagarofuran from sandalwood oils and a sesquiterpenoid $C_{15}H_{24}O$ from pinewood infested with *Lenzites trabea*. *Proc. VIIth Int. Congr. IUSSI, London*, pp. 330-331.

Ritter, F.J., I.E.M. Bruggeman, C.J. Persoons, E. Talman, A.M. van Oosten and P.E.J. Verwiel. 1977. Evaluation of social insect pheromones in pest control, with special reference to subterranean termites and Pharaoh's ants. In: *Crop Protection Agents: Their Biological Evaluation* (N.R. McFarlane, Ed.), Academic Press, New York, pp. 210-216.

Roisin, Y. and J.M. Pasteels. 1986. Reproductive mechanisms in termites: polycalism and polygyny in *Nasutitermes polygynus* and *N. costalis*. *Ins. Soc.* 33: 149-167.

Roisin, Y., J.M. Pasteels and J.C. Braekman. 1987. Soldier diterpene patterns in relation with aggressive behaviour, spacial distribution and reproduction of colonies in *Nasutitermes princeps*. *Biochem. Syst. Ecol.* 15: 253-261.

Roisin, Y., C. Everaerts, J.M. Pasteels and O. Bonnard. 1990. Caste-dependent reaction to soldier defensive secretion and chiral alarm/recruitment pheromone in *Nasutitermes princeps*. *J. Chem. Ecol.* 16: 2865-2875.

Runcie, C.D. 1987. Behavioral evidence for multicomponent trail pheromone in the termite, *Reticulitermes flavipes* (Kollar) (Isoptera: Rhinotermitidae). *J. Chem. Ecol.* 13:1967-1978.

Stuart, A.M. 1961. Mechanism of trail-laying in two species of termites. *Nature*. 189: 419.

Stuart, A.M. 1963. Studies on the communication of alarm in the termite *Zootermopsis nevadensis* (Isoptera). *Physiol. Zool.* 36: 85-96.

Stuart, A.M. 1967. Alarm, defense, and construction behavior relationships in termites (Isoptera). *Science* 156: 1123-1125.

Stuart, A.M. 1969. Social behavior and communication. In: *Biology of Termites*, vol.1 (K. Krishna and F.M. Weesner, Eds.), Academic Press, New York, pp. 193-232.

Stuart, A.M. 1970. The role of chemicals in termite communication. In: *Communication by Chemical Signals. Advances in Chemoreception*, vol.1 (J.W. Johnston, D.G. Moulton and A. Turk, Eds.), Appleton-Century-Crofts, New York, pp.79-105.

Stuart, A.M. 1975. Some aspects of pheromone involvement in the post-flight behaviour of the termite *Zootermopsis angusticollis* (Hagen) and *Reticulitermes flavipes* (Kollar). In: *Pheromones and Defensive Secretions in Social Insects* (C. Noirot, P.E. Howse and G. Le Masne, Eds.), Université de Dijon, Dijon, pp. 219-223.

Stuart, A.M. 1981. The rôle of pheromones in the initiation of foraging, recruitment and defence by the soldiers of a tropical termite, *Nasutitermes corniger* (Motschulsky). *Chem. Sens.* 6: 409-420.

Tai, A., F. Matsumura and H.C. Coppel. 1969. Chemical identification of the trail-following pheromone for a southern subterranean termite. *J. Org. Chem.* 34: 2180-2182.

Tokoro, M., M. Takahashi, K. Tsunoda and R. Yamaoka. 1989. Isolation and primary structure of the trail pheromone of the termite, *Coptotermes formosanus* Shiraki (Isoptera: Rhinotermitidae). *Wood Res.* 76: 29-38.

Tokoro, M., R. Yamaoka, K. Hayashiya, M. Takahashi and K. Nishimoto. 1990. Evidence for trail pheromone precursor in termite *Reticulitermes speratus* (Kolbe) (Rhinotermitidae: Isoptera). *J. Chem. Ecol.* 16: 2549-2557.

Tokoro, M., R. Yamaoka, K. Hayashiya, M. Takahashi and K. Tsunoda. 1991. Isolation and identification of the trail pheromone of the subterranean termite, *Reticulitermes speratus* (Kolbe) (Rhinotermitidae: Isoptera). *Wood Res.* 78: 1-14.

Traniello, J.F.A. 1981. Enemy deterrence in the recruitment strategy of a termite: Soldier-organized foraging in *Nasutitermes costalis*. *Proc. Natl. Acad. Sci. USA* 78: 1976-1979.

Traniello, J.F.A. 1982. Recruitment and orientation components in a termite pheromone. *Naturwissenschaften* 69: 343-344.

Traniello, J.F.A. and S.N. Beshers. 1985. Species-specific alarm/recruitment response in a neotropical termite. *Naturwissenschaften* 72: 491-492.

Traniello, J.F.A. and C. Busher. 1985. Chemical regulation of polyethism during foraging in the neotropical termite *Nasutitermes costalis*. *J. Chem. Ecol.* 11: 319-332.

Traniello, J.F.A., B.L. Thorne and G.D. Prestwitch. 1984. Chemical composition and efficacy of cephalic gland secretion of *Armitermes chagresi* (Isoptera: Termitidae). *J. Chem. Ecol.* 10: 531-543.

Valterová, I., J. Vrkoc, M. Lindström and T. Norin. 1992. On the natural occurence of (-)-3-carene, a component of termite defense secretions. *Naturwissenschaften* 79: 416-417.

Valterová, I., J. Vrkoc and T. Norin. 1993. The enantiomeric composition of monoterpene hydrocarbons in the defensive secretions of *Nasutitermes* termites (Isoptera): inter- and intraspecific variations. *Chemoecology* 4: 120-123.

Verron, H. and M. Barbier. 1962. L'hexène-3-ol-1, substance attractive des termites *Calotermes flavicollis* et *Microcerotermes edentatus*. *C.R. Acad. Sci. Paris* 254: 4089-4091.

Vrkoc, J., J. Krecek and I. Hrdy. 1978. Monoterpenic alarm pheromones in two *Nasutitermes* sp. *Acta Entomol. Bohem.* 75: 1-8.

Wall, M. 1971. Zur Geschlechtsbiologie der Termite *Kalotermes flavicollis* (Fabr.) (Isoptera). *Acta Tropica 28:* 17-60.

Wilson, D.S. and A.B. Clark. 1977. Above ground predator defence in the harvester termite, *Hodotermes mossambicus* (Hagen). *J. Entomol. Soc. South Africa* 40: 271-282.

9

Chemical Communication in Social Wasps

Peter J. Landolt, Robert L. Jeanne, and Hal C. Reed

Introduction

Social insects rely extensively on pheromonal communication (Wilson 1971). Members of social insect colonies must cooperate and coordinate myriad activities, including the communication of individual and colony needs, resource locations, reproduction, recognition (caste, sex, colony, species), and danger or alarm. Because many species of social wasps live in enclosed or underground nests where visual signaling is not feasible, they rely heavily on chemical communication.

This paper reviews current knowledge of pheromones in the social wasps. Pheromonal communication of relatedness, or kin recognition, is reviewed by Singer et al. (in press). Traniello and Robson (1995) review trail and territorial communication in social insects. The reader may also want to refer to recent reviews by Downing (1991) and by Jeanne (1993) emphasizing the pheromone-producing functions of wasp glands and the evolution of gland function, respectively.

Sociality in wasps is limited to the family Vespidae and the genus *Microstigmus* of the family Sphecidae (Ross and Matthews 1991). Studies of pheromones of social wasps are limited principally to a small number of genera of Vespidae, making generalizations at best tenuous. A major emphasis has long been groups well represented in temperate areas, such as *Vespula* and some *Polistes*, because of their proximity to many researchers in Europe and North America. A recent spate of studies of tropical Polistinae and of species of *Ropalidia*, however, has broadened our knowledge of social wasp behavior, including pheromonal communication.

Sex Pheromones

Sex pheromones elicit a variety of behavior patterns directly related to mating, including mate-finding and mate-selection. This behavior in insects may include the stimulation of movement or activation, attraction, arrest, or courtship responses (Shorey 1977). Although our knowledge of sexual behavior mediated by pheromones is quite limited for social wasps, there is both direct and circumstantial evidence that sex pheromones elicit attraction responses as well as arrest and courtship in some Vespidae.

Insect attractants, including sex attractant pheromones, are usually demonstrated experimentally by two approaches. Attraction may be demonstrated directly, as upwind movement in response to an odor in an airstream (Kennedy 1977), or indirectly, as capture in a trap baited with a candidate attractant. Capture in a baited trap is often considered indirect evidence that the target insect was able to locate the source of the odor. Courtship responses to pheromone are extremely diverse and are dependent on the interactions preceding mating in an individual species. Bioassays for studies of courtship pheromones are based on behavior observed in pre-mating encounters and interactions between the sexes. In social wasps, the principal courtship pheromone responses tested are mounting and copulatory attempts.

Vespinae

Sex pheromones are thought to mediate attraction and orientation behavior in Vespinae and are reported for several species (Table 9.1). For example, Ono et al. (1985) observed many males flying about the entrance to an underground nest of the hornet *Vespa mandarinia* Cameron, and suggested that males were attracted to pheromones emanating from the nest. However, sex attraction has been clearly demonstrated only in the southern yellowjacket *Vespula squamosa* (Drury). *V. squamosa* males flew upwind in a flight tunnel in response to a caged queen and to a hexane extract of queens (Reed and Landolt 1990a).

Courtship pheromones are likely to exist in many species of Vespinae. Males respond overtly and aggressively to receptive females. Groups of males sometimes form clusters trying to mate with

TABLE 9.1 Demonstrated pheromonal communication and identified pheromones in independent-founding Vespidae.

Species	Source	Structure	Ref.
ALARM PHEROMONE			
Dolichovespula media	venom		1
Vespula vulgaris	venom		2
Vespula germanica	venom		2
Vespula squamosa	venom	N-3-methylbutylacetamide	3 & 4
Vespula maculifrons	venom	N-3-methylbutylacetamide	5
Vespa crabro	venom	2-methyl-3-butene-2-ol	6
Vespa orientalis	venom		7
Provespa anomala	venom		8
Polistes carolina	venom		9
Polistes fuscatus	venom		10
Polistes exclamans	venom		10
Polybia rejecta	sting		11
Polybia occidentalis	venom		12
Ropalidia romandi	venom		13
SEX PHEROMONE			
Vespula squamosa			14
Vespa analis			15
Vespa crabro			15
Vespa mandarinia			15
Vespa simillima			15
Vespa tropica			15
Polistes exclamans			16, 17
Polistes fuscatus	venom		16,18
Belonogaster petiolata			19
QUEEN AND BROOD PHEROMONE			
Vespa orientalis		5-hexadecanolide	20, 21
Dolichovespula media			22
Vespa crabro		Z-9-pentacosene	23

TABLE 9.1 Continued

TABLE 9.1 Continued

Species	Source	Structure	Ref.
BUILDING PHEROMONE			
Vespa orientalis			24
Polistes foederatus			24
FOOTPRINT PHEROMONE			
Vespula vulgaris			25
TRAIL PHEROMONE			
Polybia sericea	5Th sternal gland		26

References: 1=Maschwitz 1984; 2=Maschwitz 1964; 3=Landolt & Heath 1987; 4=Landolt & Heath 1988; 5=Landolt et al. 1995; 6=Veith et al. 1984; 7=Ishay et al. 1967; 8=Maschwitz & Hanel 1988; 9=Jeanne 1982; 10=Post et al. 1984; 11=Overal et al. 1981; 12=Jeanne 1981a; 13=Kojima 1994a; 14=Reed & Landolt 1990a; 15=Ono & Sasaki 1987; 16=Post & Jeanne 1984; 17= Reed & Landolt 1990b; 18=Post & Jeanne 1983b; 19=Keeping et al. 1986; 20=Ikan et al. 1969; 21=Ishay et al. 1965; 22=Ishay 1973; 23=Veith & Koniger 1978; 24=Ishay & Perna 1979; 25=Butler et al. 1969; 26=Jeanne 1981.

females (*Dolichovespula sylvestris* (Scopoli), Sandeman 1938; *Vespa crabro* Christ, Batra 1980; *Vespula atropilosa* (Sladen), MacDonald et al. 1974), or in response to males that were in recent contact with receptive females (*Vespula maculifrons* (Buysson), Ross 1983; *V. germanica* (Fab.), Thomas 1960; *Vespa crabro*, Batra 1980). This behavior suggests pheromonal communication of sexual receptivity by females and contamination of courting or copulating males with female sex pheromone. Demonstrations of pheromonal roles in courtship, such as stimulation of mounting and copulatory attempts by males, have been made for species of *Vespa* (Ono and Sasaki 1987). Ono and Sasaki (1987) were able to elicit mating attempts by males with freeze-dried dead males treated with an extract of virgin queens. This response was not species-specific among *Vespa analis* Dalla Torre, *Vespa crabro*, *Vespa mandarinia*, *Vespa simillima* Smith, and *Vespa tropica* du Buysson.

Polistinae

Sex attractant pheromones also occur in the Polistinae (Table 9.1). Males of *Polistes exclamans* exhibit upwind flight in response to live conspecific females, female odor, or solvent extract of females (Reed and Landolt 1990b). Similar behavior was exhibited by females in response to males. In both sexes, there is attraction to all body tagmata of the opposite sex, indicating that the releasing pheromone is spread over the body, possibly by grooming (Reed and Landolt 1990b).

A number of species of *Polistes* have been shown to possess sex pheromones, as well as territorial and aggregation pheromones that may include a sexual function. Courtship pheromones produced by females communicate receptivity as well as sex and species identification. Post and Jeanne (1983a,b, 1984, 1985) demonstrated such pheromones in *P. fuscatus* (Fab.) and *P. exclamans* Viereck. Males of *P. fuscatus* in male-only groups made greater numbers of homosexual copulatory attempts when in the presence of excised female venom glands and sacs (or hexane extracts of female venom glands and sacs) compared to controls. This indicates that a pheromone is present in female venom that elicits copulatory responses from males (Post and Jeanne 1983b). Males in these tests also approached venom-treated filter papers, suggesting short range orientation responses. Males of *P. exclamans* attempted copulation when exposed to extracts of conspecific female venom, while males of *P. fuscatus* responded with copulatory attempts when exposed to extracts of venom of conspecific females, of *P. exclamans*, or of the eastern yellowjacket, *Vespula maculifrons* (Post and Jeanne 1984). Similar bioassays were used to demonstrate that a pheromone extracted from gynes of the polistine wasp *Belonogaster petiolata* (DeGeer) releases both antennation and copulatory attempts by males (Keeping et al. 1986).

Other types of behavior by male and female *Polistes* suggest pheromonal roles in sexual behavior, although experimental demonstrations are lacking. Males of *Polistes* species are well known to patrol and defend territories and perches in areas frequented by females. Such territoriality may involve pheromonal marking of sites, either as a territorial display to ward off other males or as a sex attractant for arriving females. Males of *P. fuscatus* defend territories at female hibernacula and nest sites, which they appear to mark with the sternal glands of the gaster through gastral dragging (Post and Jeanne 1983a). Wenzel (1987) reported both sternal rubbing (gastral dragging) and facial rubbing by males of *P. major* (Beauvois) at defended perches and patrol routes. Facial rubbing may involve deposition of pheromone from enlarged ectal mandibular glands by way of mandibular applicator brushes. Similar behavior was observed by males of *P. fuscatus, P. metricus* Say, and *Polistes dorsalis* (Fab.), that swarm at the tops of towers in Florida. Males of these species rub perch sites with their gastral sternites and mandibles (Reed and Landolt 1991). Beani and Calloni (1991a) reported male gaster-dragging and leg-rubbing in *Polistes dominulus* (Christ) that suggest scent-marking. Gaster-dragging at perch sites is also known for males of *Mischocyttarus labiatus* (Fab.) (Litte 1981) and *Mischocyttarus flavitarsis* de Saussure (Litte 1979). Although this behavior suggests deposition of pheromone from well developed exocrine glands associated with the mandibles, abdominal sterna, and legs (Landolt and Akre 1979; Jeanne et al. 1983; Beani and Calloni 1991b), there have been no demonstrations of communication by way of such marks or exocrine secretions.

To date, none of the sex pheromones of Vespidae has been structurally identified. It is conceivable that such pheromones may provide useful attractants and trap baits for pestiferous species.

Alarm Pheromones

We define alarm behavior broadly as any response to a disturbance of the colony that increases the likelihood that colony members will take defensive action. Alarm behavior may occur directly in response to an external threat to the colony, or it may occur as the result of a signal (chemical or otherwise) produced by one or more nestmates. The familiar response to a vertebrate intruder is a mass stinging attack, but not all species respond in this way. *Protopolybia fuscatus* and *Polybia emaciata*, for example, retreat inside the nest when the colony is disturbed (Jeanne 1970) and *Parachartergus colobopterus* sprays sticky venom at intruders (Jeanne and Keeping 1995). Furthermore, although pheromonal communication of alarm appears to be widespread in the social Vespidae (Table 9.1), it is not universal. The rich diversity of defensive responses among the vespids has the potential to tell us a great deal about the evolution and adaptiveness of chemical signalling in the social insects. On the chemical front, progress has been made in the search for chemical releasers of alarm behavior. We summarize both lines of research here.

Vespinae

In the first report of an alarm response to pheromone in wasps, Maschwitz (1964) found that workers of *Vespula vulgaris* (L.) and *V. germanica* left the nest entrance and attacked nearby objects when other conspecific worker wasps were held at the nest entrance and induced to spray venom. Aldiss (1983) demonstrated attack responses by *V. vulgaris* workers to crushed gasters, excised stings, and excised venom sacs, indicating the venom sac as the principal source of alarm pheromone. Aldiss (1983) also obtained a significant attack response by

V. vulgaris workers to crushed heads of conspecifics. Alarm behavior in response to venom and to crushed workers has been reported for other yellowjackets. Worker *Dolichovespula saxonica* (Fab.) responded to venom glands of conspecific workers presented at the nest entrance by leaving the nest, flying about the nest, and by attraction to the venom gland placed at the entrance (Maschwitz 1984). Venom glands of *V. squamosa* and *V. maculifrons* workers also contain an alarm pheromone that elicits departure from the nest entrance, attraction, and attack (Landolt and Heath 1987; Landolt et al. 1995).

Similar alarm behavior occurs in hornets (*Vespa, Provespa*) in response to their venom. Ishay et al. (1967) were the first to demonstrate pheromonal communication of alarm in a hornet. They elicited alarm behavior by worker *Vespa orientalis* (Fab.) in response to conspecific venom. Results of a series of experiments testing for evidence of pheromones in *Vespa crabro* in North America led Batra (1980) to conclude that this species lacks alarm pheromone. Veith et al. (1984), however, working with the same species in Europe, reported departure from the nest, defensive flights, wing buzzing, approach, and biting by workers in response to venom on filter paper placed near the nest entrance. Maschwitz and Hanel (1988) evoked attack flights from workers of *Provespa anomala* (Saussure) with squashed venom glands presented near the nest entrance. The occurrence of alarm pheromones in all species tested so far in 4 genera suggests that alarm pheromones could be universal in the Vespinae.

Alarm pheromones have been isolated and identified in only three species of social wasps, all of them vespines: *Vespa crabro, Vespula squamosa,* and *V. maculifrons*. Veith et al. (1984) identified eight compounds distilled from *V. crabro* venom sac contents. All compounds were assayed at the entrances to hornet nests, in comparison to venom sac contents. Only 2-methyl-3-butene-2-ol elicited significant wing buzzing, defense flight, and departure from the nest, when compared to wasp responses to venom sac contents. Heath and Landolt (1988) isolated and identified *N*-3-methylbutylacetamide from the venom sacs of *V. squamosa* and demonstrated that alarm and attack behavior in response to this chemical is similar to the behavioral responses elicited by extracts of conspecific venom sacs. This compound was also found in venom of *V. vulgaris* by Aldiss (1983), but was not assayed for alarm pheromone activity in that species. *N*-3-methylbutylacetamide also occurs in the venom sacs of *V. maculifrons* and elicits alarm and attack behavior in this species, although only at greater than expected dosages (Landolt et al. 1995).

Polistinae

There is now ample evidence for alarm pheromones among the swarm-founding Polistinae (Table 9.1). Jeanne (1981a) experimentally demonstrated the presence of an alarm pheromone in the venom of *Polybia occidentalis* (Olivier). The odor of venom wafted over or injected into the nest recruits large numbers of alarmed workers to the outer surface of the envelope. For several minutes these wasps remain extremely sensitive and will attack and sting any object moving near the nest. The attack and stinging response is released visually, especially by dark, moving objects. A similar response to the odor of venom has been demonstrated for *P. rejecta* (F.) (Overal et al. 1981) and there is evidence for venom-based alarm pheromones in *P. sericea* (Olivier) (Jeanne, unpublished) and *Ropalidia romandi* (le Guillou) (Kojima 1994a). Evidence for the existence of alarm pheromones, but not yet linked to venom, exists for *Protopolybia acutiscutis* (Cameron) (Naumann 1970), *Apoica pallida* (Olivier) (Schremmer 1972), and *Synoeca surinama* (L.) (Castellón 1981). *Parachartergus colobopterus* appears to be an exception in that it lacks an alarm pheromone (Jeanne and Keeping, 1995). In this cryptically-nesting species, gaster tapping on the envelope by workers seems to act as an alarm signal. Workers rarely sting in defense of the nest; instead they stand on the nest and spray mists of sticky venom toward objects moving nearby. If disturbance of the colony persists, many of the wasps temporarily flee the nest rather than defend it by stinging (Strassmann et al. 1990).

Some independent-founding polistines also possess alarm pheromones that occur in the venom. Jeanne (1982) reported that *Polistes canadensis* (L.) females attacked a moving visual model when filter paper bearing a crushed venom sac was held upwind of the nest (Jeanne 1982). *P. exclamans* and *P. fuscatus* also respond to venom glands and venom sacs with alarm and attack behavior (Post et al. 1984). As in the swarm-founding Polistinae, however, alarm pheromones are not universal in the independent-founders. Pheromonal mediation of alarm was not demonstrated in experiments testing for it in *Polistes nimpha*, *P. dominulus* (Freisling 1943), *P. biglumis* (Maschwitz 1964), *Belonogaster petiolata* (Keeping 1995), or *Mischocyttarus immarginatus* Richards (K. B. London, pers. comm.).

The progress made toward elucidating alarm behavior in a number of species, while significant, only serves to sharpen the focus on several intriguing questions. The patchy distribution of alarm pheromones among the social wasps raises questions about how many times alarm pheromones have evolved in the wasps, whether they have been lost in one or more lineages in which they are ancestral (in swarm-founding polistines with very small colonies, for example), and in what ecological contexts alarm pheromones are adaptive.

Another set of questions has to do with the evolutionary steps leading to alarm pheromones. Elsewhere, one of us has conjectured that the first step toward certain chemical signalling systems such as alarm pheromones may have been a learned association of danger with the odor of venom released in the course of stinging an intruder (Jeanne 1993). Genetic assimilation of the response could follow from this (Tierney 1986), genetically fixing it in the population. At this point, venom would still be a cue, in the sense of Seeley (1989). In some species, this may be the final step; in other words, the response may be innate, but the release of venom is still only incidental to the act of stinging. The final step would be the release of venom apart from the stinging act, i.e. as a true signal (Seeley 1989), enabling a group defense of the colony by many workers even before the first sting attempt.

Demonstration of the release of venom as a signal is not easy. In fact, no direct evidence for it has been adduced for any wasp. There is circumstantial evidence for it in *Polybia occidentalis* (Olivier) (Jeanne, unpublished data). Exhaling lightly through a straw onto a single, unalarmed worker at one edge of the nest caused her to respond with alarm (wing buzzing, gaster pumping). The alarm response then spread away from this point, within 1.5 sec reaching all the other workers standing on the nest (Figure 9.1). This experiment suggests that the first wasp released venom as an alarm signal and that each worker perceiving it responded by releasing venom in turn, causing a chain reaction.

In some species of *Vespula*, *Vespa*, and *Dolichovespula*, venom is reportedly sprayed from the sting (Maschwitz 1964; Saslavasky et al. 1973; Greene et al. 1976). However, because all such observations have been made on wasps that were restrained or were attempting to sting an impenetrable surface or the mesh of a veil, Jeanne and Keeping (1995) suggested that these examples may have been abortive sting attempts rather than the independent release of alarm pheromone. Attempts to show experimentally that an alarmed colony of *Polistes fuscatus* released venom apart from stinging were unsuccessful (Post et al. 1984).

The distinction between cue and signal is worth making. If it can be shown that there are species in which venom elicits an alarm response but is released only incidental to stinging, i.e. is only a cue and not a signal, we will have begun to tease apart the steps in the evolution of an important class of pheromones in social insects. Because of the irregular distribution of alarm pheromones and the diversity of behavioral responses to them among social wasps, these social insects have the potential to provide answers to such questions.

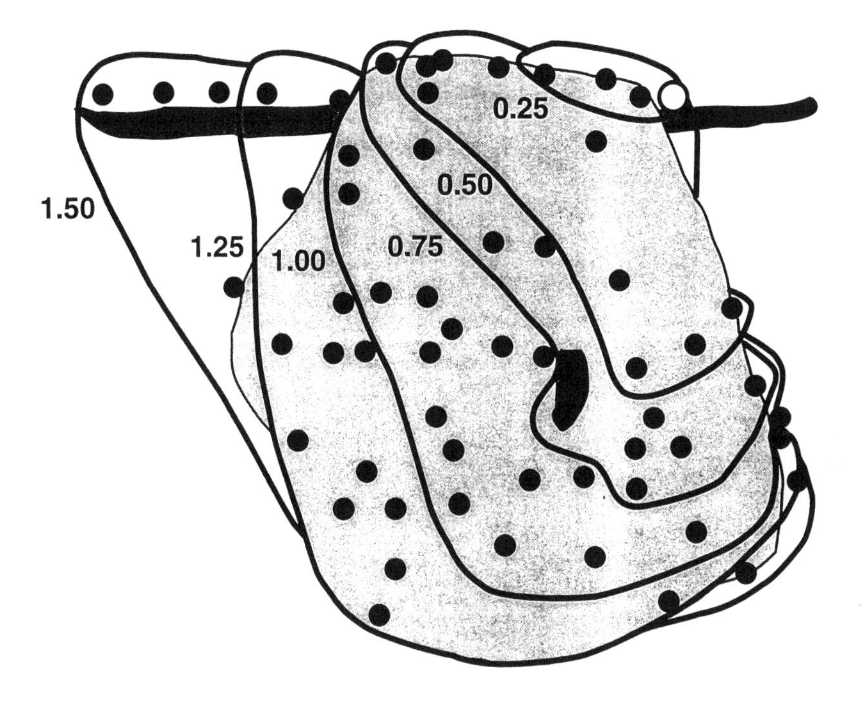

FIGURE 9.1 Spread of an alarm response over the surface of a nest of *Polybia occidentalis* attached to a horizontal twig. Each dot represents a worker. One worker (open circle, upper right) was alarmed by breathing on it lightly through a straw. Isochrone lines (0.25 s intervals) show the spread of alarm to other wasps (solid circles) on the nest. Dark gray spot near the center represents the nest entrance.

Queen Pheromones

The term queen pheromone generally refers to chemicals produced by the queen or queens of a colony to effect control of reproductive activities of workers and to stimulate certain worker-like behaviors, such as brood care, nest building and foraging. Despite evidence in a number of species of strong queen influence over worker activities and reproductive development, very little has been discovered regarding queen pheromones in the social Vespidae. In many species, queen control over workers' physiology and behavior is thought to be effected through dominance:subordinance interactions with queens (West-Eberhard 1969; Keller & Nonacs 1993).

Vespinae

The Oriental hornet, *V. orientalis*, is the only species of social wasp so far shown experimentally to possess a queen pheromone (Table 9.1). In this species, workers approach and lick the queen, forming a circle of workers around her (royal court). Similar behavior occurs in *V. crabro flavofasciata* (royal court) (Matsuura 1974) (Table 9.1). Ishay et al. (1965) elicited *V. orientalis* worker responses to an alcoholic extract of queens that were similar to those occurring in response to a real queen. Ikan et al. (1969) identified 5-hexadecanolide from the *V. orientalis* queen extract, and reported that synthetic 5-hexadecanolide elicited queen pheromone activity similar to that of the original extract, including the stimulation of the construction of queen cells at the end of the season. It is unfortunate that none of the methods used in these experiments or details of the results were provided in the reports (Ishay et al. 1965; Ikan et al. 1969). Additional work on the queen pheromone of *V. orientalis* has since focused on methods of synthesizing the enantiomers of 5-hexadecanolide (Coke and Richon 1976; Bacardit and Moreno-Manas 1983; Kikuwa and Tai 1984; Mori and Otsuka 1985).

Pheromonal control of worker physiology and activities by queens probably occurs in other Vespinae. In the Vespinae there is a general lack of evidence of the overt dominance behavior (i.e. Matsuura 1974) that occurs in *Polistes* species. There is evidence, however, of strong queen control of workers. Landolt et al. (1977) found that worker *Vespula atropilosa* separated from the queen exhibited greatly reduced foraging, nest construction, and brood care, and also showed increased ovarian development and agonistic behavior, compared to workers on a queen-right nest. Akre and Reed (1983) presented evidence that pheromone deposited on the comb by queens may have inhibited worker ovarian development in colonies of *Vespula pensylvanica* and *Vespula vulgaris*, although they did not demonstrate pheromonally-mediated behavioral responses to queens in either species.

Polistinae

Among the swarm-founding Polistinae there is circumstantial evidence of queen pheromones. Behavioral evidence suggests that workers of *Metapolybia* spp. recognize queens on the basis of odor alone and that the source of the odor is the head (West-Eberhard 1977). For some species of *Polybia*, *Protopolybia*, and *Agelaia* there is evidence suggesting that queens suppress ovary development and/or signal dominance (Naumann 1970; Simões 1977; Forsyth 1978). Such effects may occur via pheromones, but experimental support is wanting.

Trail Pheromones

Swarm-founding social wasps found new colonies by the emigration of a group of workers and queens to a new nest site, which may be as far as 100 m away from the natal nest (Jeanne 1980). Swarm emigration is characteristic of the vespine genus *Provespa* (Matsuura and Yamane 1984), probably all of the 22 New World genera in the polistine tribe Epiponini, and at least some species of *Polybioides* and *Ropalidia*. There is abundant evidence that at least some of these wasps use scent marks to lead emigrating swarm-mates to new nest-sites, although experimental demonstration of such pheromonal communication exists only for one species (Table 9.1). Surprisingly, there is no evidence that scent-marking is used by these wasps to recruit nestmates to rich food resources (Jeanne 1980).

Naumann (1970) was the first to describe swarming behavior in several epiponine genera and to suggest that swarming wasps produce and follow trails of scent marks deposited on vegetation along emigration routes to new nest sites. Some of the workers in the swarm rub or drag the venter of the gaster on visually prominent leaves and twigs along the route, suggesting that the behavior serves to deposit scent marks (Naumann 1975; Jeanne 1975). Jeanne (1981b) subsequently showed experimentally that swarms of *Polybia sericea* (Olivier) could be diverted along an artificial trail smeared with the secretion of the 5th metasomal sternal gland of females. Since in this species the secretion has a distinctive odor, detectable at spots marked by the wasps as well as at the gland opening, it is likely that this gland is the source of the trail pheromone. In *P. sericea*, at least, the pheromone is used both to assemble the swarm into a single cluster near the old nest and to mark the trail followed by nestmates to a new nest site. Naive nestmates follow the trail by searching for projecting twigs, the same cues scout wasps use in selecting sites to scent-mark. By hovering just downwind of such points or by landing on them and inspecting them, the wasps in the swarm determine whether the scent is present. If it is, they proceed farther from the cluster site in the same direction and repeat the search. If not, they search in a new direction from the cluster site.

Although not yet demonstrated experimentally, the use of chemical trails is suggested by reports of gaster-dragging behavior accompanying swarm emigration in several other genera, including *Angiopolybia*, *Agelaia*, *Charterginus*, *Leipomeles*, *Metapolybia*, *Synoeca*, *Parachartergus*, *Nectarinella*, *Polybioides*, and *Ropalidia* (Naumann 1975; Jeanne 1975; Francescato et al. 1993; Kojima 1994b; Jeanne and Keeping, unpublished observations). In an extensive survey of sternal glands in the polistine wasps, Jeanne et al. (1983) documented the existence of a well-developed sternal gland on the 5th sternite in many swarm-founding genera. What is curious, however, is that this gland is apparently absent in most of the genera for which the use of a trail pheromone is suspected (see above), raising the question of what the source of the presumed trail pheromone might be.

What were the evolutionary origins of emigration trails in the wasps? Swarm-founding behavior has probably evolved independently at least four times: once in the vespines, once each in *Ropalidia* and *Polybioides*, and at least once in the epiponine lineage. West-Eberhard (1982) posited a scenario by which swarming behavior could have evolved from independent-founding ancestors. The steps she hypothesizes begin with simple odors associated with the nest itself and progress through assembly pheromones and orientation cues to full-fledged scent-trail marking. The few details we have learned in the past ten years suggest that this scenario may be correct. There have been several reports of marked nestmates of *Polistes* and *Mischocyttarus* appearing together on newly-founded colonies within a day or two of leaving their destroyed or declining natal colony (West-Eberhard 1969; Jeanne 1972; Litte 1981). There are at least two possibilities for how the late arrivals find the newly founded nest. First, they could be using random visual search followed by contact chemoreception (nestmate recognition cues) to distinguish former nestmates from non-nestmates. Second, they could be orienting to a more specific, long-distance chemical attractant.

Evidence for the latter explanation comes from Litte's (1981) observation of females of *M. labiatus* wiping their abdomens over leaves and twigs along the route between their natal nest and the new nest they had established some meters away. Nestmates landed on and antennated these spots, suggesting that they were responding to scent marks. Queens performed the same behavior around the nest if nestmates were removed. O'Donnell (1992) observed similar behavior in *M. immarginatus* in Costa Rica, in which one or more of the cofounding females on a newly-initiated nest rubbed the gaster on leaves within a meter of the nest. Within two days at least four additional females joined the colony, after which gaster dragging was no longer seen. Although linear scent trails were evidently not laid, the behavior suggested that the marking may act as a volatile local attractant that helps potential cofoundresses orient to the nest vicinity from downwind.

The genus *Apoica*, sister group to the remaining epiponine genera, has recently been shown to possess some unusual chemical communication behavior in the context of swarm assembly following nest loss (Hunt et al. 1995). Females possess well-developed sternal glands on both the 5th and 6th metasomal sternite, each provided with a patch of stiff bristles. Rather than smearing the secretion onto leaves, however, perched females chemically 'call' nestmates by exposing the gland and allowing the chemical to volatilize into the air. It is not yet known whether this represents an intermediate step between the independent-founding condition and the substrate-borne scent marks of the other swarm founders, or whether this is a specialization, perhaps somehow adaptive in the context of the nocturnal habits of this genus.

To date, no trail pheromones have been chemically identified for any wasp species. This is due to a combination of four factors. First, bioassays for trail pheromones are difficult to develop, as compared to alarm pheromones. Second, colonies have to be induced to swarm. Then, even though it may be a day or two before the swarm emigrates, females are receptive to the pheromone only during the hour or so the swarm is actually moving. Third, swarm-founding species are limited to the tropics, distant from most chemical analysis instrumentation. This means expensive foreign travel for each round of bioassays. Finally, as Wilson and Bossert (1963) predicted and has been borne out for the ants (Hölldobler and Wilson 1990), it is likely that wasp trail pheromones will prove to be larger, more complex molecules than alarm pheromones. Despite the difficulties, however, we predict that trail pheromone chemistry will eventually prove an extremely useful character in helping to reconstruct the multiple evolutionary pathways leading to swarm-founding behavior in the wasps.

Other Pheromones

There is evidence of additional types of pheromonal communication within the Vespidae, with demonstrations of specific types of behavioral effects elicited by chemical contaminants of wasps, solvent extracts of wasps, or chemicals isolated from wasps (Table 9.1).

Worker *Vespula vulgaris* respond to a material deposited by the ambulatory traffic of conspecific workers that has been referred to as a footprint pheromone (Butler et al. 1969). Greater numbers of wasps arriving at the nest entrance landed on and crawled through glass tubes through which their nestmates had already passed as compared with untreated tubes, suggesting a chemical deposited by the passage of wasps. Such a chemical could aid workers in locating and recognizing the nest entrance. Recently, Overmyer and Jeanne (in press) have shown experimentally that foraging *V. germanica* workers took less time to approach and land on food dishes that had been visited by other wasps compared to dishes that had not been so "footprinted." Furthermore, when given a choice, significantly more wasps refused to enter clean dishes, compared to footprinted ones. These findings support the hypothesis that foragers leave a footprint pheromone and that marked sites are recognized by conspecific foragers. It is not yet known if the pheromone is colony-specific.

Francke et al. (1978) identified a set of compounds from pentane extracts of *Vespula vulgaris* workers and provided evidence that they function as repellents or aggression inhibitors among colony members. It is not known if these compounds function as recognition pheromones at the individual, colony, or species level.

Ishay (1973) provided evidence that an organic solvent-extractable material from pupal cases of *Dolichovespula media* (Retzius) stimulates the workers to warm the pupae through congregating and abdominal pumping. Z-9-pentacosene was later isolated from pupal cases of *Vespa crabro* and shown to release similar thermoregulatory behavior by workers (Veith and Koeniger 1978; Koeniger 1984).

In some species nest initiation may be triggered by a pheromone. Ishay and Perna (1979) reported that workers of *Vespa orientalis* and *Polistes foederatus* (Kohl) apparently initiate nest building in response to a chemical deposited on the substrate by resting wasps.

It is quite clear that we are far from understanding the full scope of pheromonal communication in social wasps. While it seems likely that pheromones play an integral part in all facets of the lives of these fascinating insects, particularly the organization and cohesion of social structure, we are merely speculating on all but a small number of possibilities. Only in the areas of alarm, swarming, and sexual behavior has much progress been made in unraveling the roles of pheromonal mediation of wasp behavior. Even here, we seem only to be scratching the surface, considering that so few pheromones have been identified (Table 9.1).

References Cited

Akre, R.D. and H.C. Reed. 1983. Evidence for a queen pheromone in *Vespula* (Hymenoptera: Vespidae). *Can. Entomol.* 115:371-377.

Aldiss, J.B.J.F. 1983. Chemical communication in British social wasps (Hymenoptera: Vespidae). Ph.D. Thesis, Univ. Southampton. 252 pp.

Bacardit, R. and M. Moreno-Manas. 1983. Synthesis of d lactonic pheromones of *Xylocopa hirsutissima* and *Vespa orientalis* and an allomone of some ants of genus *Camponotus*. *J. Chem. Ecol.* 9:703-714.

Batra, S.W.T. 1980. Sexual behavior and pheromones of the European hornet, *Vespa crabro germana* (Hymenoptera: Vespidae). *J. Kansas Entomol. Soc.* 53:461-469.

Beani, L. and C. Calloni. 1991a. Leg tegumental glands and male rubbing behavior at leks in *Polistes dominulus* (Hymenoptera: Vespidae). *J. Insect Behav.* 4:449-462.

Beani, L. and C. Calloni. 1991b. Male rubbing behavior and the hypothesis of pheromonal release in polistine wasps (Hymenoptera: Vespidae). *Ethol. Ecol. and Evol.* 1:51-54.

Butler, C.G., D.J.C. Fletcher and D. Watler. 1969. Nest entrance marking with pheromones by the honeybee *Apis mellifera* L. and by a wasp, *Vespula vulgaris* L. *Anim. Behav.* 17:142-147.

Castellón, E.G. 1981. Alarma e defesa no ninho de *Synoeca surinama* (L.) (Hymenoptera: Vespidae). *Acta Amazonica* 11:377-382.

Coke, J.L. and A.B. Richon. 1976. Synthesis of optically active gamma-N-hexadecalactone, the proposed pheromone from *Vespa orientalis*. *J. Org. Chem.* 41:3516-3517.

Downing, H.A. 1991. The function and evolution of exocrine glands. In: Ross, K. G. and R. W. Matthews (eds.). *The Social Biology of Wasps*. Cornell Univ. Press, Ithaca, NY pp. 540-569.

Forsyth, A.B. 1978. Studies on the Behavioral Ecology of Polygynous Social Wasps. Ph.D. dissertation, Harvard University, Cambridge, MA.

Francescato, E., S. Turillazzi and A. Dejean. 1993. Swarming behaviour in *Polybioides tabida* (Hymenoptera, Vespidae). *Actes. Coll. Insectes Soc.* 8:121-126

Freisling, J. 1943. Zur Psychologie der Feldwespe. *Z. Tierpsychol.* 5: 439-463.

Francke, W., G. Hindorf and W. Reith. 1978. Methyl-1,6-dioxaspiro[4,5] decanes as odors of *Paravespula vulgaris* (L.). *Angew. Chem. Int. Ed. Engl.* 17:862.

Greene, A., R.D. Akre, and P.J. Landolt. 1976. The aerial yellowjacket, *Dolichovespula arenaria* (Fab.): Nesting biology, reproductive production, and behavior (Hymenoptera: Vespidae). *Melanderia* 26:1-34.

Heath, R.R. and P.J. Landolt. 1988. The isolation, identification, and synthesis of the alarm pheromone of *Vespula squamosa* (Drury) (Hymenoptera: Vespidae) and associated behavior. *Experientia* 44:82-83.

Hölldobler, B. and Wilson, E.O. 1990. *The Ants*. Harvard University Press, Cambridge, MA.

Hunt, J.H., R.L. Jeanne and M.G. Keeping. 1995. Observations on *Apoica pallens*, a nocturnal Neotropical social wasp (Hymenoptera: Vespidae, Polistinae, Epiponini). *Insectes Soc.* 42:223-236.

Ikan, R., R. Gottlieb, E.D. Bergmann and J. Ishay. 1969. The pheromone of the queen of the Oriental hornet, *Vespa orientalis*. *J. Insect Physiol.* 15:1709-1712.

Ishay, J. 1973. Thermoregulation by social wasps: behavior and pheromones. *Trans. N.Y. Acad. Sciences.* 35:447-462.

Ishay, J., R. Ikan and E.D. Bergmann. 1965. The presence of pheromones in the Oriental hornet, *Vespa orientalis* F. *J. Insect Physiol.* 11:1307-1309.

Ishay, J., H. Bytinski-Salz, and A. Shulov. 1967. Contributions to the bionomics of the Oriental hornet *Vespa orientalis*. *Israel J. Entomol.* 2:45-106.

Ishay, J.S. and B. Perna. 1979. Building pheromones of *Vespa orientalis* and *Polistes foederatus*. *J. Chem. Ecol.* 5:259-272.

Jeanne, R.L. 1970. Descriptions of the nests of *Pseudochartergus fuscatus* and *Stelopolybia testacea,* with a note on a parasite of *S. testacea* (Hymenoptera, Vespidae). *Psyche* 77:54-69.

Jeanne, R.L. 1972. Social biology of the Neotropical wasp *Mischocyttarus drewseni*. *Bull. Mus. Comp. Zool.* Harvard Univ. 144:63-150.

Jeanne, R.L. 1975. Behavior during swarm movement in *Stelopolybia areata* (Hymenoptera: Vespidae). *Psyche* 82:259-264.

Jeanne, R.L. 1980. Evolution of social behavior in the Vespidae. *Ann. Rev. Entomol.* 25:371-396.

Jeanne, R.L. 1981a. Alarm recruitment, attack behavior, and the role of the alarm pheromone in *Polybia occidentalis* (Hymenoptera: Vespidae). *Behav. Ecol. Sociobiol.* 9:143-148.

Jeanne, R.L. 1981b. Chemical communication during swarm emigration in the social wasp *Polybia sericea* (Olivier). *Anim. Behav.* 29:102-113.

Jeanne, R.L. 1982. Evidence for an alarm substance in *Polistes canadensis*. *Experientia* 38:329-330.

Jeanne, R.L. 1993. The evolution of exocrine gland function in wasps. In: S. Turillazzi & M. J. West-Eberhard, eds., *Natural History and Evolution of Paper Wasps*. Proc. Confer., Florence, Italy.

Jeanne, R.L. and M.G., Keeping. 1995. Venom spraying in *Parachartergus colobopterus*: a novel defensive behavior in a social wasp (Hymenoptera: Vespidae). *J. Insect Behav.* 8:433-442.

Jeanne, R.L., H.A. Downing and D.C. Post. 1983. Morphology and function of sternal glands in polistine wasps (Hymenoptera: Vespidae). *Zoomorph.* 103:149-162.

Keeping, M.G. 1995. Absence of chemical alarm in a primitively eusocial wasp (*Belonogaster petiolata*, Hymenoptera: Vespidae). *Insectes Soc.* 42:317-320.

Keeping, M.G., D. Lipschitz and R.M. Crewe. 1986. Chemical mate recognition and release of male sexual behavior in a polybiine wasp, *Belonogaster petiolata* (DeGeer) (Hymenoptera: Vespidae). *J. Chem. Ecol.* 12:773-779.

Keller, L. and P. Nonacs. 1993. The role of queen pheromones in social insects: queen control or queen signal? *Anim. Behav.* 45:787-794.

Kennedy, J.S. 1977. Behaviorally discriminating assays of attractants and repellents. In H. H. Shorey and J. J. McKelvey, Jr. (Eds.). *Chemical Control of Insect Behavior, Theory and Application.* Wiley-Interscience, N.Y., pp. 215-229.

Kikuwa, T. and A. Tai. 1984. Synthesis of optically pure (R) and (S)-5-hexadecanolide, a proposed pheromone component of the oriental hornet. *Chem. Lett.* 1935-1936.

Koeniger, N. 1984. Brood care and recognition of pupae in the honeybee (*Apis mellifera*) and the hornet (*Vespa crabro*). In: *Insect Communication.* Royal Entomological Society of London. pp. 267-282.

Kojima, J. 1994a. Evidence for an alarm pheromone in *Ropalidia romandi* (Le Guillou) (Hymenoptera: Vespidae). *J. Aust. Entomol. Soc.* 33:45-47.

Kojima, J. 1994b. Behavior during artificially induced swarm emigration in an Old World polistine wasp, *Ropalidia romandi* (Hymenoptera: Vespidae). *J. Ethol.* 12:1-8.

Landolt, P.J. and R.D. Akre. 1979. Occurrence and location of exocrine glands in some social Vespidae (Hymenoptera). *Ann. Entomol. Soc. Amer.* 72:141-148.

Landolt, P. J. and R. R. Heath. 1987. Alarm pheromone behavior of *Vespula squamosa* (Hymenoptera: Vespidae). *Florida Entomol.* 70:222-225.

Landolt, P. J., R.D. Akre and A. Greene. 1977. Effects of colony division on *Vespula atropilosa* (Sladen) (Hymenoptera: Vespidae). *J. Kansas Entomol. Soc.* 50:135-147.

Landolt, P.J., R.R. Heath, H.C. Reed, and K. Manning. 1995. Pheromonal mediation of alarm in the eastern yellowjacket (Hymenoptera: Vespidae). *Florida Entomol.* 78:101-108.

Litte, M. 1979. *Mischocyttarus flavitarsis* in Arizona: Social and nesting biology of a polistine wasp. *Z. Tierpsychol.* 50:282-312.

Litte, M. 1981. Social biology of the polistine wasp *Mischocyttarus labiatus*: Survival in a Columbian rain forest. Smithsonian *Contr. Zool.* No. 327, 27 pp.

MacDonald, J.F., R.D. Akre and W.B. Hill. 1974. Comparative biology and behavior of *Vespula atropilosa* and *V. pensylvanica* (Hymenoptera: Vespidae). *Melanderia* 18:1-66.

Maschwitz, U. 1964. Alarm substances and alarm behavior in social Hymenoptera. *Nature* 204:324-327.

Maschwitz, U. 1984. Alarm pheromone in the long-cheeked wasp, *Dolichovespula saxonica* (Hymenoptera: Vespidae). *Dtsch. Entomol.* Z. 31:33-34.

Maschwitz, U. and H. Hanel. 1988. Biology of the southeast Asian nocturnal wasp, *Provespa anomala* (Hymenoptera: Vespidae). *Entomol. Gener.* 14:47-52.

Matsuura, M. 1974. Intracolonial polyethism in *Vespa* Part I. Behavior and its change of the foundress of *Vespa crabro flavofasciata* in early nesting stage in relation to worker emergence. *Kontyû* 42:333-350.

Matsuura, M. and S. Yamane. 1984. *Biology of the Vespine Wasps.* Springer-Verlag, Berlin.

Mori, K. and T. Otsuka. 1985. Synthesis of the enantiomers of 5-hexadecanolide, the pheromone of the queen of the oriental hornet, *Vespa orientalis*, employing enzymic resolution of (+)-2-aminotridecanoic acid as the key step. *Tetrahedron Lett.* 41:547-551.

Naumann, M.G. 1970. The Nesting Behavior of *Protopolybia pumila* in Panama (Hymenoptera: Vespidae). Ph.D. dissertation, Univ. Kansas, Lawrence, Kansas.

Naumann, M.G. 1975. Swarming behavior: Evidence for communication in social wasps. *Science* 189:642-644.

O'Donnell, S. 1992. Off-nest gastral rubbing observed in *Mischocyttarus immarginatus* (Hymenoptera: Vespidae) in Costa Rica. *Sphecos* 23:5.

Ono, M., M. Sasaki and I. Okada. 1985. Mating behavior of the giant hornet, *Vespa mandarinia* Smith and its pheromonal regulation. *Proc. XXX Int. Apic. Congr.*, Nagoya, Japan. 255-259.

Ono, M. and M. Sasaki. 1987. Sex pheromones and their cross-activities in six Japanese sympatric species of the genus *Vespa*. *Insectes Soc.* 34:252-260.

Overal, W.L., D. Simoes, and N. Gobbi. 1981. Colony defense and sting autotomy in *Polybia rejecta* (F.) (Hymenoptera: Vespidae). *Rev. Bras. Entomol.* 25:41-47.

Overmyer, S. and R.L. Jeanne. In press. Behavioral evidence for a food site marking substance in the German yellowjacket, *Vespula germanica* (Hymenoptera: Vespidae). *J. Insect Behav.*

Post, D.C. and R.L. Jeanne. 1983a. Male reproductive behavior of the social wasp *Polistes fuscatus* (Hymenoptera: Vespidae). *Z. Tierpsychol.* 62:157-171.

Post, D.C. and R.L. Jeanne. 1983b. Source of sex pheromone in the social wasp *Polistes fuscatus* (Hymenoptera: Vespidae). *J. Chem. Ecol.* 9:259-266.

Post, D.C. and R.L. Jeanne. 1984. Venom as an interspecific sex pheromone, and species recognition by a cuticular pheromone in paper wasps (*Polistes*, Hymenoptera: Vespidae). *Physiol. Entomol.* 9:65-75.

Post, D.C. and R.L. Jeanne. 1985. Sex pheromone in *Polistes fuscatus* (Hymenoptera: Vespidae). Effect of age, caste, and mating. *Insectes Soc.* 32:70-77.

Post, D.C., H.A. Downing and R. L. Jeanne. 1984. Alarm response to venom by social wasps *Polistes exclamans* and *P. fuscatus* (Hymenoptera: Vespidae). *J. Chem. Ecol.* 10:1425-1433.

Reed, H.C. and P.J. Landolt. 1990a. Queens of the southern yellowjacket, *Vespula squamosa*, produce sex attractant (Hymenoptera: Vespidae). *Florida Entomol.* 73:687-689.

Reed, H.C. and P.J. Landolt. 1990b. Sex attraction in paper wasp, *Polistes exclamans* Viereck (Hymenoptera: Vespidae), in a wind tunnel. *J. Chem. Ecol.* 16:1277-1287.

Reed, H.C. and P.J. Landolt. 1991. Swarming of paper wasp (Hymenoptera: Vespidae) sexuals at towers in Florida. *Ann. Entomol Soc. Amer.* 84:628-635.

Ross, K.G. 1983. Laboratory studies of the mating biology of the eastern yellowjacket, *Vespula maculifrons* (Hymenoptera: Vespidae). *J. Kansas Entomol. Soc.* 56:523-527.

Ross, K.G. and R.W. Matthews. 1991. *The Social Biology of Wasps.* Cornell University Press. 678 pp.

Sandeman, R.G. 1938. The swarming of the males of *Vespula sylvestris* (Scop.) around a queen. *Proc. R. Ent. Soc. London* (A)13:87-88.

Saslavasky, H., J. Ishay and R. Ikan. 1973. Alarm substances as toxicants of the Oriental hornet, *Vespa orientalis*. *Life Sci.* 12:135-144.

Schremmer, F. 1972. Beobachtungen zur Biologie von *Apoica pallida* (Olivier, 1791), einer neotropischen sozialen Faltenwespe (Hymenoptera, Vespidae). *Insectes Soc.* 19:343-357.

Seeley, T.D. 1989. Social foraging in honey bees: how nectar foragers assess their colony's nutritional status. *Behav. Ecol. Sociobiol.* 24:181-199.

Shorey, H.H. 1977. Interactions of insects with their chemical environments. In: Shorey, H. H. and J. J. McKelvey, Jr. (eds.). *Chemical Control of Insect Behavior. Theory and Application.* Wiley-Interscience, New York, pp. 1-5.

Simões, D. 1977. Etologia e Diferenciação de Casta em Algumas Vespas Sociais (Hymenoptera, Vespidae). Ph.D. dissertation, Universidade de São Paulo, Ribeirão Prêto, Brazil.

Singer, T. L., K. E. Espelie, and G. J. Gamboa. 1997. Nest and nestmate discrimination in independent-founding paper wasps. In Vander Meer, R.K., M. Breed, M. Winston, and K.E. Espelie. (Eds.). *Pheromone communication in social insects: Ants, wasps, bees and termites.* Westview Press, Boulder, CO.

Strassmann, J.E., C. R. Hughes and D.C. Queller. 1990. Colony defense in the social wasp, *Parachartergus colobopterus. Biotropica* 22:324-327.

Thomas, C. R. 1960. The European wasp (*Vespula germanica* Fab.) in New Zealand. *New Zealand Dept. Sci. and Ind. Research Information Series.* No. 27:1-70.

Tierney, A.J. 1986. The evolution of learned and innate behavior: contributions from genetics and neurobiology to a theory of behavioral evolution. *Animal Learning & Behavior.* 14:339-348.

Traniello, J. F. A. and Robson, S. K. 1995. Trail and territorial communication in social insects. In: R. Cardé and W. J. Bell (eds.). *The Chemical Ecology of Social Insects 2.* Chapman & Hall. pp. 241-286.

Veith, H.J. and N. Koeniger. 1978. Identifizierung von cis-9-Pentacosen als Auslöser für das Wärmen der Brut bei der Hornisse. *Naturwissenschaften.* 65:263.

Veith, H. J., N. Koeniger and U. Maschwitz. 1984. 2-Methyl-3-butene-2-ol, a major component of the alarm pheromone of the hornet *Vespa crabro. Naturwissenschaften* 71:328-329.

Wenzel, J.W. 1987. Male reproductive behavior and mandibular glands in *Polistes major* (Hymenoptera: Vespidae). *Insectes Soc.* 34:44-57.

West-Eberhard, M. J. 1969. The social biology of polistine wasps. *Misc. Publ. Mus. Zool., Univ. of Michigan.* No. 140, 101 p.

West-Eberhard, M. J. 1977. The establishment of reproductive dominance in social wasp colonies. *Proc. 8th Inter. Cong. Int. Union Study Soc. Insects*, pp. 223-227.

West-Eberhard, M. J. 1982. The nature and evolution of swarming in tropical social wasps (Vespidae, Polistinae, Polybiini). In: P. Jaisson (ed.), *Social Insects in the Tropics, vol. 1.* Université de Paris-Nord, Paris. pp. 97-128.

Wilson, E.O. 1971. *The Insect Societies.* Harvard University Press, Cambridge, MA.

Wilson, E.O. and Bossert, W.H. 1963. Chemical communication among animals. *Recent Progress in Hormone Research* 19:673-716.

10

Exocrine Glands and Their Products in Non-*Apis* Bees: Chemical, Functional and Evolutionary Perspectives

Abraham Hefetz

Introduction

Chemical communication has reached an advanced and sophisticated form in social insects. These are endowed with exocrine glands which produce an array of compounds, such that almost every facet of their social behavior is mediated by pheromones. The chemistry of social pheromones in Hymenoptera, as well as their glandular origins, suggests that these systems have evolved within multiple phyletic lines. As a result, neither the chemistry nor the glandular origin of a particular pheromone in the various social insects is predictable. It is therefore interesting to select particular glands, common to all Hymenoptera, and to follow the possible adaptation of their secretions for pheromonal use, with the concomitant changes in the chemicals they produce.

Bees are an excellent group for studying such evolutionary trends. They show a broad spectrum of social behavior, sometimes within a single family as in the Halictidae. Bees also have representatives with a wide diversity of ecological requirements and nesting habits. It is thus possible to follow the increase in the use of pheromones as social behavior becomes more complex, and discern the ecological constraints that may have dictated the use of an exocrine gland for the production of a particular pheromone.

The glandular system of bees has been extensively described by various authors (Cruz Landim 1967; Michener 1974; Hesselhause 1922).

Likewise, the chemistry of bees exocrine products has been extensively reviewed earlier (Blum 1981; Duffield et al. 1984; Wheeler and Duffield 1988). It is not the purpose of this chapter to list all the exocrine products identified in bees, nor to describe all the glandular systems. Instead, I will focus on several glandular systems in solitary and non-*Apis* social bees (excluding stingless bees) for which we have some knowledge on the function of their exocrine products and can draw a reasonable link between chemistry and function. Since Dufour's gland is the best studied with respect to its glandular chemistry and functional role, I will attempt to draw its possible functional evolution from structural to communicative. The chemical diversity of the gland will also serve as an example for relating exocrinology to phylogeny.

Mandibular Glands

There is very little information available concerning the function of mandibular gland secretion in solitary bees. In most of the bee species, the secretion is immediately emitted when they are netted, suggesting that the primary function of the secretion is defensive. This defensive role, in addition to deterring potential predators, may also act as a nest disinfectants (Cane et al. 1983). In addition, there are some reports that the mandibular glands may also possess a communicative function. Males of *Centris adani* mark grass in their territories with the mandibular gland constituents in order to attract females (Vinson et al. 1984). In the carpenter bee *Xylocopa sulcatipes* the mandibular gland secretion deposited on leaves within the male's territory deters other males but attracts virgin females to the site (Hefetz 1983). Female mandibular gland secretion in this species act as a sex pheromone, attracting males and inducing them to copulate (Velthuis and Gerling 1989). The terpenes constituents of the mandibular glands in the aggregative species *Colletes thoracicus* are attractive to conspecific bees, implying that the secretion comprises an aggregation pheromone (Hefetz et al. 1979).

In the primitively eusocial bee *Bombus terrestris* there are indications that the queen mandibular gland secretion contains a putative queen pheromone. It is responsible for repressing ovarian development in workers, and its decline is presumed to trigger the onset of workers reproduction. Queens from which the glands were excised were less effective in dominating reproduction (Honk et al. 1980). The glandular exudates were also able to repress the release of JH, the gonadotropin hormone in this species, in queenless workers (Röseler et al. 1981). Recent findings cast doubts on the pheromone decline hypothesis as a trigger for the onset of worker reproduction (Bloch et al. 1996). Queens after the competition phase (a phase characterized by reproduction by workers) were as effective as queens before the competition phase (that successfully repress reproduction by workers -- Duchateau and Velthuis 1988) in repressing ovarian development and the establishment of dominance hierarchy in a group of callow workers. These workers under queenless conditions establish dominance hierarchy and at least the dominant worker develops ovaries and lays eggs within 7 days of queenlessness. So far, the chemical nature of the pheromonal blend is not known. A recent study on the chemical composition of queen *B. terrestris* have shown that the mandibular glands contain over 150 compounds including hydrocarbons, alcohols, aliphatic and terpene esters and acids (Hefetz et al. 1996). It is worth noting that a homologous series of 3-hydroxy and 3-butyroxy acids were prevalent in the secretion. The same hydroxy acids were also found in queens' mandibular glands of *B. hypnorum* (Röseler, personal communication). A comparison with the caste specific mandibular glands secretion of the honeybee *Apis mellifera* may be instructive. Queens are typified by a blend of five components: the traditional 9-oxo-2-decenoic acid, two enantiomers of 9-hydroxy-2- decenoic acid, (R,E)-(-)-9HDA and (S,E)-(+)-9HDA, methyl p -hydroxybenzoate (HOB), and 4-hydroxy-3-methoxyphenyl-ethanol (HVA) (Slessor et al. 1988). Workers, on the other hand, possess two hydroxy acids, 10-hydroxy-(E)-2-decenoic acid (10-HDA), 10-hydroxydecanoic acid (10-HDAA), simple fatty acids (e. g. C_{16} and C_{18}) (4; 30), and 2-heptanone that functions as an alarm pheromone (Shearer, 1965). Considering the fact that hydroxy acids are components of the honeybees queen mandibular pheromone, it is tempting to postulate that in bumble bees 3-hydroxy acids fulfill a similar function. This, however, must await verification by explicit tests.

Labial Glands

The labial glands constitute two pairs of glands, one in the head and one in the thorax, that are intermittently developed in the bees (Cruze Landim 1967). The labial gland of the head is the only one to have been chemically studied so far. In the mason bee, *Chalicodoma siculum* , the secretion is composed of hydrocarbons that are admixed with saliva while preparing the mortar from which the brood cells are constructed, rendering it waterproof. The hydrocarbons are well embedded within the cell walls and can no longer be extracted after the mortar has dried. After constructing several cells, the bees cover the cluster with a heavy coating which provides additional protection to the brood cells against rain falling on the often exposed nests (Kronenberg and Hefetz 1984).

In male bumble bees the glands of the head are especially developed and fill most of the head cavity. The glandular secretion is used for marking posts along the male flight path to which queens and other males are attracted (Kullenberg et al. 1970; 1973). In *B. terrestris* at least, flight paths and marking spots of different males often overlap. Queens that are attracted to the pheromone land on the marked spot, where they are sought by the males and where copulation occurs. When given the choice, queens prefer spots that have been marked by several males, thus reacting to a super signal. This suggests cooperation between males to enhance the probabilities of attracting queens to a particular site (Maccagnani et al. 1994; Duchateau, personal communication). Recent studies using glass flowers to induce marking by males revealed that the source of this marking is indeed the labial glands (Bergman and Bergström, 1996).

The chemical investigation of male bumble bee labial gland secretion constitutes a pioneering work in chemical ecology of bees (Bergström et al. 1981 and references therein). It has revealed a plethora of compounds including aliphatic primary alcohols, mono- and diterpene alcohol and their corresponding acetates, ethyl ester and saturated and unsaturated hydrocarbons. When the complete bouquet is considered, the secretion is species-specific, but the various subgenera can be characterized by the common occurrence of the major constituents. In the subgenus *Bombus* (sensus stricto) fatty acid esters (ethyl esters or acetates) constitute the major products, whereas the subgenus *Pyrobombus* possesses primarily mono- sesqui- and diterpenes and the subgenus *Alpinobombus* is characterized by possessing mostly straight chain alcohols. Nevertheless, considering the limited species that have been analyzed to date, there are species which deviate from the general trends. For example *B. terrestris* is outstanding among the members of the subgenus *Bombus* in possessing the sesquiterpene alcohol 2,3-dihydro-6-*trans*-farnesol and the diterpene alcohol geranyl-citronellol. This demonstrated diversity within a phylogenetic group is not surprising. The secretion is involved in flight path marking and attracts virgin queens. Since many of the species are sympatric, at least in part of their distribution, the selection for diversification should be strong. The existence of chemotypes in *Bombus lucorum* labial gland secretion provides another example demonstrating the selection for diversity. Populations of this species consist of two forms, "blond" and "dark", based on hair coloration, that are also chemically distinct. The secretion of the "blond" males is dominated by ethyl tetradec-*cis*-9-enoate, whereas "dark" males possess ethyl dodecanoate as the major component (Bergström et al. 1973). Since the secretion is involved in the attraction of females to the marking site, these differences may reflect populations that are reproductively isolated. Unfortunately, there are no data pertaining to covariance between females preference and color morph to substantiate this hypothesis.

The composition of queen labial gland secretions was recently analyzed in *B. terrestris*(Hefetz et al. 1996). They are highly developed and similar in appearance to males, but their chemistry is gender specific. The comparison should be taken with some reservation since the males analyzed were from the Swedish population of *B. terrestris* whereas the queens analyzed were from the Israeli population. The secretion of males is composed mostly of 2,3-dihydro-6-*trans*-farnesol with lesser amounts of geranylcitronellol and ethyl dodecanoate and traces of 2,3 dihydrofarnesyl acetate, ethyl decanoate and ethyl tetradecanoate. Queens have in common the three ethyl esters, but lack completely the sesquiterpene or diterpene alcohols or dihydrofarnesyl acetate. Instead, the major group of compounds in the queen's labial gland consist of high molecular weight, wax type esters derived from dodecanol and a homologous series of saturated and unsaturated acids ranging from C_8 to C_{18}. Interestingly, geranylcitronellol that was found in large amounts in the male's labial glands, occurs in trace amount in the queen's Dufour's gland.

There are some similarities between the labial gland secretions of queens of *B. hypnorum* (Genin et al. 1984) and *B. terrestris* (Hefetz et al. 1996), but the bouquet is species specific. Many of the ethyl and methyl esters and the series of geranylcitronellate esters are in common. *Bombus hypnorum* possesses, in addition, esters of geranyl-geraniol and straight chain fatty acids as well as terpene alcohols, whereas *B. terrestris* possess aliphatic alcohols, 2-alkanones and aldehydes that are absent in *B. hypnorum.*

Dufour's Gland

The best studied gland in the non-*Apis* bees is the Dufour's gland. This gland is excellent for demonstrating evolutionary trends in function and the chemical composition of exocrine glands. It is highly developed in bees, and our knowledge of its chemistry enables a comprehensive treatment with a comparative perspective.

It seems that the primary function of the Dufour's gland in bees is to line the brood cells with a hydrophobic lining. All the ground nesting bees investigated so far utilize Dufour's exudates, modified or unmodified, to line their brood cells. Comparative chemical analyses in the various ground nesting species revealed that while all use Dufour's gland, the chemical means by which this hydrophobicity is attained is diverse. Out of the many cases that have been investigated so far (reviewed by Hefetz 1987), the following three examples best demonstrate this diversity.

The first example is *Colletes thoracicus* (Colletidae) in which, characteristic of this genus, the hydrophobic lining is in the form of a membrane that envelops the brood provisions (Hefetz et al. 1979). This membrane is extremely resistant to organic solvents of various polarities, and cell lining that were dug out from 2-3 years old nests remained almost intact. The Dufour's gland secretion of this species is composed of 8 compounds, dominated by macrocyclic lactones including octadecanolide, eicosanolide and docosanolide. A solvent wash of this membrane contained all Dufour's gland constituents, but the proportion of the lactones was much reduced, suggesting that they had been chemically transformed. Further chemical analyses by direct probe mass spectrometry and using GC/MS for the analysis of the products of HCl methanolysis demonstrated that the membrane is a polymer composed of ω–hydroxy fatty acids with chain lengths corresponding to Dufour's gland lactones. Thus the lactones that were present in the gland were transformed to a laminester that provided the hydrophobic lining. However not all the lactones were transformed, leaving a coat of free lactones that gave the cell its characteristic musky fragrance.

A similar polymer was also described from *Hyleus bisinuatus*, a species that belongs to a different subfamily (Hyliinae) within the Colletidae, suggesting that this phenomenon is general to the family. Macrocyclic lactones also characterize the Dufour's gland secretion of many Halictidae (Hefetz et al. 1978; Hefetz and Graur 1988; Duffield et al. 1981). Although, Dufour's gland lactones were found in the cell lining and at the nest entrance of *Evylaeus (Lasioglossum) malachurum* (Hefetz et al. 1986), it is unknown whether they are partially transformed to a polymer during this process. It is noteworthy that Dufour's gland secretion of many halictine bees is also very rich in long chained straight alkanes that can provide a waterproof lining. Ontogenetic studies with *L. malachurum* seem to corroborate the hypothesis that the lactones are involved in nesting behavior. Unmated queens possess predominantly isopentenyl esters in the Dufour's gland secretion, whereas in actively nesting queens or workers the major constituents are the lactones (Ayasse et al. 1993). Since nest foundation in the Halictidae is by haplometrosis, the queen needs to provide the cell lining from her own resources until the first worker caste emerges.

The hydrophobic lining of *Anthophora abrupta* (Anthophoridae) is produced by a drastically different chemical mechanism which is intriguing (Norden et al. 1980). When the brood cells of this ground nesting bee are opened an acrid odor characteristic of short chain fatty acids is emitted from the rather heavy white cell coating. Unlike in *Colletes,* this lining is degradable and is consumed by the larva after it had finished the supplied bee bread. A comparative chemical analysis of Dufour's gland exudates and cell lining revealed that the gland contains triglycerides composed of two short fatty acyl groups, including acetic, butyric and hexanoic moieties, and one residue of acyl palmitate. Free palmitic acid was also detected in the glandular exudate. Analysis of the cell lining, on the other hand, disclosed the presence of dipalmitin and traces of the free short chained fatty acid. Thus, subsequent to the application onto the walls of the brood cell, the secretion is trans-esterified to form dipalmitin while releasing large amounts of free short chained fatty acids.

The cell lining of *Proxylocopa olivieri* constitutes a third example for the diverse evolutionary solutions to providing a waterproof coating for the brood cells. As in the two former species, the source of the cell lining in *P. olivieri* is the Dufour's gland, but its chemistry is different. It is composed of a series of hydrocarbons ranging from C_{23} to C_{29}, that are deposited without any changes into the brood cell walls, creating a glossy hydrophobic cover for the cell walls (Kronenberg and Hefetz 1984).

Although we do not as yet have sufficient information along a phylogenetic continuum, the diverse chemical solutions for creating a hydrophobic cell lining, suggest an independent evolution of cell linings in bees. Despite this chemical divergence, the common use of Dufour's gland for this purpose in ground dwelling bees is adaptive since a large part of nest digging and smoothing of the cells is performed with the aid of the pygidium, very close to the aperture of this gland (Batra 1972; 1984). However, despite the constraints imposed by the physical characteristics of the lining, i.e., high hydrophobicity and high molecular weight, the Dufour's gland evolved different biosynthetic pathways for obtaining its chemical precursors. Corroborating this hypothesis of independent evolution is the use of different glandular sources for cell lining when the nesting ecology of the bee species is different. In the mason bee, *C. Siculum,* the labial glands have adaptively evolved as the secretory source (Kronenberg and Hefetz 1984), whereas in the wood nesting bee, *Xylocopa pubescens,* the intersegmental glands became the source of the hydrophobic secretion (Gerling et al. 1989). This evolutionary scenario is in accordance with the prevailing hypothesis that in the Hymenoptera existing exocrine glands were used differentially for different purposes as the need arose.

Functional Evolution of Dufour's Gland Secretion

The complexity of the Dufour's gland secretion in many of the bees studied suggests that it has evolved a communicative function in addition to cell lining. In all of the bees investigated to date the secretion comprises a multicomponent blend, increasing in complexity in species exhibiting some social behavior. For example, it possesses over 24 components in *Evylaeus (Lasioglossum) malachurum* (Hefetz et al. 1986), and over 40 compounds in the various bumble bee species (Tengö et al. 1991). Even in the strictly solitary species such as *Eucera palestinae*, there are 22 components in the secretion (Shimron et al. 1985). If the secretion was intended solely for supplying a hydrophobic lining, one or a few compounds should be sufficient. One explanation for having a multicomponent system is that it helps to lower the melting points of the compounds so that they can remain liquid while stored in the exocrine glands. This may explain, at least in part, why in *Anthophora* the dipalmitate applied to the cell lining is stored as a triglyceride with two short fatty acyl moieties, and also explains the storage of the lactone precursors of the polymeric lining in *Colletes*. Another possible explanation is that while some of the compounds may function for cell lining, additional compounds with a lower molecular weight may serve as a solvent. Neither of these explanations fully justify the chemical complexity exhibited by the gland. On the other hand, both the chemical diversity and the complexity of the secretion can be explained if we consider that it has an additional, communicative, role.

One of the postulated communicative functions of Dufour's gland secretion is to chemically mark the precise location of the nest. Although bees are highly visual animals, landmarks may often change, especially for the ground dwelling bees, thus hampering the bees' ability to find the nest entrance. This is especially important in bees that nest in dense aggregations. The use of chemical signals to locate the nest can be readily observed in the field after changing the visual cues in the surrounding of the nest, or by covering the nest entrance. The bees generally alight in the vicinity of the nest, and search the entrance while touching the ground with their antennae (Hefetz unpublished observations).

Bees are also often observed marking the nest entrance. Upon exiting the nest a bee frequently pauses for a few seconds at the nest entrance before exiting and can sometimes be observed rubbing its abdomen around the walls of the entrance hole. In *L. malachurum* the marking is performed by all nest members resulting in a heavy coating. Chemical analysis of this coating demonstrated the presence of Dufour's gland components including lactones and hydrocarbons (Hefetz 1990). Similar analysis done with sand surrounding the nest entrance of *Colletes halophilus* corroborated the identity of Dufour's gland secretion as the source of the marking (Hefetz, A., O'Toole C., and Birch, M. unpublished data).

Behavioral experiments to substantiate this hypothesis were done in two species, *Eucera palestinae* (Shimron et al. 1985), and *E. malachurum*. (Hefetz 1990; Ayasse, 1990). In both cases the experimental paradigm was to confuse the bees by changing the visual cues, followed by applying various secretions to the nest entrance. The tests with *E. palestinae* have shown that the bees did not have any problem in locating their nest after a short antennation in the vicinity of the entrance. On the other hand, application of a pooled sample of Dufour's gland secretion, or even of a single alien glandular secretion, resulted in prolonged hesitation at the nest entrance. The bees did not hesitate before entering the nest if the material applied was synthetic citral, eliminating the possibility that the bees were merely reacting to a novel scent at their nest entrance. Finally, by careful dissection, the Dufour's gland of several bees was removed without killing them, after which the bees were replaced in their original nest. These bees reacted by hesitation, as before, to application of Dufour's secretion of an alien bee, but were indifferent to the application of their own glandular secretion (Shimron et al. 1985).

These experiments demonstrated that Dufour's gland secretion is used for nest entrance marking and that the bees can at least discriminate self from alien odor. Do bees also recognize individual odors? This seems to be the case in the large carpenter bee *Xylocopa pubscens* (Hefetz 1992), although it is not yet clear whether Dufour's gland secretion is the source of the nest marking secretion in this species. These bees nest in wood, and can be induced to nest in canes as well. When the entrances between canes were exchanged the bees attempted to enter the wrong nest, sometimes for several minutes. Some of the bees, however, finally found their original nest and entered it. After two or three times the bees became familiar with their new nest entrance odor and no longer hesitated before entering it. If at this stage this alien nest entrance was replaced by a second, alien nest entrance, the bees hesitated again and searched actively for the original nest entrance. This suggests that the bees can recognize and learn individual odor blends.

The fine discrimination between odors at the nest entrances of neighboring bees, especially in aggregative species (as in the case of *L. zephyrum* and *L. malachurum*), requires the secretions to consist of complex blends. It is further postulated that in the social species, in which several individuals occupy a single nest, the discriminatory ability should be even finer. It is therefore predicted that the pheromonal blend in the social species is more complex than in the solitary bees. Based on the empirical variance in the relative abundance of the secretory constituents of Dufour's gland in various halictine species, an assessment was made of the number of compounds needed for discrimination between sympatric species, between different nests in an aggregation and between nestmates, whether full or half sibs (Hefetz and Graur 1988). As predicted, an increasing number of components was needed to discriminate between bees depending on the level of specificity required. Discrimination between conspecific solitary species can be achieved with just a few components, but discrimination between nestmates and alien bees requires a very complex blend. The analysis also confirmed (based on the few species for which the social behavior is known) that in the social species the secretion is much more complex than in the solitary species.

In bumble bees, Dufour's gland secretion is possibly used to mark trails extending from the center of the nest to the outside. Evidence of trail marking was first obtained in *B. terrestris*, by letting the bees mark over a paper and then cutting it and changing the angle of the observed trail. The majority of the bees followed the pre-marked trail despite the changes in its angle with respect to the nest entrance (Cederberg 1977). In a similar set of experiments, trail marking and trail following was also substantiated for *B. hypnorum*. Based on the chemical congruency between the trail and the glandular exudate it was concluded that in this species Dufour's gland is the source of the trail pheromone (Hefetz et al. 1993). Formation of trails extending from the nest were also recently demonstrated by field observation on *B. transversalis* in the tropical forest of the Amazonian Basin (Cameron and Whitefield, 1996). These trails are not elaborate food trails as exhibited in ants, but can be regarded as an extension of the nest entrance marking to the vicinity of the nest.

Based on the above description we can postulate a scenario for the evolution of Dufour's gland function. However, before doing so it is important to state that this evolution need not necessarily follow a phylogenetic continuum, nor need there be intermediate steps. The ancestral glandular function, according to this hypothesis, was for cell lining and possibly for strengthening the nest main tunnel in the ground dwelling bees. It was then adapted for self communication in the form of a nest entrance marker. In the communal and social ground dwelling bees it evolved further for intraspecific communication. Finally, entrance marking was extended to a short trail as found in bumble bees. This evolutionary trend could have occurred in the primitive bee families as well as in the advance bees in an independent manner, but in all cases cell lining is assumed to be the ancestral function from which the communicative function derived.

The evolutionary trend in Dufour's gland function described above, does not exclude the possibility that in selected species the glandular secretion may have additional functions. A detailed account of the role of the gland in halictine bees has been summarized by Smith and Breed (1995). The macrocyclic lactone constituents of the gland were implied to act as a sex pheromone in *Lasioglossum zephyrum* (Smith et al. 1985). In contrast, in *L. malachurum* the amount of lactones is highly correlated with nesting behavior. They constitute the major fraction of the secretion in workers or mated queen, but are present only in small quantities in unmated queens. Unmated queens possess large amounts of isopentenyl esters that diminish in quantity after mating (Ayasse 1990; Ayasse et al. 1993). Although there is no explicit evidence that in *L. malachurum* the sex pheromone is composed of isopentenyl esters, it is implied from the ontogenetic studies. Behavioral assays further demonstrated that males react very specifically to the sex signals emitted by females (Smith and Ayasse 1987). Whether this kin based discrimination is also reflected in the chemical composition of Dufour's gland is still unknown. Indirect support for the involvement of Dufour's gland secretion in kin discrimination comes from the correlation between chemical similarity of its constituents and genetic relatedness. Based on the relative proportions of the various components (including all classes of compounds), it was demonstrated that in *L. malachurum* nestmates are more similar to each other than to alien conspecific (Hefetz et al. 1986). Similar results were obtained for *L. zephyrum* using the macrocyclic lactones only (Smith and Wenzel 1988). Lactones were also reported to be involved in the competition over nesting between queens of *L. malachurum*. Larger queens had higher amounts of lactones in the gland and were more likely to win the contest. Unfortunately, most of the above-presented evidence on the involvement of the glandular secretion in kin based behavior is only circumstantial. Direct experiments to test this hypothesis and to resolve the question of whether the whole secretion or only its lactones constituents are biologically active, are still to be conducted.

In the large carpenter bee *Xylocopa virginica texana*, the Dufour's gland secretion is applied on flowers after they have been visited and presumably depleted of nectar (Frankie and Vinson 1977; Vinson et al. 1978). Similar marking was also reported for *X. sulcatipes* (Eisikowitch 1987). This marking constitutes a "negative signal" that informs the visiting bee of the resource status of the flower she is about to visit, resulting in avoidance behavior. The bee approaches the flower, but instead of landing she hovers briefly then shifts direction to encounter another flower. Similar, pheromone induced, avoidance behavior was also postulated in honey bees, where 2-heptanone, a worker mandibular gland component is implied as the putative pheromone ((Nuñez, 1982). In contrast, "positive" flower marking of rewarding resources (artificial flowers) was reported in bumble bees (Cameron 1981 in *B. vosnesenskii*; Schmitt and Bertch, 1990 in *B. terrestris*). Rewarding flowers are visited and probed more frequently than non-rewarding flowers, and when given the choice, the bees recognize and prefer the rewarding flowers. Although they may land on the non-rewarding flowers they rarely probe them. Chemical analysis of the platform of the artificial flowers revealed that the source of the secretion in *B. terrestris* are the tarsal glands, and specifically the hydrocarbon constituents of these glands (Schmitt, 1990). Although these reports seem contradictory, it is possible that bees utilize both "positive" and "negative" markings of flowers, probably in relation to their social organization and foraging ecology. Moreover, the use of different glands as the source of marking suggests that this behavior has evolved independently in these bees and, as often occurs in the Hymenoptera, any of the existing exocrine glands could have been selected as the source of the pheromone.

Recent studies in our laboratory have revealed that in the honey bee *Apis mellifera* the secretory composition is caste specific (T. Katzav-Gozansky et al. in preparation). The gland in queens is hypertrophied and contains as much as 12 times more material than workers (17.6 μg/gland vs 1.4 μg/gland respectively). Qualitatively, the exudate of workers is composed of a homologous series of odd *n*-alkanes ranging from C_{23} to C_{31}, whereas the glandular exudate of queens is endowed, in addition, with long chain esters. This caste specificity does not seem to be rigid, since under queenless conditions egg laying workers, but not foragers, produce the queen-like esters in roughly the same proportion as found in queens. Production of these esters in egg laying workers is not minor since they comprise up to 30% of the secretory composition. The dependence of ester production on the social environment of the hive suggests that Dufour's gland secretion may be another source of a queen signal.

Chemosystematic Perspectives of Dufour's Gland Chemical Diversity

The chemical diversity expressed in Dufour's gland of bees and its function in the bees' biology may also be used for chemosystematics (Cane 1983 a,b), and moreover provides an insight into the ecological constraints that may have shaped these secretions. However the reconstruction of bee phylogeny on the basis of chemical data is seriously hampered by the incomplete information currently available on the chemistry of the glandular secretions and especially the lack of knowledge regarding the biosynthesis of these compounds. The following two examples demonstrate the necessity for chemical data for understanding the evolutionary processes pertaining to Dufour's gland secretions in bees. They are also used to emphasize the shortcomings of the system.

The first example is the co-occurence of the unusual triglycerides in disparate bee families. Their common presence in *Anthophora abrupta* (Anthophoridae) (Norden et al. 1980) and *Empheropsis* (Cane and Carlson 1984), both of which belong to the Anthophoridae, may not be surprising, but they were also found in several species of *Megachile,* bees from a different phylogenetic line and with a very different nesting ecology. Unlike in the ground dwelling anthophorid bees, the secretion in *Megachile* is not used for cell lining, but as a maternal food supplement to the bee bread (Cane and Carlson 1984). This represents another example of convergent evolution (if we consider *Anthophora abrupta* as an example of the genus) with specific adaptations to different nesting ecologies. The twig nesting megachilids do not possess a cell lining (or if they do, it originates from a different exocrine gland) and use Dufour's gland secretion as a food supplement only. The ground nesting anthophorids, on the other hand, follow the general pattern of using Dufour's gland secretion for cell lining, extending its function (due to the particular chemistry) as food supplement. While this comparison may tell us about the phylogenetic relationship between the Anthophoridae and Megachilidae, considering the triglycerides in both groups as homologous without knowledge of the biosynthetic routes by which they are produced may prove erroneous.

Another example in which phylogeny and ecology may have shaped the chemical nature of Dufour's gland secretion is in the case of the large carpenter bees in the genus *Xylocopa*. Six species have been studied to-date, revealing essentially three patterns that characterize the composition of the secretion. *Xylocopa micans* (Williams et al. 1983), *X. pubescens* and *X. valga* (Gerling et al. 1989) produce exclusively alkanes and alkenes. In *X. sulcatipes,* on the other hand, the secretion is dominated by ethyl and methyl esters with only minor amounts of two hydrocarbons (Kronenberg and Hefetz 1984). Two other species, *X. virginica* (Williams et al. 1983) and *X. iris* (Gerling et al. 1989) represent an intermediate stage by possessing a mixture of ethyl esters and hydrocarbons. *Proxylocopa,* a possible progenitor of the genus

Xylocopa (Hurd, 1958), synthesizes exclusively hydrocarbons (e.g., *P. olivieri;* Kronenberg and Hefetz 1984).

The chemical data on this limited number of species, matches the classical scheme of the evolution of *Xylocopa* suggesting that as the bees started to nest in dry wood they lost the need for cell lining (Iwata 1976), and concomitantly the function of the gland diverted. For example, in *X. virginica texana* the blend of methyl esters and hydrocarbons present in the gland is used for marking visited flowers (Frankie and Vinson 1977; Vinson et al. 1978). Similar marking was also reported for *X. Sulcatipes* (Eisikowitch 1987), the secretion of which is dominated by methyl and ethyl esters. The latter compounds are not suitable for nest lining, but can serve for communication. Therefore, the gradual switch in the chemistry may have been associated with the changes in glandular function. The ground nesting progenitor of the genus, currently represented by *Proxylocopa,* used hydrocarbons for cell lining. As the genus *Xylocopa* evolved, some species, like *X. pubescens,* retained the hydrocarbon factory while other, e.g., *X. sulcatipes* switched to the production of ethyl esters with an intermediate stage portrayed by *X. micans*. Whether these changes are associated with the change in function along the phyletic line is still obscure. We lack knowledge on the flower visiting behavior of the intermediate species, particularly clear evidence on the phylogenetic relationships among the species.

A different evolutionary sequence may be deduced from recent studies of the in-nest behavior in *X. sulcatipes* and *X. pubescens,* which revealed the presence of cell lining. By using a soft X-ray technique, it was observed that the bees extensively brush the abdomen on the cell walls with the consequent appearance of an opaque X-ray line presumably representing a cell lining. Extraction of the cell wall revealed a series of long chain hydrocarbons that apparently originate from the intersegmental glands (Gerling et al. 1989; Hefetz, unpublished observation). This finding, if true for additional species in the genus, makes the alternative evolutionary sequence just as logical. The switch from ground nesting to dry wood nesting habits in the ancestral *Xylocopa* type, was accompanied by a change in the glandular source used for cell lining, reflecting the independent evolution of the need for lining in bees, and its glandular source. The copious Dufour's gland secretion in these species was used for self-communication in flower marking. Chemical changes in selected, but not all, species accompanied this change in function. Presumably, it was the retention of hydrocarbon biosynthesis in the postulated ancestral xylocopine that enabled *Proxylocopa* to switch back to ground nesting, re-adapting Dufour's gland secretions for cell lining. In this reconstructed phylogeny, *Proxylocopa* is a derived genus rather than the progenitor of *Xylocopa*. Undoubtedly reconstructed phylogenies based on chemical data are at best speculative. There is an urgent need to construct phylogenetic trees by other means which are not so greatly affected by the ecology of the species. With the advance in molecular systematics it will be possible to test such evolutionary hypotheses and moreover, to understand how evolutionary history and ecology have shaped the chemistry and function of exocrine products.

Dufour's gland composition in bumble bees seems to be highly diverse in comparison to the other bee families (Tengö et al. 1991). However, before assigning any biological significance to this diversity, we have to bear in mind that recent analyses of exocrine products are much more complete due to the increased sensitivities of the analytical instrumentation (mainly highly sensitive gas chromatograph/mass spectrometers). These advanced analytical procedures enable proper identification of even trace compounds, making comparisons with secretions analyzed a decade or two ago much more complicated. Over 150 compounds were identified in the 9 palearctic bumble bee species investigated, representing at least six classes of compounds including hydrocarbons, esters (from methyl to decyl; acetates and long wax type esters), terpenoid esters, alcohols, ketones, and fatty acids. The vast majority of the constituents are hydrocarbons, but the additional oxygenated compounds denote species specificity. Some of the species (e.g., *B. lapidarius, B. ruderarius, and B. pratorum*) lack oxygenated compounds altogether. The group included in the subgenus *Bombus sensus stricto* (*B. terrestris, B. magnus, B. lucorum,* and *B. cryptarum*) are characterized by a multitude of methyl and ethyl esters, while *B. vorticosus* (subgenus *Sibiricobombus*) is fortified with a series of acetates. The secretion may also be fortified with idiosyncratic compounds. For example, *Bombus cryptarum* and *B. terrestris* are unique in having terpenoid ester, while *B. terrestris* was the only species that possessed pentadecan-2-ol, its corresponding pentadecan-2-one and heptadecan-2-one. Within the species *B. terrestris* there seems to be population differences. Bees collected in Israel contain both geranylcitronellol and farnesyl hexanoate, whereas bees collected in Leuven, Belgium had only geranylcitronellol. The British subspecies *B. t. audax*, on the other hand, had farnesyl hexanoate but not geranylcitronellol (Oldham et al. 1994). Noteworthy is the congruency between the composition of Dufour's gland and cuticular lipids reported in five species of bumble bees (Oldham et al. 1994). This congruency was not complete in all species, and was most apparent when hydrocarbons were compared. For example, the terpenoid alcohols and esters found in both subspecies of *B. terrestris* are not present in their cuticular wash.

Since hydrocarbons are common constituents of cuticular lipids, it is hard to assess whether they originate from Dufour's gland.

The Dufour's gland secretion of *Bombus hypnorum* was studied more extensively, exemplifying the possible diversity in its chemical composition (Hefetz et al. 1993). Bees from seven nests collected in the island of Öland, Sweden and the adjacent mainland were analyzed revealing large quantitative differences among nests. In three of the nests, alkanes and alkenes were the dominant compounds with only very minute traces of ethyl esters and acetates. The other nests had progressively higher amounts of the acetates, and in one nest acetates formed the major groups accompanied by similar proportions of alkanes, alkenes and ethyl esters.

Despite the shortcomings of the chemosystematics described above, some general conclusions on chemical diversity and bee phylogeny can be drawn, as has been done successfully in the Halictidae (Cane 1983a) and in the Andrenidae (Cane 1983b). [But see also Hefetz (1987) for reservations]. Assuming that common compounds that are found in remotely related bee families are produced by the same biosynthetic pathway, it is possible either that the same pathway has evolved several times, or that this biosynthetic capability was never lost but became expressed differentially in the various bee species. Unlike other exocrine glands, the compounds produced by Dufour's gland are rather simple, and most of them can be readily derived from existing metabolic pools. Nevertheless, the likelihood of a specific biosynthetic pathway (even if we consider only the few final steps needed to produce the final product from an existing metabolic pathway) evolving several times in a homologous gland is rather small. It is more likely that this biosynthetic ability was present, albeit suppressed, throughout the evolution of the bees. Such differential expression could result in compounds that range from undetectable quantities to large amounts. Although some of the biosynthetic abilities of the gland, such as the production of macrocyclic lactones in the short tongued bees, may have been lost in evolution, it would not be surprising if some of the compounds considered as idiosyncratic of these bees will also be found in the long-tongued bees, as more species are investigated.

Conclusions

Our knowledge of the chemistry of exocrine secretions produced by solitary and social bees far exceeds our understanding of their function. One of the major obstacles is the need to conduct the critical behavioral assays in the field, and the restriction due to the seasonality of the bees' activity. This may be the reason why there are so few comprehensive studies pertaining to social communication with non-*Apis* bees. An additional difficulty is the immense diversity of bees and their habitats, which seems to be an advantage for evolutionary studies, but becomes a handicap when we want to generate evolutionary hypotheses. As we have seen, many of the glandular exudates are very complex blends of compounds which, coupled with the need to conduct field bioassays within a short nesting period, has made the elucidation of the active components almost impossible. This is probably the reason why most of the behavioral studies were done using the glandular exudates, rather than their synthetic substitutes. Other studies concentrated on chemical analyses and rely on statistical differences in the composition to draw behavioral and evolutionary conclusions. While in many of the solitary species the situation is not likely to change, I see great prospects in the renewed interest in bumble bee chemical communication. The use of bumble bees for commercial pollination, and the developments in mass production of several species throughout the world, will undoubtedly enhance all aspects of their research, including chemical communication. I see also similar prospectives for development of many of the halictine species and some of the *Xylocopa* species. They are good pollinators of crops, which may provide an impetus for efforts to develop techniques for their mass production, and since they exhibit progressive levels of sociality they may provide us with essential clues to the understanding of social communication among bees.

References

Ayasse, M. 1990. Odor based interindividual and nest recognition in the sweat bee *Lasioglossum malachurum* (Hymenoptera: Halictidae). In: *Social insects and their environment. Proc. 11th Inter. Cong. IUSSI,.* (G.K. Veeresh, B. Malik, and C.A. Viraktamath, eds.), Oxford & IBH Publishing Co. PVT. LTD, New Delhi. pp. 515-516

Ayasse, M., W. Engels, A. Hefetz, J. Tengo, G. Lübke and W. Francke. 1993. Ontogenetic patterns of volatiles identified in Dufour's gland extracts from queens and workers of the primitively eusocial halictine bee, *Lasioglossum malachurum* (Hymenoptera: Halictidae). *Insectes Sociaux* 40: 41-58.

Batra, S. W. T. 1984. Solitary bees. *Sci. Amer.* 86-93.

Batra, S. W. T. 1972. Some properties of the nest-building secretions of *Nomia, Anthophora, Hylaeus* and other bees. *J. Kans. Entomol. Soc.* 45: 208-218.

Bergman, P. and Bergström, G. 1996. Chemical communication in bumblebee premating behavior. ISCE 13th Annual Meeting, Prague, Czech Republic 43-44.

Bergström G., B. Kullenberg and S. Stallberg-Stenhagen. 1973. Studies on natural odoriferous compounds VII. Recognition of two forms of *Bombus lucorum* L. (Hymenoptera, Apidae) by analysis of the volatile marking secretion from individual males. *Chem. Scr.* 4: 174-182.

Bergström, G., B.G. Svensson, M. Appelgren and I. Groth. 1981. Complexity of bumble bee marking pheromones: Biochemical, ecological and systematical interpretations. In: *Biosystematics of Social Insects*. (P.E. Howse and J.L. Clément, Eds.) Academic Press, London. pp. 175-183

Bloch, G., D.W. Borst, Z.Y. Huang, G.E. Robinson and A. Hefetz. 1996. Effects of social conditions on JH-mediated reproductive development in *Bombus terrestris* workers. *Physiol. Ent.* 21: 257-267.

Blum, M.S. 1981. *Chemical Defenses of Arthropods*. Academic Press, New York.

Cameron, S.A. 1981. Chemical signals in Bumble bee foraging. *Behav. Ecol. Sociobiol.* 9: 257-260.

Cameron, S. A. and J.B. Whitfield. 1996. Use of walking trails by bees. *Nature* 379: 125.

Cane, J.H. 1983. Chemical evolution and chemosystematics of the Dufour's gland secretions of the lactone-producing bees (Hymenoptera: Colletidae, Halictidae and Oxeidae). *Evolution* 37: 657-674.

Cane, J.H. 1983. Preliminary chemosystematic of the Andrenidae and exocrine lipid evolution of the short-tongued bees (Hymenoptera: Apoidea). *Syst. Zool.* 32: 417-430.

Cane, J. H. and R.G. Carleson. 1984. Dufour's gland triglyceride from *Anthophora, Emphoropsis* (Anthophoridae) and *Megachile* (Megachilidae) bees (Hymenoptera: Apoidea). *Comp. Biochem Physiol.* 78B: 769-772.

Cane, J. H., S. Gerdin and G. Wife, G. 1983. Mandibular gland secretions of solitary bees (Hymenoptera:Apoidea): Potential for nest cell disinfection. *J. Kans. Entomol. Soc.* 56: 199-204.

Cederberg, B. 1977. Evidence for trail marking in *Bombus terrestris* workers (Hymenoptera, Apidae). *Zoon* 5: 143-146.

Cruz Landim, C. d. 1967. Estudo comparativo de algumas glandulas das abelhas (Hymenoptera, Apoidea) e respectivas implicacoes evolutivas. *Arq. Zool.* 15: 177-290.

Duchateau, M.J. and H.H.W. Velthuis. 1988. Development and reproductive strategies in *Bombus terrestris* colonies. *Behavior* 107: 186-207.

Duffield, R.M., A. Fernandes, C. Lamb, J.W. Wheeler and G.C. Eickwort. 1981. Macrocyclic lactones and isopentenyl esters in the Dufour's gland secretion of halictine bees (Hymenoptera: Halictidae). *J. Chem. Ecol.* 7: 319-331.

Duffield, R.M., J.W. Wheeler and G.C. Eickwort. 1984. Sociochemicals of bees. In: *Chemical ecology of insects.* (W.J. Bell and R.T. Carde, Eds.), Chapman and Hall, London. pp. 387-428

Eisikowitch, D. 1987. *Claotropis procera* (Ait.) Ait. F. (Asclepiaceae) and *Xylocopa* spp.: a study of interrelationship. *Insects-Plants, Proc.6th Symp. Insect-Plant Relat., PAU1986.* Junk., Dordrecht, The Netherlands. pp. 341-345

Frankie, G.W. and S.B. Vinson. 1977. Scent marking of passion flowers in Texas by females *Xylocopa virginica texana* (Hymenoptera: Anthphoridae. *J. Kans. Entomol. Soc.* 50: 613-625.

Genin, E., R. Jullien, F. Perez, C. Fonta and C. Masson. 1984. Preliminary results on the chemical mediators of the bumble bee *Bombus hypnorum*. *Comp. Rend. Acad. Sci. III* 299: 297-302.

Gerling, D., H.H.W. Velthuis and A. Hefetz. 1989. Bionomics of the large carpenter bees of the genus *Xylocopa*. *Ann. Rev. Entomol.* 34: 163-190.

Hefetz, A. 1983. Function of secretion of mandibular gland of male in territorial behavior of *Xylocopa sulcatipes* (Hymenoptera:Anthophoridae). *J. Chem. Ecol.* 9: 923-931.

Hefetz, A. 1990. Individual badges and specific messages in multicomponent pheromones of bees (Hymenoptera: Apidae). *Entomol. Gener.* 15: 103-113.

Hefetz, A. 1992. Individual scent marking of the nest entrance as a mechanism for nest recognition in *Xylocopa pubescens* (Hymenoptera, Anthophoridae). *J. Insect Behav.* 5: 763-772.

Hefetz, A. 1987. The role of Dufour's gland secretions in bees. *Physiol. Entomol.* 12: 243-253.

Hefetz, A., G. Bergström and J. Tengo. 1986. Species, individual and kin specific blends in Dufour's gland secretions of halictine bees: Chemical evidence. *J. Chem. Ecol.* 12: 197-208.

Hefetz, A., M.S. Blum, G.C. Eickwort and J.W. Wheeler. 1978. Chemistry of the Dufour's gland secretion of halictine bees. *Comp. Biochem. Physiol.* 61B: 129-132.

Hefetz, A., H.M. Fales and S.W.T. Batra. 1979. Natural polyesters: Dufour's gland macrocyclic lactones from brood cell laminesters in *Colletes* bees. *Science* 204: 415-417.

Hefetz, A. and D. Graur. 1988. The significance of multicomponent pheromones in denoting specific compositions. *Biochem. Syst. Ecol.* 16: 557-566.

Hefetz, A., T. Taghizadeh and W. Francke. 1996. The exocrinology of the queen bumble bee *Bombus terrestris* (Hymenoptera: Apidae, Bombini). *Z. Naturforsch.* 51: 409-422.

Hefetz, A., J. Tengo, G. Lübke and W. Francke. 1993. Inter-colonial and intra-colonial variations in Dufour's gland secretion in the bumble bee species *Bombus hypnorum* (Hymenoptera: Apidae). In: *Sensory Systems of Arthropods.* (K. Weise, F.G. Gribakin and G. Renninger, Eds.), Birkhauser Verlag, Basel. pp. 469-480

Heselhause, F. 1922. Die Hautdrusen der Aiden und verwandterr Formen. *Zool. Jahr. Abt. Anat. Ont. Tiere* 43: 369-464.

Honk van, C.G.J., H.H.W. Velthuis, P.F. Roseler and M.E. Malotaux. 1980. The mandibular glands of *Bombus terrestris* queens as a source of queen pheromone. *Entomol. Exp. Appl.* 28: 191-198.

Hurd, P.D.J. 1958. Observation of the nesting habits of some New World carpenter bees with remarks on their importance in the problem of species formation (Hymenoptera, Apoidea). *Ann. Entomol. Soc. Amer.* 51: 365-375.

Iwata, K. 1976. *Evolution of Instinct. Comparative ethology of Hymenoptera.* Amerind Publishing Co. Pvt. Ltd., New Delhi.
Kronenberg, S. and A. Hefetz. 1984. Comparative analysis of Dufour's gland secretion of two carpenter bees (Xylocopinae: Anthophoridae) with different nesting habits. *Comp. Biochem. Physiol.* 79B: 421-425.
Kullenberg, B., G. Bergström, B. Bringer, B. Carlberg and B. Cederberg. 1973. Observation of scent marking by *Bombus* Latr. and *Psithyrus* Lep. males (Hym. Apidae) and localization of site of production of the secretion. *Zoon. Suppl.* 1: 23-30.
Kullenberg, B., G. Bergström and S. Stallberg-Stenhagen. 1970. Volatile components of the cephalic marking secretion of male bumble bees. *Acta Chem. Scand.* 24: 1481-1483.
Maccagnani, B., H.H.W. Velthuis and M.J. Duchateau. 1994. Mating behavior in *Bombus terrestris* L. (Hymenoptera, Apidae). *Les Insectes Sociaux; Abstract presented in the 12th IUSSI congress.* pp. 460
Michener, C.D. 1974. *The social behavior of the bees. A comparative study.* Harvard University Press, Cambridge, Massachusetts.
Norden, B., S.W.T. Batra, H.F. Fales, A. Hefetz and J.C. Shaw. 1980. *Anthophora* bees; Unusual glycerides from maternal Dufour's gland serve as larval food and cell lining. *Science* 207: 1095-1097.
Nuñez, J.A. 1982. Honeybee foraging strategies at a food source in relation to its distance from the hive and the rate of sugar flow. *J. Apic. Research* 21: 139-150.
Oldham, N.J., J. Billen and E.D. Morgan. 1994. On the similarity of the Dufour gland secretion and the cuticular hydrocarbons of some bumblebees. *Physiol. Entomol.* 19: 115-123.
Roseler, P.F., I. Roseler and C.G.J. van Honk. 1981. Evidence for inhibition of corpora allata in workers of *Bombus terrestris* by a pheromone from the queen's mandibular gland. *Experientia* 37: 348-351.
Schmitt U. And A. Bertsch. 1990. Do foraging bumblebees scent-mark food sources and does it matter? *Oecologia,* 82: 137-144.
Schmitt, U. 1990. Hydrocarbons in tarsal glands of *Bombus terrestris. Experientia,* 46: 1080-1082.
Shearer, D.A. and R. Boch. 1965. 2-Heptanone in the mandibular gland secretion of the honey-bee. *Nature* 206: 530.
Shimron, O., A. Hefetz and J. Tengo. 1985. Structural and communicative functions of Dufour's gland secretion in *Eucera palestinae* (Hymenoptera: Anthophoridae). *Insect Biochem.* 15: 635-638.
Slessor, K.N., L. Kaminski, G.G.S. King, J.H. Borden and M.L. Winston. 1988. Semiochemical basis of the retinue response to queen honey bees. *Nature* 332: 354-356.
Smith, B.H. and M. Ayasse. 1987. Kin-based male mating preferences in two species of halictine bee. *Behav. Ecol. Sociobiol.* 20: 313-318.
Smith, B.H. and M.D. Breed. 1995. The chemical basis for nestmate recognition and mate discrimination in social insects. *Chemical ecology of insects 2*. Carde, R. T. and Bell, W. J. (Eds) Chapman and Hall, New York. pp. 287-317

Smith, B.H., R.G. Carlson and J. Frazier. 1985. Identification and bioassay of macrocyclic lactone sex pheromone of the halictine bee *Lasioglossum zephyrum*. *J. Chem. Ecol.* 11: 1447-1456.

Smith, B. H. and J.W. Wenzel. 1988. Pheromonal covariation and kinship in social bee *Lasioglossum zephyrum* (Hymenoptera: Halictidae). *J. Chem. Ecol.* 14: 87-94.

Tengö, J., A. Hefetz, A. Bertsch, U. Schmitt, G. Lübke and W. Francke. 1991. Species specificity and complexity of Dufour's gland secretion of bumble bees. *Comp. Biochem. Physiol. B* 99: 641-646.

Velthuis, H. H. W. and D. Gerling. 1980. Observations on territoriality and mating behavior of the carpenter bee *Xylocopa sulcatipes*. *Ent. Exp. Appl.* 28: 82-91.

Vinson, S. B., G.W. Frankie, M.S. Blum and J.W. Wheeler. 1978. Isolation, Identification and function of Dufour's gland secretion of *Xylocopa virginica texana* (Hymenoptera: Anthophoridae. *J. Chem. Ecol.* 4: 315-323.

Vinson, S. B., H.J. Williams, G.W. Frankie, J.W. Wheeler, M.S. Blum and R.E. Coville. 1984. Mandibular glands of male *Centris adani*, (Hymenoptera: Anthophoridae): Their Morphology Chemical constituents and function in scent marking and territorial behavior. *J. Chem. Ecol.* 8: 319-327.

Wheeler, J.W. and R.M. Duffield. 1988. Pheromones of Hymenoptera and Isoptera. In: *CRC Handbook of natural pesticides*, (E.D. Morgan and N.B. Mandava, Eds.), Vol. IV: Pheromones CRC Press Inc., Boca Raton, Florida. pp. 59-206

Williams, H. J., G.W. Elzen, M.R. Strand, and S.B. Vinson. 1983. Chemistry of Dufour's gland secretions of *Xylocopa virginica texana* and *Xylocopa micans* (Hymenoptera: Anthophoridae) - A comparison and re-evaluation of previous work. *Comp. Biochem. Physiol.* 74B: 759-761.

11

Mass Action in Honey Bees: Alarm, Swarming and the Role of Releaser Pheromones

Justin O. Schmidt

Introduction

Honey bees are the grandmasters of chemical communication in the insect world. Not only are more pheromones described for honey bees than any other species, but the diversity of functions for honey bee pheromones is unsurpassed. A wide variety of pheromonal functions occur within the confines of the colony's nest -- pheromones ranging in function from alerting the workers to the presence of the queen and her reproductive state, to communicating to adults information about the quantity, age, state, health, and gender of eggs and larvae, to marking prepared cells for egg oviposition by the queen. Outside their immediate nest environment honey bees use equally impressive pheromones for chemical communication. These are all releaser pheromones, pheromones whose action is immediate -- in contrast to many of the within-hive primer pheromones such as queen pheromone which act to cause later behavioral or physiological changes (Wilson and Bossert, 1963). Releaser pheromones include the honey bee sex pheromone, (*E*)-9-oxo-dec-2-enoic acid (Gary, 1962), and worker pheromones used to communicate information such as the recency of visitation of floral resources, the presence of a valuable water source, the presence of potential intruders and threats to the colony, and the presence and suitability of a cavity as a potential nest site for a reproductive or absconding swarm. This chapter will focus only on those releaser pheromones that coordinate and elicit mass actions --

mass attack of potential predators of the colony and the mass movement of bees during the swarming processes.

Alarm Pheromone from the Sting Apparatus: the Pheromone of Defense

To human observers the most obvious and unmistakable honey bee pheromone is the alarm pheromone emanating from the sting. This master pheromone of defense was also the first honey bee pheromone to be recognized. Huber (1814) noted that a colony of bees dramatically increased its defensive attacks following the planting of the first sting in a beekeeper, and that the autotomized sting released an odor which caused the increased attacks. This sting pheromone, which is present in worker bees about the time of guarding behavior or older (Boch and Shearer, 1966; Whiffler et al. 1988), quickly alerts nest mates that a potentially dangerous intruder is present. The pheromone is located mainly in the setose membrane surrounding the sting shaft base (Ghent and Gary, 1962; Mauchamp and Grandperrin, 1982) and is readily released from autotomized stings imbedded in the integument of attacked animals (Figure 11.1). From this position, the pheromone is admirably situated to communicate to nest mates "here is the intruder to attack" and that "this part of the intruders integument is vulnerable to attack". In addition to its role of targeting an intruder, alarm pheromone can be released by exposing the sting chamber to alert and orient other workers to potential danger (Maschwitz, 1964).

Defensive Behaviors and Factors Eliciting Defense

Sting alarm pheromone does not elicit attack behavior simply on its own. Rather, it operates as one component in an array of factors that govern a complex of hierarchial defensive behaviors. The defensive behavior of honey bees near a colony often involves four distinctive progressive behaviors: (1) diffuse aggressiveness, or alerting; (2) directed aggressiveness, or activating and recruiting; (3) recognition, or orienta-tion toward the target; and (4) effective aggressiveness, or direct attack (Lecomte, 1961; Collins et al. 1980). In the first behavior, a bee is alerted by visual, chemical, or mechanical cues that a potential threat is nearby. The bee, usually a guard bee, raises to a high stance, waves it antennae rapidly, partially extends it wings, and appears to be seeking information about the nature of the possible threat (Ghent and Gary, 1962). Once a potential threat is perceived, the alerted bee in the second behavior usually buzzes its wings, opens its sting chamber to release alarm pheromone, and rushes into the

FIGURE 11.1 Autotomized honey bee sting simultaneously releasing alarm pheromone and injecting venom into the victim.

colony to recruit other bees. In response, a mass of highly activated bees comes out of the colony entrance, all ready to defend (Maschwitz, 1964). Without visual signals from a potential predator, the progression of defensive behaviors stops at this point (Maschwitz, 1964). If, however, a visual stimulus, especially from a moving potential predator, is present, the activated bees progress to the third stage and fly out in search of the disturbance and orient toward than source of disturbance. In the final stage, the defending bees attack the potential predator by attempting to sting it, by buzzing vigorously in fur/feathers, by biting and pulling hairs, or by mechanically colliding with the object. The defensive behaviors can terminate at any stage. Some bees will actively orient (stage 3) but not progress to attack, and other bees might express some of the final attack behaviors, such as colliding with the target or pulling hair, but without stinging. Different cues, or types of cues, can be responsible for inducing different stages of defensive behavior. If the threatening cues are sufficiently strong, for example a moving mammal or bird that breaths directly into the colony entrance, stages 1-3 may be omitted, or at least passed through so quickly as to be missed by an observer, and a direct vigorous stinging attack is initiated.

The complex nature of honey bee defensive behavior has stimulated the development of a wide variety of bioassays to measure defensiveness (Table 11.1). These assays often dramatically differ from one another, measure different aspects or stages of defensive behavior, and may be difficult to compare each to one another. Nevertheless, these assays have greatly facilitated an understanding of the factors that elicit defensive behaviors by honey bees. A listing of external factors that stimulate defensive behavior around a colony is provided in Table 11.2. Internal factors, such as age of bees, hunger of individual bees, or other within colony factors are not listed in the table and cursory listing of genetic factors is provided because of the widespread interest in this topic.

Perhaps the strongest single factor in stimulating defensive behavior is motion of an object near the colony. Motion can cause all stages of defense from alerting to stinging. Moving targets are more stimulating than stationary targets, and faster moving targets induce increased stimulation relative to more slowly moving targets (Free, 1961). In the case of separate targets presented simultaneously, dark colored targets are attacked and stung more frequently than light colored targets. Attackers can also distinguish between dark and light colors on the same target and prefer to sting the darker areas. In a recent stinging attack on a dog with yellow and grey-black patches, the dark areas were stung almost 2.5 times more than the yellow areas (Schmidt, 1996). Contrast between the target and background facilitates discovery of the target, but contrast is not as important as color (Free, 1961). Objects having properties similar to actual potential predators stimulate more vigorous attack than those less representative of predators: three dimensional shapes are attacked more than equal area flat objects; objects about the size of typical vertebrate predators are attacked more vigorously than smaller or larger objects; furry or rough textured objects are attacked more vigorously than smooth objects; and warm objects are attacked more than ambient or cool objects (Table 11.2).

A large variety of olfactory cues stimulate defensive behaviors in honey bees. The presence of a previous sting in an object dramatically increases the strength of the attack. Likewise, the presence of alarm pheromone, or components of alarm pheromone intensify the attack. Pheromone, and not mechanical buzzing by activated bees, is responsible for the alarming response (Maschwitz, 1964). By far, the greatest olfactory stimulus for attack behavior is breath directed into a colony entrance by a warm-blooded potential predator. In most cases, such breath causes an immediate and explosive outpouring and attack of defending bees (Schmidt and Spangler, unpublished). Factors in

TABLE 11.1. Assays for honey bee alarm or defensive behavior.

Assay	Stimulating factor	References
Activation of guard bees	Stings/extracts placed in hive	Ghent and Gary, 1962
	Stings, extracts, etc. in hive	Maschwitz, 1964
	Pheromone and/or shot marble	Collins and Kubasek, 1978
	Pheromone components	Free et al., 1989
Attraction of colony bees	Stings/extracts placed hive	Ghent and Gary, 1962
	Stings/extracts, etc., hive entrance	Maschwitz, 1964
	Pheromone components, analogues	Boch and Shearer, 1971
	Breath	Boch and Rothenbuhler, 1974
	Pheromone disks near colony	Koeniger et al., 1979
Movement/buzzing of caged bees	Pheromone and components	Collins and Rothenbuhler, 1978 , Collins, 1981
Stinging of still objects	Pheromone and/or shot marble	Collins and Kubasek, 1982
	Cotton wool ball at hive entrance	Free, 1961
	Stings/extracts above open hive	Ghent and Gary, 1962
Stinging of moving objects	Black leather ball	Stort, 1974
	Pheromones and/or shot marble	Collins and Kubasek, 1982
	Cotton wool balls, many variables	Free, 1961
	Black fabric balls, many variables	Maschwitz, 1964
Collisions with black target	Human breath and disturbance	Spangler et al., 1990
Non-stinging attack	Human breath, black fabric target	DeGrandi-Hoffman et al. unpub
Bees flying in air	Removal of hive top	Boch and Rothenbuhler, 1974
	Pheromone and moving target	Collins et al., 1988
Wing beat frequency		Spangler, 1993
Sound production	Pheromone components, chemicals	Lefebvre and Beattie, 1991
Stings in investigator	Nest dissection	Schneider and McNally, 1992
Inhibition of Nasonov scenting	Pheromone and components	Free et al., 1983
Inhibition of Nasonov attraction	Pheromone components	Free et al., 1983
Metabolic energy	Isopentyl acetate	Moritz et al., 1985
Electroantennogram response	Eicosenol	Pickett et al., 1982
Electrical threshold to stinging	Electric voltage	Kolmes and Fergusson-Kolmes,1989
Abdominal contractions	Pheromone	Tel Zur and Lensky, 1995

breath that are important include warmth, and high humidity, but breath odors also appear to have an effect. The combination of high temperature, high humidity, and respiratory odors in breath causes maximal response. Other odors that stimulate defensive behavior include animal scents and (human) sweat, though these odors cause considerably less response than breath or pheromone. Indeed, the ability of human sweat to elicit attack is unclear (Table 11.2), and, at most, has only a very minor effect. Part of the confusion about the effects of sweat may derive from the fact that sweating people are usually also hot, and hot targets elicit more attacks than cold targets. Odor of smoke on targets has no effect, but smoke puffed into a bee colony reduces the effect of alarm pheromone and reduces guarding and flight activity.

Many investigators have devoted considerable effort toward determining the influence of weather and climate on the defensive behavior of honey bees. The strongest correlation between weather and defensiveness seems to be the presence of an impending storm. Just before a storm, bee behavior and defensiveness change dramatically, with a resultant major increased in all stages of defensiveness (Schua, 1952; Lecomte, 1961). In some studies high air temperature is a factor increasing defensiveness, while in others, little or no effect is evident (Table 11.2). The differences might reflect that bees in hot climates seem not to be affected much by temperature, whereas bees in colder climates might exhibit some effect. Overall, the effect is probably small, and likely is evident mainly as a contribution to other factors. Similar conclusions apply to other weather factors such as relative humidity and barometric pressure (Table 11.2). The greatest effect of weather on defensiveness of bees is to be seen when all factors are trending toward their maximal effect -- that is, during hot, humid, sunny, low wind days in cold climates (Southwick and Moritz, 1987)

The effect of colony factors in governing defensiveness has received less attention than other factors. Colony size, nest cavity size, comb area, and other factors probably contribute to defensiveness in ways that would appear logical -- factors representing greater resources possessed by the bees and greater rewards for successful predators will cause greater defensiveness. Another intrinsic colony factor, genetics, clearly plays a major role in the defensiveness of honey bee colonies. Both genetic variability among colonies of one type of bee population and between different populations (e.g. Africanized and European bees) have been clearly demonstrated (Table 11.2). Nestmate recognition vis-á-vis defensive behaviors toward other conspecifics and the division of defensive roles within colonies will not be addressed here (see Section on Nestmate Recognition in this volume for further information).

TABLE 11.2 Factors eliciting colony defensive behaviors in honey bees[1].

Factor	Defensive effect or trend	Reference	Comments
Motion of object	Increases alerting, orienting & attack	Lecomte, 1951, 1961	
		Free 1961	Movement > none; fast motion > slow
		Maschwitz, 1964	
		Collins and Kubasek, 1982	
Color of object	Dark color greater than light color	Lecomte, 1951	Comparisons among colored targets
		Free, 1961	"
		Maschwitz, 1964	"
		Schmidt, 1996	Stings to colored areas of dog
Contrast with background	Contrasting object > low contrast	Free, 1961	Color more important than contrast
		Maschwitz, 1964	
Shape of object	3 dimensional shape > 2 dimensional	Maschwitz, 1964	Sphere more active than equal area patch
Texture of object	Rough/furry > smooth	Free, 1961	
		Maschwitz, 1964	
Size of object	Maximum activity for medium size	Lecomte, 1961	Sizes smaller than a bee not attacked
		Maschwitz, 1964	Area of about 64 cm^2 is maximum
Temperature of object	Warm > cold	Maschwitz, 1964	35° target attacked >> 24° target, but initial visits to target same

TABLE 11.2 (Continued)

Factor	Defensive effect or trend	Reference	Comments
Previous sting in object	Sting present > no sting	Huber, 1814	
		Free, 1961	
		Lecomte, 1961	
		Ghent and Gary, 1962	
		Maschwitz, 1964	Bees without stingers induce no alarm
Alarm pheromone	Alarm pheromone increases defense	Ghent and Gary, 1962	
	Other odors do not cause alarm	Maschwitz, 1964	Acids, human sweat, floral: no effect
	Increases attraction of nest bees	Koeniger et al., 1979	Increases attraction and stinging by Apis dorsata, florea, cerana, and mellifera
	Isopentyl acetate increases metabolism	Moritz et al., 1985	Oxygen use increases on exposure
Buzzing of agitated bee	No effect	Maschwitz, 1964	Without sting pheromones buzzing guards could not elicit recruitment
Breath	Breath >> ordinary air	Maschwitz, 1964	
		Schmidt and Spangler, unpub.	
	Breath > animal scent	Schmidt and Boyer-Hassen, 1996	Stings to nasal vs other areas of dog
Warmth	Warmth > cool	Maschwitz, 1964	Warm air > cool air
Moisture	Moist > dry	Maschwitz, 1964	Moist air > dry air
Odors	Breath odors > odorless	Schmidt and Spangler, unpub.	Breath with odors adsorbed less active than ordinary human breath
Animal scent	Scent > none	Free, 1961	Scent of voles or shrews on targets

TABLE 11.2 (Continued)

Factor	Defensive effect or trend	Reference	Comments
Human sweat	Sweat > none	Schua, 1951	No distance attraction; but bees cling to object
	Sweat > none	Lecomte, 1961	Increases attacks
	Sweat > none	Free, 1961	Stings increased
	No effect	Maschwitz, 1964	Repeated of Free, 1961; no effect
Smoke	No effect on target	Free, 1961	Neutralized alarm pheromone effect
	Reduced guarding and flights	Newton, 1969	Reduced effects of alarm pheromone
Weather	Storms >> defensiveness	Schua, 1952	
		Lecomte, 1961	
Temperature of air	Little effect	Lecomte, 1961	
	No difference	Stort, 1971	
	High > low	Rothenbuhler, 1974	1 colony of 3 greater at 28° than 24°
	No effect	Brandeburgo, 1979 (cited by Schneider and McNally, 1992)	
	High > low	Collins, 1981	35° >30° in caged bees
	Depends on season and location	Brandeburgo et al. 1982	No effect in hot area, + effect in cool
	High > low	Southwick and Moritz, 1987	Stings: cold climate bees
	High > low	Villa, 1988	Stings: 34° > 17°
	No difference	Schneider and McNally, 1992	Number of stings; hot climate
Relative humidity	High > low	Collins, 1981	High increases intensity, caged bees
	Probably high > low	Brandeburgo et al., 1982	Hot and humid location, positive effect;

TABLE 11.2 (Continued)

Factor	Defensive effect or trend	Reference	Comments
			cooler and drier location, no effect
	High > low	Southwick and Moritz, 1987	Stings: cold climate
	No difference	Schneider and McNally, 1992	Stings: hot climate
Barometric pressure	High > low	Southwick and Moritz, 1987	Stings: cold climate
	No difference	Schneider and McNally, 1992	Stings: hot climate
Colony size	Large colonies > small colonies	Schneider and McNally, 1992	Stings in investigators
Nest cavity size	Colonies in large cavities more prone to sting	Schneider and McNally, 1992	Stings in investigators
Colony comb area	Colonies with large comb area more prone to sting	Schneider and McNally, 1992	Stings in investigators
Genetics of colonies	Defensiveness based on genetics	Boch and Rothenbuhler, 1974	Differences among European bees
	Africanized bees > European bees	Stort, 1975	
	Africanized bees > European bees	Collins et al., 1988	
	Africanized bees > European bees	Villa, 1988	

[1]Emphasis on major and historic studies; only limited representative citations to work on pheromones and genetics.

Chemistry and Biological Activity of Alarm Pheromones

The first chemical component to be identified in alarm pheromone of honey bees was isopentyl acetate (IPA) (Boch et al. 1962). However, based on its low ability to elicit responses in bioassays relative to stings, the authors concluded that IPA was only one component in the blend. Morse et al. (1967) showed that all four then commonly recognized species of honey bees, *Apis mellifera, A. dorsata, A. cerana*, and *A. florea*, contained IPA, results which were later expanded by Koeniger et al. (1979) to show that, in addition to IPA, all four species contained 1-octyl acetate, and that *Apis dorsata* and *A. florea* additionally contained sizable quantities of 2-decen-1-yl acetate. These findings confirmed that the presence of alarm pheromones is a trait shared by all species of *Apis*. Moreover, the blends of components in the alarm pheromones were generally similar among the familiar four species (*A. mellifera, A. cerana, A. dorsata, A. florea*) but differed dramatically in the giant Himalayan honey bee, *A. laboriosa* (M. S. Blum, pers. communication).

The alarm pheromones of *A. mellifera* consist of an amazing array of oxygenated volatile compounds (Table 11.3). The 30+ major components fall into several chemical classes. The two predominant classes are the alcohols, consisting of simple n-alcohols plus hydrocarbons with alcohol moieties on carbon 2, and esters. Most of the esters are acetates. Many of the esters are derivatives of alcohols as indicated in Table 11.3 by connecting lines. Some of the esters such as 2-nonyl acetate appear to be further derivatized to acetates in which the alcohol linkage is shifted to the primary position to make nonyl acetate, and that molecule is further derivatized with the addition of an n -ß unsaturation to yield 1-non-2-enyl acetate. In addition to alcohols and esters, the alarm pheromone contains small quantities of acids and aromatics (Table 11.3).

The quantities and ratios of components in the alarm pheromone blends vary considerably among populations of honey bees. Africanized honey bees possessed an overall greater quantity of alarm pheromone than European bees and contained especially more 2-nonanol, hexyl acetate, butyl acetate, octyl acetate and octanol. They contained the same amount of IPA and less isopentanol than European bees (Collins et al. 1989). Large differences in the blends of adjoining and introgressing populations of the cape honey bee (*Apis mellifera capensis*) and the African honey bee (*A. m. scutellata*) were also discovered (Hepburn et al. 1994). Likewise, differences in alarm pheromone blends sufficiently large to distinguish among populations and subpopulations of North African and Spanish honey bees were reported (Hepburn and Radloff,

TABLE 11.3 Honey bee alarm pheromone components. Components with stinging or activating activity: active, **bold**; inactive, *italic*; those in parentheses likely are tissue contaminants and not actual alarm pheromone components; rest either not tested, very minor activity, or yielded conflicting results. Compounds biosynthetically related indicated by arrows.[1]

Alcohols (rel. amount)[2]	Esters (rel. amount)[2]	Acids (rel. amount)[2]
Butanol (t) →	**Butyl acetate** (+)	
↘	Butyl butyrate (+)	
Isopentanol (+) →	**Isopentyl acetate** (3+)	*3,3-Dimethylacrylic acid* (t)
↘	Isopentyl butyrate (+)	
	Hexyl acetate (2+)	
	Isohexyl acetate (t)	
Heptanol (+)		
2-Heptanol (+) →	**2-Heptyl acetate** (2+)	
Octanol (2+) →	*Octyl acetate* (2+)	*Octanoic acid* (+)
	↓ ?	
	1-Oct-2-enyl acetate (2+)	
2-Octenol (2+)		
2-Nonanol (2+-4+) →	*2-Nonyl acetate* (3+)	
	↓ ?	
	Nonyl acetate (2+)	
	↓ ?	
	1-Non-2-enyl acetate (2+)	
	Decyl acetate (+-3+)	Aromatics
	2-Decenyl acetate (+)	*Phenol* (t)
2-Undecanol (t)		Cresol (t)
(9-octadecenol (3+))		*Benzyl alcohol* (+)
(Octadecanol (2+))		*Benzoic acid* (+)
Δ^x-Nonadecenol (2+)		**Benzyl acetate** (2+)
(Z)-11-Eicosenol (5+) → ?	*Eicosyl acetate* (2+)	

[1]Data from Blum et al., 1978; Collins and Blum, 1982, 1983; Pickett et al., 1982; Free et al., 1983, 1989; Blum and Fales, 1988; Hepburn and Radloff, 1996.

[2]Relative quantities increasing from t (trace) to a maximum of 5+; major components are 3+ or greater.

1996). In general these studies indicate that honey bees from Africa appear to have secretions enriched with 2-nonanol and decyl acetate relative to bees derived from European populations. Too little is known about the differences in alarm pheromone chemistry among populations of other species to make useful statements. One observed difference between the open nesting species (*A. dorsata* and *A. florea*) and the cavity nesters (*A. mellifera* and *A. cerana*) is that the open nesters contain considerably greater quantities of the large, slowly vaporizing 2-decenyl acetate than the cavity nesters. The duration of the alarm reaction is considerably extended in these two species relative to the cavity nesters, leading the authors to suggest that open nesters require better defenses, and that the slowly evaporating 2-decenyl acetate probably serves in that role (Koeniger et al. 1979).

The exact roles of each of the plethora of components in honey bee alarm pheromones is unknown and behavioral testing has been only rudimentary. The compounds listed in bold in Table 11.3 are those for which some individual activity in assays has been shown. Many compounds have been shown to be inactive, or have very low activities on their own. These components are indicated in italics in Table 11.3. Components that have not been tested, or for which conflicting results have been reported are shown in standard type. Assays of octanol and (Z)-11-eicosenol yielded conflicting results with low activity for octanol (Collins and Blum, 1983) and activity for eicosenol (Pickett et al. 1982) as compared to lack of stinging activity of either component by Free et al. (1989).

The unusual compound (Z)-11-eicosen-1-ol differs from the other alarm components in chemistry and possible function. It is much larger and less volatile, and is present in greater quantity, than the other alarm pheromone components (Table 11.3). Its site of production is unclear, though in *Apis mellifera* it appears to be produced in conjunction with the other alarm pheromone components (Pickett et al. 1982) and is present in the setose membrane area of the sting (Schmidt et al. 1997). It is absent in the alarm pheromone of *Apis laboriosa* (M.S. Blum, pers. communication), whereas in *A. cerana,* eicosenol is present in prodigious quantities in the venom within the venom reservoir (Schmidt et al. 1997). The role of eicosenol is unclear. Free et al. (1989) reported it lacked activity in their stinging assay, whereas Pickett et al. (1982) obtained increased stinging in targets impregnated with eicosenol relative to controls. Eicosenol in combination with IPA was also more active than IPA alone and the combined two components were as active as actual stings (Pickett et al. 1982). Clearly, further assays are needed to resolve the question of the activity of eicosenol in *Apis mellifera*. In *A. cerana*, the function of the massive quantities of eicosenol also is not obvious. The bees do not appear to spray venom into the air surrounding intruders, thereby releasing alarm pheromone, as is common among some species of social wasps (Pers. observations). Two possibilities for the operation of eicosenol are its volatilization from tiny droplets of venom present on the sting tip during alarm behavior which thereby signals other bees, or its vaporization from venom that leaked during the stinging process onto an assailant's skin near the sting implantation site and from there it operates as a long term marker.

Mechanism of Action of Alarm Pheromone

The exact mechanism of action of alarm pheromone is unclear. Its mechanism is, however, considerably more complicated and depends more on other factors than does the mechanism of typical sex pheromones. In honey bee sex pheromone, only one component, 2-oxydecenoic acid, is required for total activity (Gary, 1962). Also, the requirement for the effective operation of 2-oxydecenoic acid is apparently simply that it be released approximately four meters above the ground (Butler and Simpson, 1967). In contrast, no one component of alarm pheromone elicits a response nearly as great as an intact sting (Boch et al. 1962; Free and Simpson, 1968), yet a great many of the individual components induce some alarm reaction (Collins and Blum, 1982; Free et al. 1983). These observations suggest that alarm pheromones operate in fundamentally very different manners than sex pheromones. The best known sex pheromones are those of moths, which are characterized by having very specific blends of components, all or most of which are essential, and all of which must be present in very close to the natural ratio for full attraction to occur (Bell et al. 1995). If slight deviation in sex pheromone blend ratio occurs, or if a component is added or deleted from the blend, the pheromone either dramatically decreases in its activity, or becomes outright repellent. In the case of alarm pheromones, many individual components are capable of eliciting at least some response; indeed, many components not even present in the alarm pheromone (but which are chemically similar) can induce alarm (Boch and Shearer, 1971). Moreover, unlike sex pheromones where variations in the ratios of components within an interbreeding population are small (Bell et al. 1995), the ratios of honey bee alarm components vary considerably (Hepburn and Radloff, 1996). The ratios vary even more greatly among allopatric populations of honey bees. Finally, the alarm pheromones of several honey bee species are similar and induce cross-specific alarm reaction (Koeniger et al. 1979). Thus, unlike sex pheromones, honey bee alarm pheromones simply do not serve as species, or population, isolating mechanisms.

Why should alarm and sex pheromones, both releaser pheromones, be so different? The answer may lie in the function of the two pheromones. Sex pheromones have dual roles of enabling males and females of the same species to find each other, and of preventing matings with a different species from occurring. The second role -- the prevention of mating with a member of another species -- is the driving force for the specificity and precision of the sex pheromonal system. Simply stated, mistakes are too costly in evolutionary terms to allow imprecision in the sexual communication system. Alarm pheromones, on the other hand, serve a dramatically different function. Alarm pheromones function to arouse colony mates to awareness of a potential threat to the colony, and then to direct those aroused colony mates to the target and vulnerable areas on that target. For alarm pheromone to function properly, it needs to be very quick acting and to induce sufficient arousal to be effective. Alarm pheromones need not be nearly as precise in their communicative message as sex pheromones: a mistake simply is not nearly as costly. The worst that would happen from a communication mistake with alarm pheromone is that the colony would be aroused more slowly to a defensive mode, or it that would be falsely or prematurely aroused by an overly active pheromone. In evolutionary terms, both of these mistakes are much less costly than a mistake with sex pheromones. For an alarm pheromonal mistake, the cost mainly would be loss of energy consequent to activity in defending against an unimportant situation or to loss of foraging effort to collect resources. The other cost of a mistake would be a slight increased risk of successful predation. Either consequence is relatively minor.

Given the role of alarm pheromones to generally alert and arouse the colony to risks from potential predators, how then does the pheromone operate, and why so many components? Positive answers cannot be given, but plausible mechanisms can be suggested. Alarm pheromones of honey bees have chemical similarities to alarm pheromone components of some ant species (Blum, 1969). Perhaps, because ants can be major threats to honey bees, bees evolved an alarm response to chemicals indicative of ant presence. Such an alarm response would also be useful against larger vertebrate predators. But vertebrate predators present no ant-like chemical cues. If honey bees evolved one or more alarm pheromone components similar to those they already responded to, these would be evolutionarily favored defenses. Because of the general nature of the chemicals responsible for the alarm, the addition of one or more components to the blend likely would enhance the response of the bees to the alarm blend. Thus, in theory, the system is not driven as much by precision in signal, as by detectability and overall quantity of the signal. This would explain two factors: why the signal can be so complex and why the presence or absence of individual components is not crucial to the communication system. In essence, the alarm pheromone system probably operates as a loosely additive system -- many components contribute to enhancing the effectiveness of the whole, but none are essential to, or governing of, the communication.

Nasonov, the Orientation Pheromone Outside the Hive

Nasonov pheromone is the primary pheromone used by bees for orientation outside the environment. Originally described in 1883 by the Russian scientist Nassanoff, the pheromone is released by individual bees that have discovered the route to the hive entrance, and serves as an airborne chemical trail to lead other workers to the entrance (Sladen, 1901). The pheromone is produced in the Nasonov glands located in the dorsal intersegmental membrane between tergites 6 and 7 of worker bees and has a pleasant aroma of citrus. It consists of (*E,E*)-farnesol plus six monoterpenes which are various isomerizations and oxidative levels of the basic geraniol-nerol structure (Figure 11. 2). Nasonov pheromone can be considered a master orientation pheromone because in all of its roles, it reorients bees that are either disoriented or

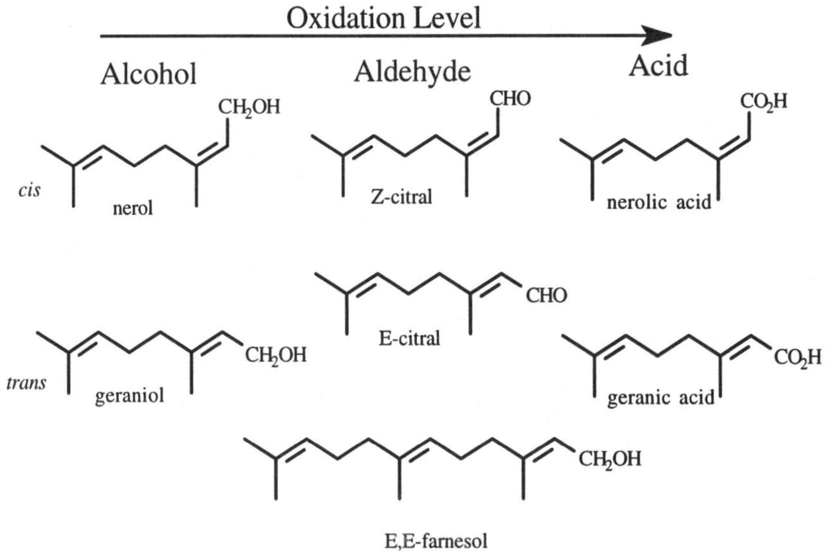

FIGURE 11.2 Chemical composition of Nasonov pheromone illustrating how all components are simply geometric isomers, different oxidative levels, or the addition of an isoprene unit to the basic geraniol structure.

lack directional knowledge. Three classical situations in which Nasonov pheromone is released include: 1) when a nest entrance lost by calamity or beekeeper manipulation is discovered (Figure 11.3); 2) at low odorant sugar syrup feeders or water; and 3) when workers discover a queen that was unable to continue with a swarm in flight and has landed (Free, 1987). In all of these situations, orientation information is communicated to bees in need of that information. When a nest entrance is blocked and a new one is discovered, the bees outside the colony at the time will lack the knowledge or chemical cues necessary to find the entrance. Bees that discover the entrance quickly communicate that information to arriving bees, often by forming strings of fanning and Nasonov-releasing bees that lead from the location of the old entrance to the new one.

Water and nectar foragers usually use odor to locate their sources of liquid (Williams et al. 1981). If the water, or artificial sources of sugar syrup are not aromatic, bees that discover the sources will scent mark with Nasonov to orient other foragers to the location (Free and Williams, 1970). But this use of Nasonov likely represents simply an effective means for a forager to add odor to an odorless source, rather than acting as a specific pheromonal signal (Wells et al. 1993).

The mature queen honey bee is the poorest flying member of a honey bee colony. During actual movement of a swarm through the air the queen, especially if she is a heavy queen from a European bee colony, might drop out of the flying swarm and land. In response to a landed queen, the first bees discovering her and those arriving shortly

FIGURE 11.3 Nasonov pheromone being released by bees that discovered a nest entrance.

afterwards cluster around her and release Nasonov pheromone to attract the rest of the swarm members (Avitabile et al. 1975).

Honey Bee Reproduction and Nasonov Pheromone

The culmination of the honey bee reproductive process is the production of a swarm which issues from the parent colony and seeks a new nest site. The process begins with the initiation of queen cell formation in preparation for producing queens to replace the old queen and/or to leave with swarms, and ends when the airborne swarm successfully reaches the chosen nest cavity and has moved in. Nasonov pheromone plays key roles in the outside activities throughout virtually all of this period. Several days before a swarm departs the natal nest, scout bees leave the nest and search for new nest sites. Historically these were usually cavities in trees or holes in rocks or the ground, but, with the advent of human activities, bees have an expanded variety of potential nest cavities in man-made structures or refuse. Scouts that discover potentially suitable nest cavities carefully examine the cavities for size, structural suitability, location, and microenvironmental conditions (Seeley, 1977). If the cavity is suitable, the scout will return to recruit other scouts to investigate the cavity (Lindauer, 1951). Sometime after this process begins a swarm will emerge from the nest and cluster nearby, often on a tree branch. From this bivouacking swarm cluster, scouting activity intensifies. Scouts that have discovered a good nest cavity will continue to return from the swarm to the cavity and often will scent at the new nest site with Nasonov to recruit other scouts. At the swarm cluster, the scout often will further scent with Nasonov. Typically, various scouts will have found different cavities and be recruiting others to investigate those cavities. This process continues with scouts and recruits investigating the various cavity choices until a "consensus" begins to emerge (Lindauer, 1955). The "winning" cavity now is visited almost continuously by numerous scouts and Nasonov scenting activity is strong. During this latter stage of recruiting the bees exhibit a peculiar behavior of bouncing off the inner sides of the cavity and generating a sound like "popcorn popping" that can be clearly heard when the cavity is a hollow resonant swarm trap (Schmidt et al. 1989; Schmidt, unpublished observations). Just before a swarm breaks bivouac and flies to the chosen cavity, almost all the scouts depart the cavity (Seeley et al. 1979), leaving it so quiet that a nearby human who has been observing the process is left with an "eerie feeling". The swarm breaks cluster, ascends into the air as a diffuse swirling mass of bees flying in large circular arcs, with bees moving in both directions and slowly begins to move toward the chosen cavity. At first, all the bees including the queen, move slowly as this diffuse mass in the direction of the new nest site. Once moving, the swarm increases speed and condenses into the shape of a comet or airborne "slug". Nasonov plays a key role during this time. It is being released in high concentration by bees already at the nest cavity and apparently also by knowledgeable scouts that rapidly fly back and forth through the moving swarm leading it in the correct direction (Lindauer, 1951; Avitabile et al. 1975; Seeley et al. 1979). When the swarm approaches the nest cavity, it slows and begins swirling in diffuse circles, somewhat reminiscent of the departing swarm. Before this, some bees had flown ahead of the moving swarm, landed at the cavity entrance, and begun releasing Nasonov pheromone and fanning (Seeley et al. 1979). These scenting bees are joined by the first arriving bees from the swarm which also land near the cavity entrance and heavily Nasonov scent and fan, as do many others that have landed on nearby vegetation (Figure 11.4). Airborne bees, including the queen, quickly cluster at the cavity and begin moving into the entrance. The whole landing and moving in process occurs quickly, often within a 10 minute period.

Nasonov pheromone serves key communication roles in two other situations relating to the swarming process. First, it is the crucial pheromone synergized by queen pheromone to induce swarm clustering and to maintain the cluster once formed (Morse and Boch, 1971; Seeley

Figure 11.4 First arriving bees that landed on vegetation and are releasing Nasonov pheromone to attract and orient airborne bees to settle and form a swarm cluster. Arrows point to exposed Nasonov glands.

et al. 1979; Free et al. 1981). Second, as alluded to earlier, it is essential for discovery of a queen downed during flight by flying workers (Mautz et al. 1972). Without these roles of Nasonov pheromone, swarms would experience difficulty maintaining a functional colony group during the swarm cluster phase and many swarms would perish from loss of their reproductive queen.

Chemical Properties and Biological Assays of Nasonov Pheromone Communication During the Swarming Process

Nasonov pheromone is a multi-component blend and serves in multiple communication situations. Are all the components of the pheromone essential for its activity? Do the various separate components have different roles in different situations? These questions are difficult to answer, and answers are complicated by the fact that the results of bioassays in one situation cannot be interpreted to apply to all situations. For example, assays involving the role of Nasonov components in its communicative effectiveness for water collecting bees cannot be assumed to describe correctly the role of the components in the context of the swarming process. In the case of communicating information that influences swarm clustering and selection of nest cavities, clear differences in the components are evident. Free and coworkers extensively investigated these functional idiosyncracies by use of moving "roundabouts" containing lures with different pheromone components and blends. They determined that swarm clustering behavior is highly dependent upon the pheromone blend. Individual components in decreasing order of importance were: nerolic acid, geranic acid, (*E*)-citral, (*Z*)-citral, geraniol, nerol, and (*E,E*)-farnesol. Mixtures were clearly better than individual components with nerolic acid, (E)-citral, and geraniol essential for good swarm clustering (Free et al. 1981). The natural blend of Nasonov pheromone (geraniol: nerolic acid: (*E,E*)-farnesol: geranic acid: (*E*)-citral: (*Z*)-citral: nerol 100:78:44:11: 1.1:0.6:0.4, Pickett et al. 1980) was not required for optimal activity; in fact a blend of equal portions of geraniol, citrals (in normal synthetic ratio), and nerolic + geranic acids (in normal synthetic ratio) was as effective as the natural blend in formation of clusters (Free et al. 1982). In this assay nerol and farnesol were unimportant, perhaps even detrimental, and could be eliminated without reduction of activity (Free et al. 1981).

In assays involving attraction of natural swarms to artificial nest cavities (swarm traps) Schmidt and Thoenes (1987a,b) found that equal amounts of citrals, geraniol, and nerolic + geranic acids sealed in slow release polyethylene lures were highly effective in attracting swarms. Pheromone concentration can affect the communication signal and consequent behavior of the receiver as illustrated by altered behaviors including repellency of otherwise attractive alarm pheromones in bees and ants (Blum, 1969), or disruption of upwind orientation of moths flying to sex pheromone (Roelofs, 1978). Swarm attraction to Nasonov pheromone in the swarm trap assay occurs over a wide range of concentrations, with the general observation that within the release rates tested, the higher the concentration, the greater the attractancy. In direct choice tests, traps containing 1 lure with 10 mg synthetic Nasonov pheromone attracted only 1 swarm, contrasted to 13 swarms attracted by traps with 5 tubes of 10 mg pheromone each, and 32 swarms attracted by traps with 25 tubes of 10 mg pheromone each (Schmidt and Thoenes, 1990). At the highest concentration tested, odor of pheromone was readily detectable by human observers standing downwind of a trap.

Very little is known about the Nasonov pheromones of species other than *Apis mellifera*. *Apis cerana* in Malaysia appears to contain at least (*E*) and (*Z*) citrals and geraniol, and homeless bees of this population tend to cluster around citral lures (Abdullah et al. 1990). *Apis cerana japonica* appears to have a different Nasonov blend with linalool and related compounds as the major constituents (Sasagawa et al. 1996). In our preliminary studies we have found that Indonesian species in the *cerana* species group are attracted to small swarm traps baited with equal proportions of geraniol, citrals, and nerolic + geranic acids (Schmidt et al. unpublished).

The relative importance of the individual components of Nasonov pheromone has not been thoroughly determined in the context of swarm attraction to nest cavities. The acids, nerolic and geranic, are, however, important for full attractiveness of the pheromone. In choice tests involving side-by-side identical swarm traps, only 1 swarm was attracted to a trap containing equal portions of geraniol and the citrals, compared to 17 attracted to traps containing equal portions of geraniol, citrals, and nerolic + geranic acids (Schmidt and Thoenes, 1992).

Pheromonal communication cannot be separated from the colony's needs, and from the particular environment in which the pheromone is released. Often one is faced with the difficulty of determining if a pheromone is actually playing a role via communication or if the behavior would have occurred independently of the pheromone. In most releaser pheromone research this question can be easily answered by careful observation, but in the case of pheromonal communication in rare events, such as swarms choosing a nest cavity, observation alone is not adequate. To determine if Nasonov pheromone or other properties of the swarm trap itself are governing attractiveness of swarm traps to swarms, side by side swarm traps were set up in Guanacaste, Costa Rica. One of the traps contained the usual synthetic pheromone blend of equal amounts of citrals, geraniol, and nerolic + geranic acid, another trap contained the pheromone without the acid components, and the third trap contained no pheromone. Tropical bees were selected because they are generally believed to be less choosy than temperate bees (Schmidt and Hurley, 1995). No swarms chose the pheromone-free traps, 1 swarm chose traps with the citral-geraniol pheromone blend, and 8 swarms chose traps containing the usual complete pheromone blend (Schmidt and Thoenes, 1992). These results revealed that tropical bees not only are attracted to cavities containing synthetic Nasonov pheromone in preference to non-scented cavities, but also that they are preferentially attracted to more complete pheromone blends than to incomplete blends.

The argument could be made that the traps themselves were sufficient for all the attraction observed and that the pheromone only "tricked" the bees during the last few meters of flight to move into the pheromone-containing traps instead of into the control traps. To eliminate this possibility, a crossover experimental design was used in which half of the trap locations contained traps with Nasonov pheromone and half were pheromone-free controls. On a weekly basis the pheromone lures were removed from the traps containing the lures and placed in the traps without them. This protocol controls for both temporal and spacial variables because at any given time a swarm can only discover and select or reject a cavity, but cannot chose between cavities with and without pheromone. Only 4 swarms chose swarm traps without pheromone, compared to 19 that selected traps with pheromone (Schmidt, 1994). Since the traps were identical and in constant locations, with only the pheromone being moved, the results show that Nasonov pheromone enhances trap selection by an almost five-fold factor compared to controls.

Even the crossover experiment testing the attraction of Nasonov pheromone to swarms cannot rule out the possibility that Nasonov pheromone is simply acting as an olfactory cue to enable scouts to locate a cavity. Wells et al. (1993) discovered that in the case of foragers seeking sugar solutions, Nasonov components were not necessarily acting as a pheromone, but simply as an added odor. As evidence for this, they reported that their Nasonov mixture (citrals: geraniol: nerol of 2:2:1) was not more attractive to scouts than the common odors bay, anise, cajeput, or clove. To test if a similar effect could be happening in the case of bees seeking nest cavities, I tested Nasonov versus clove oil and Nasonov versus linalool in crossover experiments. In these experiments,

Nasonov attracted about four times as many swarms as the other scents, which, in turn, attracted no more swarms than previously recorded in pheromone-free traps (Schmidt, unpublished). Thus, although these tests are not definitive proof that Nasonov is a pheromone of swarming, they do show that Nasonov is not simply "an odor" and provide strong evidence that it is, in fact, acting in a pheromonal role.

Mechanism of Communication by Synthetic Nasonov Pheromone in Swarm Trap Assays

The basic question is "how does synthetic Nasonov pheromone cause swarms to select swarm traps for nest cavities"? The answer cannot be given with certainty, but the probable mechanism is understood. Synthetic Nasonov pheromone helps scout bees engaged in nest cavity seeking behavior to discover the swarm trap. The mechanism is simple -- the artificial pheromone communicates to searching scouts that "another scout has already discovered a good potential nest cavity". The pheromone achieves this communication by generating a long downwind volume of air that contains the pheromone. A scout bee that flies into this pheromone-laden air will switch to an upwind anemotactic (zig-zag) flight behavior to locate the pheromone source and the entrance to the cavity. This system increases the discovery rate of potential nest cavities by scouts because entrances to suitable nest cavities are rare, small, and often hidden from direct vision. By providing a chemical signal to aid in nest location, the pheromone also reduces the time spent by the scout in investigating unpromising or dangerous objects that appear like cavity entrances.

Scout bees use Nasonov pheromone alone in the manner described above, and not queen pheromone for discovering nest cavities. Evidence for this comes from experiments using combinations of both queen pheromone and Nasonov pheromone in traps. Nasonov pheromone alone increases trap attractancy by five-fold (Schmidt, 1994), but could queen pheromone enhance the attraction? Using a crossover design experiment, half of the test locations received separate traps with Nasonov pheromone alone, queen pheromone alone, and no pheromone, and the other half of the locations received separate traps with Nasonov pheromone + queen pheromone, queen pheromone alone, and no pheromone. The results revealed that neither the untreated (one swarm) nor queen pheromone alone (0 swarms) traps were attractive to swarms, whereas Nasonov containing traps were highly attractive (21 swarms in Nasonov + queen traps; 24 swarms in Nasonov alone traps). The conclusion from this experiment is that Nasonov and Nasonov alone acts as the long range attractant for scout bees (and hence swarms) and that queen pheromone has no long range attractancy either alone or in combination with Nasonov pheromone. Thus, no synergism is occurring between queen pheromone and Nasonov pheromone (Schmidt et al. 1993), a conclusion supported by a previous experiment with a different design (Villa and Schmidt, 1992).

In another manipulation all locations simultaneously contained separate traps with Nasonov + queen pheromone, Nasonov pheromone alone, queen pheromone alone, and no pheromone. In this experiment 17 swarms selected the traps with Nasonov + queen pheromone versus 2 swarms that selected traps with Nasonov alone (and no swarms chose the other two). Although Nasonov acts alone for long distance communication, at short range queen pheromone acts by communicating to the arriving bees that "their queen is here" and, hence, they select the trap with the queen + Nasonov instead of the one with Nasonov alone (Schmidt et al. 1993). In this situation, the two pheromones operate differently -- Nasonov pheromone acts primarily at long distance where it serves to communicate the location of a nest cavity, and queen pheromone acts at short distance where it communicates the location of the queen, which when combined with Nasonov overrides the solo Nasonov effect.

The Pheromone-Environmental Relationship in Nest Seeking Behavior

Nasonov pheromone is the messenger that a potentially suitable nest cavity is nearby; the cavity environment and the cavity itself determine the acceptability of the cavity to the scouts and swarm. In large part, the two factors operate independently. For centuries, and probably millennia, honey bee observers knew that swarms do not randomly chose nest sites and that bees are selective about their nest cavities. In more recent times quantitative observations and experiments have elucidated many cavity properties that are important to swarm bees selecting a nest (Table 11.4). Swarms typically prefer cavities with internal volumes in the range of 20-60 liters, though sometimes smaller or larger cavities are inhabited. Bees from temperate populations tend to prefer larger cavities than bees from tropical populations, and their size preferences are stronger than those of their tropical relatives. Bees also prefer sturdy, well insulated wooden cavities over unsound or inferior appearing cavities such as cardboard, plastic, combinations of the two, or polystyrene. Bees will, however, accept poorer quality cavities when the choices are few. Based on the limited available data, swarms appear not to be overly influenced by cavity shape, exposure (sunny versus deep shade) and,

TABLE 11.4 Properties of cavities affecting nest cavity selection by honey bee swarms.

Cavity Property	Finding	Reference	Comments
Volume	Feral colonies (NY): 20-60 l, median 45 l	Seeley and Morse, 1976	n=21 colonies
	Feral colonies (Africa): 34 ± 5 l	McNally and Schneider, 1996	n=113 colonies
	Feral colonies (Mex): 70% in 10-30 l range	Ratnieks et al., 1991	n=16 colonies
	Tests (NY): pref. in 25-40 l range	Seeley, 1977	n=small, 70 l acceptable
	Tests (AZ): 31 l preferred to 13.5	Schmidt and Hurley, 1995	68 in 31 l, 22 in 13.5 l
	Tests (Costa Rica): random, 31, 24, 13.5 l	Schmidt and Hurley, 1995	9 in 31 l, 8 in 24 l, 14 in 13.5 l
Material	Tests (AZ): wood pulp over cardboard	Schmidt and Thoenes, 1987a	35 in 48 pulp, 0 in 16 cardboard
	Tests (AZ): pulp over cardboard+plastic	Schmidt et al., 1989	12 in pulp, 0 in cardboard+plastic
	Wood pulp pref. to plastic tubs	Rubink, pers. communication	
Elevation	Tests (NY): 5 m over 1 m	Seeley and Morse, 1978	only six swarms total
	Tests (AZ): 3 m over 1 m	Schmidt and Thoenes, 1987b	25 at 3 m, 7 at 1 m
Shape	Tests (NY): random, cube vs elongate box	Seeley and Morse, 1978	6 to cube, 3 to box
	Tests (AZ): random, pot vs elongate box	Schmidt and Thoenes, 1992	13 to pot, 7 to box
Exposure	Tests (NY): random, exposed vs not exp.	Seeley and Morse, 1978	8 exposed; 2 unexp, p=ns, χ^2
Dryness	Tests (NY): insufficient data	Seeley and Morse, 1978	3 to wet, 2 to dry boxes
Draftiness	Tests (NY): random (?)	Seeley and Morse, 1978	4 drafty, 3 sound boxes

TABLE 11.4 (Continued).

Cavity Property	Finding	Reference	Comments
Odors	Tests (NY): no pref. for old comb	Visscher et al., 1985	10 comb, 5 no comb; p=ns, χ^2
	Tests (NY): old comb attractive	Ratnieks, 1988	6 to old combs, 0 to foundation
	Tests (AZ): old comb attractive	Schmidt, unpublished	11 to old comb, 0 to no comb
	Tests (AZ): propolis attractive	Schmidt, unpublished	11 to propolis, 3 to no propolis
Entrance size	Feral colonies (NY): mode of 10-20 cm^2	Seeley and Morse, 1976	n=33 colonies
	Feral colonies (CT): most 10-70 cm^2 range	Avitabile et al., 1978	n=108 trees
	Feral colonies (CA): median = 46 cm^2	Gambino et al. 1990	n=69
	Tests (NY): 12.5 over 75 cm^2	Seeley and Morse, 1978	6 small, 0 large; $p<.05$, χ^2
	Tests (FL): 8 over 31 cm^2	Morse et al., 1993	26 small, 0 large
Entrance location	Feral colonies (NY): bottom preferred	Seeley and Morse, 1976	17 bottom, 5 mid, 7 top
	Feral colonies (Africa): top preferred	McNally and Schneider, 1996	61 top, 10 mid, 20 bottom
	Tests (NY): random	Seeley and Morse, 1978	8 bottom, 2 top; p=ns, χ^2
Entrance orientation	Feral colonies (NY): random	Seeley and Morse, 1976	n=41 entrances
	Feral colonies (CT): southwest preference	Avitabile et al., 1978	n=106 primary entrances
	Feral colonies (CA): random	Gambino et al., 1990	n=70 entrances
	Feral colonies (Vict. Aust.): random	Oldroyd et al., 1994	n=27 entrances
	Feral colonies (Africa): away from sun	McNally and Schneider, 1996	n=105 entrances
	Tests (NY): random[1]	Seeley and Morse, 1978	124 artificial cavities, 74 swarms

[1]Recalculation of the original data reveals no directional preference among the 8 compass directions ($\chi^2_7=5.88$; $P\approx.5$); visual inspection of the data suggests an apparent preference for orientations toward SE, S, and SW, but the lowest directional frequency (W) is adjacent to the highest directional frequency (SW).

perhaps, dryness or draftiness (Table 11.4). They do, however, prefer cavities elevated 3 or 5 m above ground to those only 1 m above ground.

Odors sometimes appear to influence cavity selection. Some man-made odors are repellent (Witherell and Lewis, 1986; Schmidt, unpublished), whereas odors indicating that another honey bee colony had previously inhabited the cavity are attractive. Such "colony" odors are present in old bee comb or propolis (Table 11.4).

Cavity entrance parameters also influence nest site selection by honey bee swarms (Table 11.4). In feral bee colonies nest entrances typically range in area from 10-70 cm^2, though smaller and larger entrances are observed. In choice tests, swarms prefer entrances of 8 or 12.5 cm^2 to those of 31 or 75 cm^2. All of these observations relate to bees from temperate areas. Differences likely will be observed for tropical bees: indeed, tropical bees often appear to prefer larger entrances, presumably to facilitate cooling (Schmidt, personal observations). Preference for entrance location appears to depend upon the environment and the population of bees living in the environment. In temperate areas, bees prefer entrances located near the bottom of the cavity; in tropical areas, bees prefer entrances located near the top of the cavity. These choices reflect the environmental conditions -- in cold climates bottom entrances conserve warmth, and in hot climates top entrances allow the heat to rise out the entrance. Much inconsistency appears in the data relating to possible swarm preference for entrance direction. Most reports (Table 11.4) indicate random entrance orientation, that is, no preference, but the two largest studies indicated preference for entrances facing toward the afternoon sun and warmth in cold climates, and away from the hot afternoon sun in hot climates (Avitabile et al. 1978; McNally and Schneider, 1996). These numerous investigations indicate that swarms do not typically express a preference for nest entrance orientation, but that under extreme conditions in hot or cold environments a slight preference might be observed.

Based on the studies listed in Table 11.4, it is evident that honey bee swarms have distinct, often strong, preferences for nest cavities, and that they are capable of discriminating among cavities. Observations of colonies nesting in unusual, even maladaptive, locations do not indicate that bees are not choosy. Rather, these observations suggest that bees necessarily have a wide latitude in their tolerances and that, when given a set of poor choices for nest sites, they will accept the "least bad" available site. There is no *a priori* reason to expect bees of different populations to have identical cavity preferences. Indeed, much anecdotal and experimental evidence suggests that each bee population has its own idiosyncratic set of preferences and tolerances for nest cavity parameters. The mere fact that swarms do have strong as well as subtle preferences for cavity properties is clear evidence that swarms can, and do, critically evaluate potential new nests.

Nasonov pheromone cannot override the environment. It alone cannot entice a swarm into a nest cavity. A poor potential nest cavity will not be accepted even if it contains Nasonov pheromone. Because pheromone is present, scouts will inspect the cavity, but they will carry the selection process no further. To attract a swarm, the cavity fundamentally must be acceptable; and the more ideal in terms of the criteria important to bees, the more likely the cavity is to be accepted. Pheromone is the messenger, the cavity is the message.

Conclusion

Large social insect colonies as typified by honey bees have a number of advantages, as well as disadvantages, over insects that are solitary or live in small colonies. Through the use of mass action releaser pheromones, the large number of individuals in these colonies can mobilize and effectively solve some of their most serious problems. Predation on the colony resources and population by large mammalian or avian predators represents a major survival threat. Honey bee alarm pheromone, by mobilizing masses of individuals -- sometimes as many as half the adult population -- into defensive attacks against predators orchestrate perhaps the best insect defense on earth. When it comes time for a honey bee colony to reproductively divide or move, its population size presents other problems. Nasonov pheromone is the mass action releaser pheromone that coordinates tens of thousands of individuals to achieve an orderly movement to a new nest cavity. Alarm and Nasonov pheromones share several chemical features in common, features that differ from those of typical sex or other pheromones. Both pheromones consist of large blends of volatile compounds and, in both cases, neither the blend ratio, nor even some of the components themselves are essential or required for activity. These pheromones appear to function mainly to alert, activate, and direct, rather than to communicate specific information of a highly precision nature. Thus, unlike sex pheromones, small errors on the receiving individual's part would not generally have a major negative impact. In all of biology there are probably no better systems for communication outside the nest of a social species than honey bee alarm and Nasonov pheromones, and few examples exist of such exquisite synergisms of behavior and chemical communication.

References Cited

Abdullah, N., A. Hamzah, J. Ramli and M. Mardan. 1990. Identification of Nasonov pheromones and the effects of synthetic pheromones on the clustering activity of the asiatic honeybee (*Apis cerana*). *Pertanika* 13:189-194.

Avitabile, A., R. A. Morse and R. Boch. 1975. Swarming honey bees guided by pheromones. *Ann. Entomol. Soc. Amer. 68:*1079-1082.

Avitabile, A., D. P. Stafstrom and K. J. Donovan. 1978. Natural nest sites of honeybee colonies in trees in Connecticut, USA. *J. Apic. Res. 17:*222-226.

Bell, W. J., L. R. Kipp and R. D. Collins. 1995.The role of chemo-orientation in search behavior. *In*: R.T. Cardé and W.J. Bell, Eds. *Chemical Ecology of Insects* 2, pp. 105-152, Chapman & Hall, New York.

Blum, M. S. 1969. Alarm pheromones. *Annu. Rev. Entomol. 14:*57-80.

Blum, M. S. and H. M. Fales. 1988. Chemical releasers of alarm behavior in the honey bee: informational 'plethora' of the sting apparatus signal. *In:* G.R. Needham, R.E. Page, M. Delfinado-Baker and C.E. Bowman, Eds., *Africanized Honey Bees and Bee Mites*, pp. 141-148, Wiley, publ., New York.

Blum, M. S., H. M. Fales, K. W. Tucker and A. M. Collins. 1978. Chemistry of the sting apparatus of the worker honeybee. *J. Apic. Res. 17:*218-221.

Boch, R. and W. C. Rothenbuhler. 1974. Defensive behaviour and production of alarm pheromone in honeybees. *J. Apic. Res.* 13: 217-221.

Boch, R. and D. A. Shearer. 1966. Iso-pentyl acetate in stings of honeybees of different ages. *J. Apic. Res. 5:*65-70.

Boch, R. and D. A. Shearer. 1971. Chemical releasers of alarm behaviour in the honey-bee, *Apis mellifera. J. Insect Physiol. 17:*2277-2285.

Boch, R., D. A. Shearer and B. C. Stone. 1962. Identification of iso-amyl acetate as an active component in the sting pheromone of the honey bee. *Nature 195:*1018-1020.

Brandeburgo, M. A. M. 1979. Estudo da influencia do clima da agressividade da abelha africanizada. Masters Thesis, Faculdade de Medicina de Ribeirao Preto, Univ. de Sao Paulo.

Brandeburgo, M. M., L. S. Goncalves and W. E. Kerr. 1982. Effects of Brazilian climatic conditions upon the aggressiveness of Africanized colonies of honeybees. *In*: P. Jaisson, Ed. *Social Insects in the Tropics,* pp. 255-280, *Univ. Paris-Nord*, Paris.

Butler, C. G. and J. Simpson. 1967. Pheromones of the queen honeybee *Apis mellifera* L.) which enable her workers to follow her when swarming. *Proc. R. Entomol. Soc. Lond. (A) 42:*149-154.

Collins, A. M. 1981. Effects of temperature and humidity on honeybee response to alarm pheromones. *J. Apic. Res. 20:*13-18.

Collins, A. M. and M. S. Blum. 1982. Bioassay of compounds derived from the honeybee sting. *J. Chem. Ecol. 8:*463-470.

Collins, A. M. and M. S. Blum. 1983. Alarm responses caused by newly identified compounds derived from the honeybee sting. *J. Chem Ecol. 9:*57-65.

Collins, A. M. and K. J. Kubasek. 1982. Field test of honey bee (Hymenoptera: Apidae) colony defensive behavior. *Ann. Entomol. Soc. Amer. 75*:383-387.

Collins, A. M. and W. C. Rothenbuhler. 1978. Laboratory test of the response to an alarm chemical, isopentyl acetate, by *Apis mellifera. Ann. Entomol. Soc. Amer. 71*:906-909.

Collins, A. M., T. E. Rinderer, K. W. Tucker, H. A. Sylvester and J. J. Lackett. 1980. A model of honeybee defensive behaviour. *J. Apic. Res. 19*:224-231.

Collins, A. M., T. E. Rinderer and K. W. Tucker. 1988. Colony defence of two honeybee types and their hybrid 1. naturally mated queens. *J. Apic. Res. 27*:137-140.

Collins, A. M., T. E. Rinderer, H. V. Daly, J. R. Harbo and D. Pesante. 1989. Alarm pheromone production by two honeybee (*Apis mellifera*) types. *J. Chem. Ecol. 15*:1747-1756.

Free, J. B. 1961. The stimuli releasing the stinging response of honeybees. *Anim. Behaviour 9*:193-196.

Free, J. B. 1987. *Pheromones of Social Bees*. Ithaca, NY: Cornell Univ. Press.

Free, J. B. and J. Simpson. 1968. The alerting pheromones of the honeybee. *Z. Vergl. Physiol. 61*:361-365.

Free, J. B. and I. H. Williams. 1970. Exposure of the Nasonov gland by honeybees (*Apis mellifera*) collecting water. *Behaviour 37*:286-290.

Free, J. B., A. W. Ferguson and J. A. Pickett. 1981. Evaluation of the various components of the Nasonov pheromone used by clustering honeybees. *Physiol. Entomol. 6*:263-268.

Free, J. B., A. W. Ferguson, J. A. Pickett and I. H. Williams. 1982. Use of unpurified Nasonov pheromone components to attract clustering honeybees. *J. Apic. Res. 21*:26-29.

Free, J. B., A. W. Ferguson, J. R. Simpkins and B. N. Al-Sa'ad. 1983. Effect of honeybee Nasonov and alarm pheromone components on behaviour at the nest entrance. *J. Apic. Res.* 22:214-223.

Free, J. B., A. W. Ferguson and J. R. Simpkins. 1989. Honeybee responses to chemical components from the worker sting apparatus and mandibular glands in field tests. *J. Apic. Res. 28*:7-21.

Gambino, P., K. Hoelmer, and H. V. Daly. 1990. Nest sites of feral honey bees in California, USA. *Apidologie 21*:35-45.

Gary, N. E. 1962. Chemical mating attractants in the queen honey bee. *Science 136*:773-774.

Ghent, R. L. and N. E. Gary. 1962. A chemical alarm releaser in honey bee stings (*Apis mellifera* L.). Psyche 69:1-6.

Hepburn, H. R. and S. E. Radloff. 1996. Morphometric and pheromonal analyses of *Apis mellifera* L along a transect from the Sahara to the Pyrenees. *Apidologie* 27:35-45.

Hepburn, H. R., G. E. Jones and R. Kirby. 1994. Introgression between *Apis mellifera capensis* Escholts and *Apis mellifera scutellata* Lepeletier: the sting pheromones. *Apidologie* 25:557-565.

Huber, F. 1814. *New Observations Upon Bees*. Translated from French by Dadant (C.P.) Ed. 1926. Hamilton, Illinois: American Bee Journal.

Koeniger, N., J. Weiss, and U. Maschwitz. 1979. Alarm pheromones of the sting in the genus *Apis*. *J. Insect Physiol.* *25*:467-476.

Kolmes, S. A. and L. A. Fergusson-Kolmes. 1989. Measurements of stinging behaviour in individual worker honeybees (*Apis mellifera* L.). *J. Apic. Res.* *28*:71-78.

Lecomte, J. 1951. Les facteurs de l'agressivite chez l'abeille. *C. R. Acad. Sci.* *232*:1376-1378.

Lecomte, J. 1961. Le comportement agressif des ouvrieres d'*Apis mellifica* L. *Ann. Abeille* *4*:165-270.

Lefebvre, M. G. and A.J. Beattie. 1991. Sound responses of honey bees to six chemical stimuli. *J. Apic. Res.* *30*:156-161.

Lindauer, M. 1951. Bienentänze in der Schwarmtraube. *Naturwissenschaften 38:* 509-513.

Lindauer, M. 1955. Schwarmbienen auf Wohnungssuche. *Z. Vergl. Physiol.* *37*:263-324.

Mauchamp, B. and D. Grandperrin. 1982. Chromatographie en phase gazeuse des composes volatils des glandes a pheromones des abeilles: methodes d'analyse directe. *Apidologie 13*:29-37.

Maschwitz, U. 1964. Gefahrenalarmstoffe und Gefahrenalarmierung bei sozialen Hymenopteren. *Z. Vergl. Physiol.* *47*:596-655.

Mautz, D., R. Boch and R. A. Morse. 1972. Queen finding by swarming honey bees. *Ann. Entomol. Soc. Amer.* *65*:440-443.

McNally, L. C. and S. S. Schneider. 1996. Spatial distribution and nesting biology of colonies of the African honey bee *Apis mellifera scutellata*, in Botswana, Africa (Hymenoptera: Apidae). *Environ. Entomol.* 25: 643-652.

Moritz, R. F. A., E. E. Southwick and M. Breh. 1985. A metabolic test for the quantitative analysis of alarm behavior of honeybees (*Apis mellifera* L.). *J. Exp. Zool.* *235*:1-5.

Morse, R. A. and R. Boch. 1971. Pheromone concert in swarming honey bees (Hymenoptera: Apidae) *Ann. Entomol. Soc. Amer.* *64*:1414-1417.

Morse, R. A., D. A. Shearer, R. Boch and A. W. Benton. 1967. Observations on alarm substances in the genus *Apis*. *J. Apic. Res.* *6*:113-118.

Morse, R. A., J. N. Layne, P. K. Visscher and F. Ratnieks. 1993. Selection of nest cavity volume and entrance size by honey bees in Florida. *Florida Sci.* *56*:163-167.

Newton, D. C. 1969. Behavioural response of honeybees to colony disturbance by smoke. II. guards and foragers. *J. Apic. Res.* *8*:79-82.

Oldroyd, B. P., S. H. Lawler and R. H. Crozier. 1994. Do feral honey bees (*Apis mellifera*) and regent parrots (*Polytelis anthopeplus*) compete for nest sites? *Aust. J. Ecol.* *19*:444-450.

Pickett, J. A., I. H. Williams, A. P. Martin and M. C. Smith. 1980. Nasonov pheromone of the honey bee, *Apis mellifera* L. (Hymenoptera: Apidae) Part I. chemical characterization. *J. Chem. Ecol.* *6*:425-434.

Pickett, J. A., I. H. Williams and A. P. Martin. 1982. (Z)-11-Eicosen-1-ol, an important new pheromonal component from the sting of the honey bee, *Apis mellifera* L. (Hymenoptera, Apidae). *J. Chem. Ecol. 8:*163-175.

Ratnieks, F. L. W. 1988. Improved bait hives. *Amer. Bee J. 128:*125-127.

Ratnieks, F. L. W., M. A. Piery, and I. Cuadriello. 1991. The natural nest and nest density of the Africanized honey bee (Hymenoptera, Apidae) near Tapachula, Chiapas, Mexico. *Can. Entomol. 123:*353-359.

Roelofs, W. L. 1978. Threshold hypothesis for pheromone perception. *J. Chem. Ecol. 4:*685-699.

Rothenbuhler, W. C. 1974. Further analysis of committee's data on the Brazilian bee. *Amer. Bee J. 114:*128.

Sawagawa, H., S. Matsuyama, R. Yamaoka and M. Sasaki. 1996. The oriental orchid lures the Japanese honey bee with Nasonov pheromone mimics. *Proc. 20Th Intl. Cong. Entomol., Firenze, Italy.* P. 439.

Schmidt, J. O. 1994. Attraction of reproductive honey bee swarms to artificial nests by Nasonov pheromone. *J. Chem. Ecol. 20:*1053-1056.

Schmidt, J. O. 1996. Chemical and visual cues used by defending honey bees to recognize mammalian predators. *Proc. 20 Intl. Congr. Entomol, Firenze, Italy*. p. 417.

Schmidt, J. O. and L. V. Boyer-Hassen. 1996. When Africanized bees attack: what you and your clients should know. *Veterinary Med. 91:*923-928.

Schmidt, J. O. and R. Hurley. 1995. Selection of nest cavities by Africanized and European honey bees. *Apidologie 26:*467-475.

Schmidt, J. O. and S. C. Thoenes. 1987a. Honey bee swarm capture with pheromone-containing trap boxes. *Amer. Bee J. 127:*435-438.

Schmidt, J. O. and S. C. Thoenes. 1987b. Swarm traps for survey and control of Africanized honey bees. *Bull. Entomol. Soc. Amer. 33:*155-158.

Schmidt, J. O. and S. C. Thoenes. 1990. Honey bee (Hymenoptera: Apidae) preferences among artificial nest cavities. *Ann. Entomol. Soc. Amer. 83:*271-274.

Schmidt, J. O. and S. C. Thoenes. 1992. Criteria for nest site selection in honey bees (Hymenoptera: Apidae): preferences between pheromone attractants and cavity shapes. *Environ. Entomol. 21:*1130-1133.

Schmidt, J. O., S. C. Thoenes and R. Hurley. 1989. Swarm traps. *Amer. Bee J. 129:*468-471.

Schmidt, J. O., K. N. Slessor and M. L. Winston. 1993. Roles of Nasonov and queen pheromones in attraction of honeybee swarms. *Naturwissenschaften 80:*573-575.

Schmidt, J. O., E. D. Morgan, N. J. Oldham, R. R. Do Nascimento and R. R. Dani. 1997. (Z)-11-Eicosen-1-ol, a major component of *Apis cerana* (Hymenoptera: Apidae) venom. *J. Chem. Ecol.* (in press).

Schneider, S. S. and L. C. McNally. 1992. Colony defense in the African honey bee in Africa (Hymenoptera: Apidae). *Environ. Entomol. 21:*1362-1370.

Schua, L. 1952. Untersuchungen über den Einfluss meteorologischer Elemente auf das Berhalten der Honigbienen (*Apis mellifica*). *Z. Vergl. Physiol. 34:*258-277.

Seeley, T. 1977. Measurement of nest cavity volume by the honey bee (*Apis mellifera*). *Behav. Ecol. Sociobiol. 2*:201-227.

Seeley, T. D. and R. A. Morse. 1976. The nest of the honey bee (*Apis mellifera* L.). *Insect. Soc. 23*:495-512.

Seeley, T. D. and R. A. Morse. 1978. Nest site selection by the honey bee, *Apis mellifera*. *Insect. Soc. 25*:323-337.

Seeley, T. D., R. A. Morse and P. K. Visscher. 1979. The natural history of the flight of honey bee swarms. *Psyche 86*:103-113.

Sladen, F. W. L. 1901. Scent-producing organ in the abdomen of the bee. *Gleanings Bee Cult. 29*:639-640.

Southwick, E. E. and R. F. A. Moritz. 1987. Effects of meteorological factors on defensive behaviour of honey bees. *Int. J. Biometeor. 31*:259-265.

Spangler, H. G. 1994. Are wingbeat frequencies of honey bees an indicator of populations or behavior? *Amer. Bee J. 134*:53-55.

Spangler, H. G., J. O. Schmidt, S. C. Thoenes and E. H. Erickson. 1990. Automated testing of the temperament of Africanized honey bees -- a progress report. *Amer. Bee J. 130*:731-733.

Stort, A. C. 1971. Estudo genetico da agressividade de *Apis mellifera*. PhD dissertation, Faculdade de Filospfia, Ciencias Letras, Araraquara, Brazil.

Stort, A. C. 1974. Genetic study of aggressiveness of two subspecies of *Apis mellifera* in Brazil 1. some tests to measure aggressiveness. *J. Apic. Res. 13*:33-38.

Stort, A. C. 1975. Genetical study of the aggressiveness of two subspecies of *Apis mellifera* in Brasil. V. number of stings in the leather ball. *J. Kansas Entomol. Soc. 48*:381-387.

Tel-Zur, D. and Y. Lensky. 1995. Bioassay and apparatus for measuring the stinging response of an isolated worker honey-bee (*Apis mellifera* L. var. *ligustica* Spin.). *Comp. Biochem. Physiol. 110A*:281-288.

Villa, J. D. 1988. Defensive behaviour of Africanized and European honeybees at two elevations in Colombia. *J. Apic. Res. 27*:141-145.

Villa, J. D. and J. O. Schmidt. 1992. Does queen pheromone increase swarm capture in hives baited with Nasonov pheromone? *J. Apic. Res. 31*:165-167.

Visscher, P. K., R. A. Morse and T. D. Seeley. 1985. Honey bees choosing a home prefer previously occupied cavities. *Insect. Soc. 32*:217-220.

Wells, P., H. Wells, V. Vu, N. Vadehra, C. Lee, R. Han, K. Han and L. Chang. 1993. Does honey bee Nasonov pheromone attract foragers? *Bull. Southern Calif. Acad. Sci. 92*:70-77.

Whiffler, L. A., M. U. H. Drusedau, R. M. Crewe and H. R. Hepburn. 1988. Defensive behaviour and the division of labour in the African honeybee (*Apis mellifera scutellata*). *J. Comp. Physiol. A 163*:401-411.

Williams, I. H., J. A. Pickett and A. P. Martin. 1981. The Nasonov pheromone of the honeybee *Apis mellifera* L., (Hymenoptera, Apidae). Part II. Bioassay of the components using foragers. *J. Chem. Ecol. 7*:225-237.

Wilson, E. O. and W. H. Bossert. 1963. Chemical communication among animals. *Recent Progr. Hormone Res. 19*:673-716.

Witherell, P. C. and J. E. Lewis. 1986. Studies on the effectiveness of bait hives and lures to attract honey bee swarms -- a survey tool for use in Africanized honey bee eradication programs. *Amer. Bee J. 126:*353-361.

PART FOUR

Social Insect Primer Pheromones

12

Primer Pheromones in Ants

Edward L. Vargo

Introduction

Primer pheromones undoubtedly play a central proximate role in establishing and maintaining the sophisticated social structure of ant colonies. Such basic processes as caste determination and reproductive development appear to be regulated by pheromones, produced primarily by queens (Wilson 1971; Hölldobler and Wilson 1990). Despite their importance in shaping the structure and function of ant colonies, no ant primer pheromones have been identified and little specific information exists about them. The main obstacle in studying ant primer pheromones has been the lack of sensitive, reliable bioassays. As with research on any semiochemicals, the key to sustained progress is an effective bioassay. In contrast to the effects of releaser pheromones, e.g., sex attractants, trail following and alarm behavior, which in many cases can be assayed in a matter of minutes, primer pheromones have much more subtle physiological effects which may not be apparent for several days or weeks.

Traditionally, several lines of evidence have suggested the occurrence of primer pheromones. In many instances, the presence of a queen is clearly associated with some inhibitory effect on development or reproduction, e.g., the development of new queens (gynes) or reproduction by workers. This together with the lack of obvious aggression or other behavioral displays by queens that could serve as cues has implicated the presence of pheromones. However, the presence of a queen has other correlates besides possible chemical cues that could be involved in producing primer effects, e.g., tactile cues and source of eggs which affects the worker to larva ratio. Therefore, demonstration of the involvement of queen primer pheromones requires carefully designed experiments to exclude other possible non chemical cues.

Strong evidence for the involvement of pheromones comes from showing activity with queen corpses, excluding any role for behavioral cues, together with proper controls to exclude possible tactile cues associated with a queen. This level of evidence in ants has only been approached in a handful of cases (Carr 1962; Fletcher and Blum 1981a; Vargo and Fletcher 1986a; Vargo 1988, 1992; Vargo and Passera 1991). Even stronger evidence involves the demonstration of biological activity with extracts of queens. This level has only recently been achieved in studies of the fire ant *Solenopsis invicta* (unpubl. data). Intermediate between these two levels is showing that queen corpses lose their activity after rinsing with an organic solvent (Fletcher and Blum 1981a; Vargo and Passera 1991), which treatment presumably removes chemical but not tactile cues from the body. Finally, irrefutable evidence for the involvement of primer pheromones comes from the identification of one or more biologically active compounds and the demonstration of activity with synthetic material, a level that has yet to be achieved for any ant.

Even if we lack much detailed information about ant primer pheromones, there is considerable circumstantial evidence from many species giving us a good indication of the roles they play in ant colonies. There appear to be two main effects of primer pheromones in ant colonies: (1) on larval development, where they influence caste determination in female larvae; and (2) on reproductive activity of colony members, where they inhibit ovary development and/or oviposition. Much of this circumstantial evidence has been reviewed elsewhere (Brian 1979, 1980; Passera 1984; Fletcher and Ross 1985; Hölldobler and Wilson 1990). Here I review the literature on the better studied species, focusing on those studies with the strongest evidence for the involvement of primer pheromones. I end by considering the role of primer pheromones in the fire ant, *S. invicta*, a species that has emerged as a model system for studies of primer pheromones in social insects.

Role of Primer Pheromones in Caste Determination

Gyne-Worker Determination

Wheeler (1986) has developed a general caste determination model for ants (Figure 12.1). In most species, it seems that female larvae experience some period of bipotentiality, where they have the ability to develop as either gynes or workers. In some species, a period of

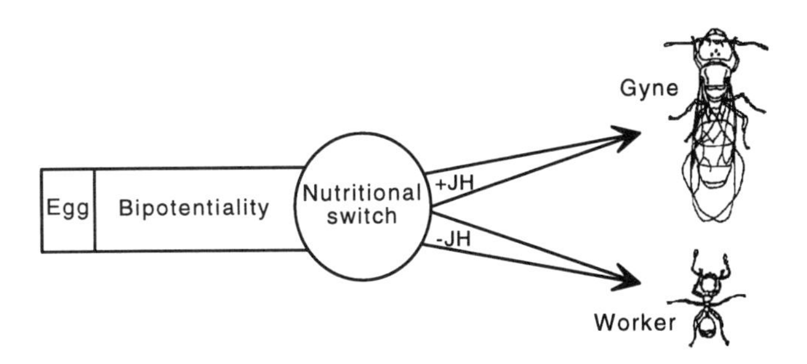

FIGURE 12.1 A general model of caste determination in ants as developed by Wheeler (1986). Female larvae are bipotent early on before undergoing a nutritional switch which acts through the endocrine system to determine their developmental fate. With sufficient nutrition, JH levels rise, leading to the expression of the gyne developmental program. Without sufficient nutrition, larvae develop as workers.

overwintering is required for larvae to be bipotent. At some point in larval development, there is a nutritional switch that determines the developmental fate of larvae. If the nutritional switch elevates juvenile hormone (JH) titers above a certain threshold, then the gyne program is expressed, otherwise the larvae develop into workers. In many species, it appears that a queen primer pheromone acts on this nutritional switch. Precisely how it acts is not known, but what evidence we have suggests that it influences the feeding behavior of workers. When the pheromonal signal is strong, workers are more likely to restrict food to larvae during this critical period resulting in low JH titers and expression of the worker developmental program. When the pheromonal signal is weak or absent, workers are more likely to feed larvae sufficient quantity and/or quality of food, resulting in elevated JH titers and expression of the gyne program. Although queen primer pheromones appear to influence caste determination indirectly by affecting the feeding behavior of larvae, the possibility of a direct effect on larval physiology can not be ruled out. An influence of queen pheromones on the feeding behavior of workers may be considered a releaser effect, but because their ultimate effect is on the developmental fate of larvae, they are more properly viewed as primer pheromones. The best evidence for the involvement of primer pheromones in gyne-worker determination comes from work on two myrmicines (*Myrmica rubra* and *Solenopsis invicta*), a formicine (*Plagiolepis pygmaea*) and a dolichoderine (*Linepithema humile*).

Queen pheromones in many species also can induce workers to execute gyne-determined larvae (e.g., Vargo and Fletcher 1986a; Vargo and

Passera 1991). Although not a primer effect, the execution of sexualized larvae may play a role in determining whether gynes are permitted to develop in the colony.

Over a 30 year period Mike Brian and his colleagues studied caste determination in *Myrmica*, especially *M. rubra*. Figure 12.2 summarizes the process of caste determination in *M. rubra*. Queens of this species influence the development of female larvae in two ways. They stimulate rapid growth of small larvae resulting in their metamorphosis into workers, and they cause workers to feed large larvae less and to bite them forcing their development as workers (Brian and Hibble 1963a, 1964; Brian 1973a, b). Some eggs laid by queens in summer develop relatively slowly and fail to reach the first critical size; these larvae overwinter in the nest. Only the overwintered larvae are bipotent. Those larvae that develop relatively quickly and reach the first critical size do not overwinter and develop into workers. The presence of the queen causes workers to feed more food to early third instar larvae, speeding up their development and metamorphosis resulting in worker determination (Brian 1973a, 1975).

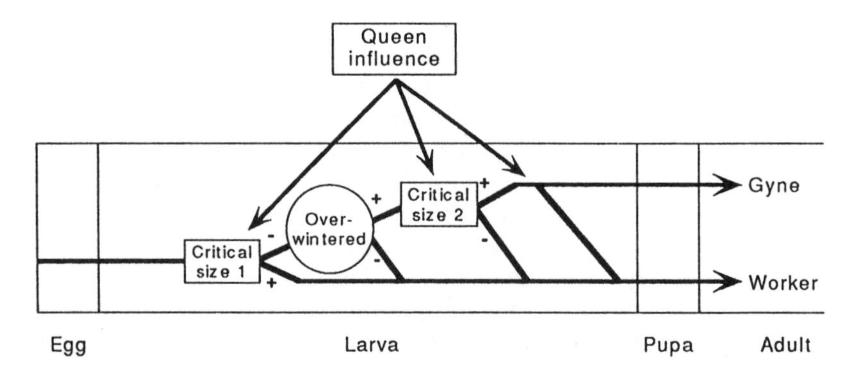

FIGURE 12.2. The process of caste determination in *Myrmica rubra*. Queen influence, probably through primer pheromones, can affect caste determination at three separate points in larval development. In the summer and fall, queens cause workers to feed larvae in a manner that speeds up their development, pushing larvae over the critical size beyond which they complete their development before the onset of winter. Only overwintered larvae are bipotent. In spring, queens cause workers to restrict food to larvae which, by presumably keeping JH titers low, prevents the larvae from reaching the second critical size necessary for expression of the gyne program. Finally, the larger larvae can be bitten and starved by workers, forcing them down the worker line of development. Based on Brian and Hibble (1963a, 1964) and Brian (1973a, b, 1975).

Larvae that have overwintered are bipotent, but they still have to reach a second critical size to become gynes. During this second critical period, workers respond to the presence of the queen by restricting protein fed to larvae, feeding larvae a dilute regurgitant rather than whole pieces of prey (Brian 1973b). Without adequate protein, larvae are unable to attain the second critical size necessary for expression of the gyne phenotype, presumably because JH titers remain too low. Larvae that successfully reach the second critical size are not yet assured of development as gynes, because workers can respond to queen influence by biting larvae and starving them (Brian 1973a, b), forcing them down the path of worker development.

The most convincing demonstration that queen pheromones are involved in caste determination in *M. rubra* comes from the work of Carr (1962), who found that the daily addition of queen corpses to small groups of workers and larvae significantly reduced the number of gynes produced compared to controls. Although her results are suggestive of a pheromonal effect, they suffer from small sample sizes (only 3 replicates each of experimentals and controls) and the use of inadequate controls; corpses of workers were used in controls instead of corpses of virgin queens which have the same form as functional queens. Carr (1962) also assessed the ability of extracts of queens obtained by various methods to influence larval growth but reported that none of these extracts had a detectable effect.

Brian and Hibble (1963b) found that whole mandibular glands of *M. rubra* queens placed in paraffin oil caused increased growth of small larvae and that whole extracts of queen heads in ethanol inhibited the production of gynes in one trial but not another. While these results tend to suggest that the mandibular gland is the source of the queen primer pheromone, they are far from conclusive. The head extracts were tested in a highly artificial manner. In the experimental trials they were applied in all of the following ways simultaneously: by painting on larvae and workers and by putting them into food and on filter paper. There were only two trials in which the effects of queen mandibular glands were tested: this was done by placing the oil containing the glands into the brood chamber of laboratory nests. In each trial, there were just three replicates consisting of 10 workers and 5 larvae each, and the mean gain in larval weight in one of the experimental replicates of each trial was as low or lower than that of any of the controls, yet the authors concluded there was a significant effect.

To obtain information on the chemical characteristics of possible pheromone components, Brian and Blum (1969) tested various fractions of extracted *M. rubra* queen heads on larval growth, and reported that the free fatty acid fraction, when painted on larvae, depressed growth rate and reduced survival of both male and female larvae. However, controls were not treated with carrier solvent in most trials and the numbers of larvae used were unspecified in many of the trials, making it difficult to compare across trials.

To sum up the results on *M. rubra*, clearly an effect of the queen on gyne determination is well-documented. There is suggestive evidence that a queen pheromone is involved and that the source of the pheromone is in the head, but more rigorous studies are needed to confirm this. The effect of the queen on gyne determination may be universal in the genus *Myrmica*; Elmes and Wardlaw (1983) found that in four other species queens inhibit large overwintered larvae from developing into gynes.

In the formicine, *Plagiolepis pygmaea*, Passera (1969) showed that queens inhibit the rearing of gynes. Passera (1980) later obtained circumstantial evidence for mediation by pheromones. When workers were allowed restricted access to parts of a living queen, the smaller the surface area contacted by workers, the weaker the inhibition. However, the author was unable to detect any inhibitory influence using queen corpses.

Working on the Argentine ant, *Linepithema humile* (=*Iridomyrmex humilis*), Vargo and Passera (1991) obtained strong evidence for the involvement of primer pheromones in inhibiting gyne development. Using small standardized colony fragments, the addition of queen corpses effectively inhibited the development of gynes (Figure 12.3), but the presence of living virgin queens did not.

The inhibitory influence was removed when corpses were washed in pentane, lending further support to the involvement of pheromones that can be extracted with an organic solvent. The main effect of the primer pheromone appears to be achieved by preventing the sexualization of bipotent larvae, probably by affecting the brood-rearing behavior of workers. In addition, these authors found that queens also may cause the execution of female larvae after they've become sexualized; the addition of living queens or queen corpses to previously queenless colony fragments containing sexualized larvae caused workers to attack and execute gyne larvae. Moreover, queens sometimes participated in attacking gyne larvae.

Regulation of gyne production by a queen primer pheromone in *L. humile* fits well with the timing of gyne production in the field (Figure 12.4). For most of the year, queen number remains high in colonies of this polygyne species (about 15 queens per 1,000 workers) and gyne production does not occur during this time (Markin 1970; Benois 1973; Keller et al. 1989). Gynes are produced only in spring after a

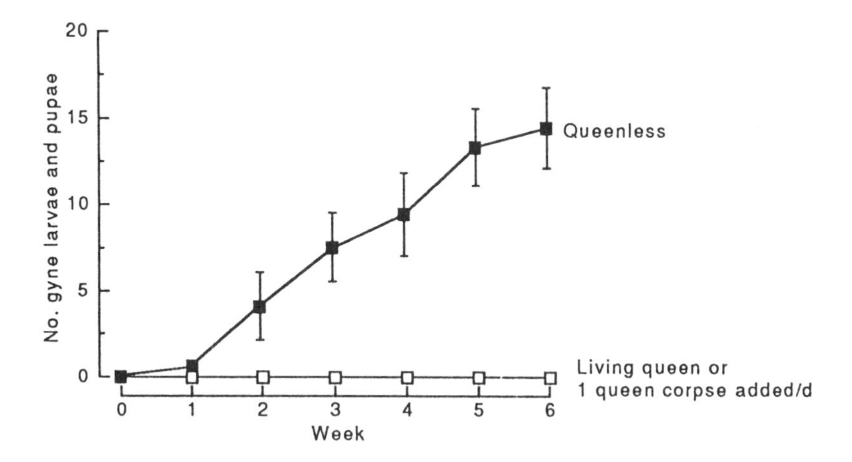

FIGURE 12.3. Strong evidence that queen primer pheromones influence caste determination in the Argentine ant, *Linepithema humile*. Queen corpses were as effective as living queens in inhibiting gyne development. Corpses of virgin queens were not needed as a control in this experiment, because an earlier experiment had shown that living virgin queens were not inhibitory (Vargo and Passera 1991). Shown are means ± SE (n = 5 in each case). (Redrawn with permission from Vargo and Passera 1991).

massive execution of queens by workers (Markin 1970; Keller et al. 1989). This act of regicide seems to lower the level of queen primer pheromone permitting some female larvae to develop as gynes. Gynes mature and mate in the nest within days of eclosion (Markin 1970; Passera et al. 1988; Keller et al. 1989), thereby restoring queen number to its previously high level. Because these young queens begin producing the primer pheromone soon after mating (Vargo and Passera 1991), the inhibitory influence is reinstated, preventing additional gynes from developing.

In addition, season affects the potential of female larvae to sexualize (Figure 12.4). Vargo and Passera (1992) showed that although some female larvae produced throughout the year are capable of sexualization, those that overwinter in the nest have the highest tendency to develop as gynes. Furthermore, workers under spring conditions have a greater tendency to sexualize competent larvae than they do at other times of the year (Vargo and Passera 1992). Thus, the period of greatest sexualization potential coincides with a massive reduction in queen pheromone levels leading to the short burst of annual gyne production.

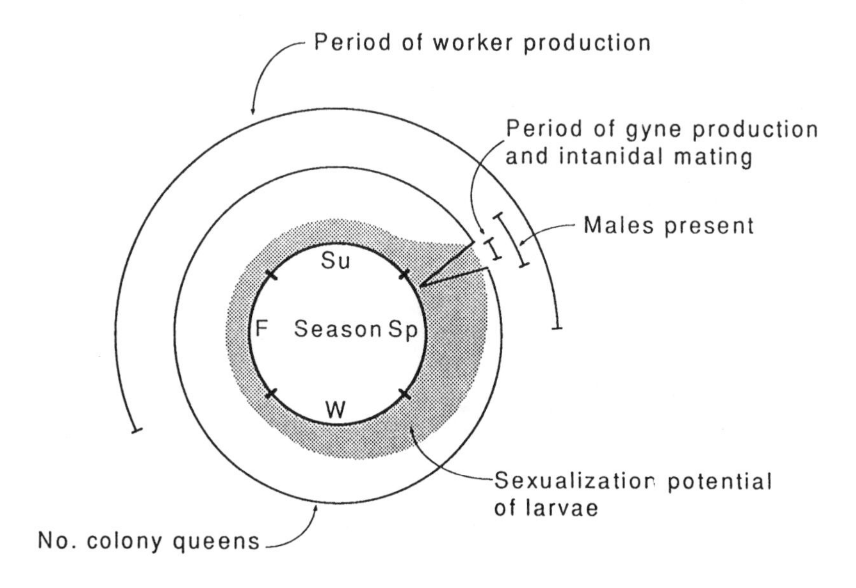

FIGURE 12.4. Model for the role of queen primer pheromones and overwintering in regulating gyne production in natural populations of *L. humile*. Based on studies done in the field (Markin 1970; Benois 1973; Keller et al. 1989) and in the laboratory (Passera et al. 1988; Vargo and Passera 1991, 1992). See text for details.

Major-Minor Worker Determination

The production of major workers in *Pheidole*, a myrmicine genus characterized by the presence of a dimorphic worker caste, may be regulated by primer pheromones. Working with *P. morrisi* Gregg (1942) found that the presence of majors (also called soldiers) inhibited the development of additional majors; experimental colonies containing all majors or all minors adjusted major worker production to restore caste ratios to natural proportions. Working with *P. pallidula*, Passera (1974) showed that an experimental colony made up of 50-100% adult majors inhibited major worker production, whereas one comprised of 0-20% majors did not. Johnston and Wilson (1985) found that individual colonies of *P. dentata* collected in the field varied in their caste ratios, and that each colony restored the ratios to their original values following experimental manipulation. Gregg (1942) proposed that a contact pheromone produced by major workers inhibits worker-determined larvae from developing into majors.

Wheeler and Nijhout (1984), working with *P. bicarinata*, also found that the presence of major workers inhibited the production of majors, and that the degree of suppression increased with increasing contact between larvae and adult majors. These authors showed that treatment of larvae with relatively high doses of the JH analog, methoprene, could override the inhibition. They proposed that majors produce an inhibitory contact pheromone that acts by increasing the threshold titer of JH necessary to induce the major worker phenotype during critical periods in larval development. While this model of major worker determination in *Pheidole* is intriguing, solid evidence for the existence of a primer pheromone produced by major workers is still lacking, and precisely how it might alter the JH titer threshold of developing larvae is unknown. If future studies corroborate pheromonal inhibition of major worker production, this would appear to be the only known case of an ant primer pheromone produced by the worker caste.

Inhibition of Worker Reproduction

In the presence of a functional queen, workers from many species representing nearly all subfamilies of ants are prevented from laying reproductive eggs (Choe 1988). The general lack of aggressive behavior between the queen and workers in these species suggests the involvement of queen-produced primer pheromones. Yet, despite their apparent widespread occurrence, we have only weak evidence for the existence of primer pheromones that inhibit worker oviposition and there is little detailed information about their possible mode of action. The best studied species all belong to the Formicinae.

In *Plagiolepis pygmaea*, Passera (1969) showed that queens inhibit workers from laying reproductive eggs (male-producing) but not trophic eggs. Passera (1980) found that live queens dipped in acetone lost some of their ability to inhibit workers from laying reproductive eggs, a finding in support of pheromonal inhibition. However, no inhibition was achieved by queen corpses, so that strong evidence for pheromonal mediation is still lacking.

Working with the weaver ants, *Oecophylla longinoda* and*O. smaragdina*, Hölldobler and Wilson (1983) found in the laboratory that males were produced in queenless colony fragments, whereas no males were produced in the presence of the queen. These authors observed workers laying eggs in queenright colonies and also found no difference in degree of ovary development in workers from queenless and queenright colonies. They therefore concluded that like *P. pygmaea* the queen inhibits workers from laying reproductive eggs but not trophic eggs. Hölldobler and Wilson (1983) found that workers are prevented from laying reproductive eggs in colony fragments for up to six months by a single queen corpse, strongly implicating a queen pheromone.

However, only four colony fragments of *O. longinoda* and one of *O. smaragdina* were given queen corpses, and controls did not receive corpses of any kind. Additional studies employing larger sample sizes and better controls are needed before firm conclusions can be reached concerning pheromonal mediation of worker reproduction in *Oecophylla*. Hölldobler and Wilson (1983) made some interesting observations regarding the possible source of a queen primer pheromone: workers retain and tend the heads and abdomens of queens which are richly endowed with exocrine glands, suggesting that these body parts may contain the source of a primer pheromone. Whatever the nature of the queen inhibitory signal, it must be exceedingly strong; these arboreal species have multi-nest colonies that may contain up to 500,000 workers distributed over 17 trees, yet the single queen inhibits the workers from laying reproductive eggs (Hölldobler and Wilson 1977).

Effects of primer pheromones on the reproductive system of workers may be more subtle than oviposition. Working with *Camponotus festinatus*, Martinez and Wheeler (1991) found that the presence of the queen suppressed vitellogenin titers but did not totally prevent the development of vitellogenic oocytes. Young (2 week old), minor workers that contained a developed oocyte did not have vitellogenin concentrations significantly higher than same-aged minors who lacked a developed ovary. In general, workers maintained with a queen had very low vitellogenin titers, whereas queenless workers showed increasing vitellogenin concentrations, reaching levels 15 times greater than queenright workers after 7-10 weeks. In addition, the presence of larvae inhibited vitellogenin titers, especially in minor workers. Nevertheless, these authors found that despite the accumulation of vitellogenin in queenless workers, even above the levels detected in some reproductively active queens, vitellogenin uptake by the ovaries did not occur so that there was no oocyte development in most of the individuals. In this species, it is not known whether workers with developed ovaries may oviposit and produce some males, or whether there might be some other function of ovary development and high vitellogenin titers, such as nutrient storage (Wheeler and Martinez 1992).

The Role of Primer Pheromones in the Fire Ant *Solenopsis Invicta*

Beginning with the studies of Fletcher and Blum (1981a, b) on pheromonal inhibition of ovary development in virgin queens, the study of queen primer pheromones in *S. invicta* has made considerable progress. While inadequate sample sizes in studies of other species are probably due to the difficulties of obtaining sufficient material, this is not a problem with *S. invicta*. This pest species occurs throughout the southeastern and south central U.S., where it builds conspicuous mounds that can reach densities of 600 or more per hectare (Porter et al. 1992). The abundance of this ant, together with the development of methods to collect and handle many colonies rapidly (e.g., Banks et al. 1981) and the wealth of background information on this species (e.g., Lofgren et al. 1975; Vinson and Greenberg 1985) has set the stage for fruitful studies of queen primer pheromones. With the development of sensitive bioassays in which queen corpses could be tested for the presence of chemical cues, three distinct effects of queen primer pheromones have been demonstrated in *S. invicta*: inhibition of ovary development and dealation (wing shedding) in virgin queens (Fletcher and Blum 1981a, b); inhibition of the production of new queens through an effect on caste determination (Vargo and Fletcher 1986b; Vargo 1988); and suppression of egg production by mature queens in polygyne colonies (Vargo 1992). In addition, there has been much recent progress in elucidating the mode of action of these pheromones, especially the pheromone affecting reproductive development in winged virgin queens.

Caste Determination

Under both field and laboratory conditions, Vargo and Fletcher (1986a, b, 1987) found that colonies with many functional queens produced far fewer sexuals than did colonies containing a single or no queen. This negative effect of queen number on the production of sexuals was demonstrated to involve a primer pheromone using small colony fragments that permitted the testing of groups of queen corpses for pheromonal activity (Vargo and Fletcher 1986a). Subsequently, I developed (Vargo 1988) a relatively sensitive bioassay in which a single queen corpse could be tested. With this technique, pheromonal activity can be detected in only two to three days, making it an exceptionally sensitive and rapid bioassay for a primer pheromone (Figure 12.5). When placed in small cages that prevented direct contact by workers but through which volatile compounds could pass, queen corpses were not inhibitory. This suggests that the pheromone is relatively non-volatile and workers must come into contact with queens to pick up the active compounds which are then dispersed through the colony by contact. I also found (Vargo 1988) that queen gasters were as effective as whole corpses, whereas heads and thoraces together exhibited only minor activity, suggesting that the glandular source of the pheromone is in the abdomen.

The pheromone appears to act indirectly by affecting the behavior of workers toward larvae in a sex-specific manner. In response to the pheromone, workers presumably restrict the quantity and/or quality of food given to female larvae resulting in their development as workers; in contrast, they kill young male larvae, which have no alternative line of development. In addition to preventing the sexualization of female larvae, workers also may respond to the pheromone by killing late last (fourth) instar larvae of both sexes after they have become sexualized (Vargo and Fletcher 1986a).

Pheromonal Inhibition of Reproduction by Virgin Queens

Workers of the genus *Solenopsis* are sterile, but virgin queens are capable of shedding their wings and reproducing in the nest if the mother queen is absent. With the development of a bioassay in which

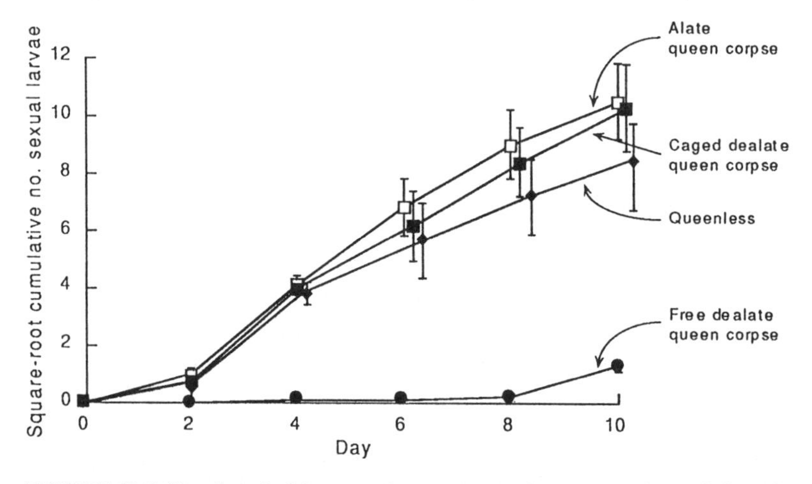

FIGURE 12.5. Results of a bioassay for a primer pheromone that inhibits the production of sexuals, both male and female, in the fire ant *Solenopsis invicta*. This experiment shows the non-volatile nature of the pheromone. Free dealate (egg-laying) queen corpses, which workers could contact, were highly inhibitory. However, when dealate queen corpses were placed in double screen cages, which prevented the workers from contacting the bodies, but did not prevent diffusion of volatile compounds from the bodies to the nest, they lost their inhibitory power. Thus the pheromone appears to be relatively non volatile, requiring that it be picked up from the queens' bodies and dispersed through the colony by worker contact. Shown are means ± SE (n = 12 in each case). There was a significant difference among treatments for days 4-10 (all $F_{3,40} \geq 17.7$, $P < 0.0001$). (Redrawn with permission from Vargo 1988).

queen corpses could be tested, Fletcher and Blum (1981a, b) showed that the queen produces a primer pheromone that inhibits reproductive development in virgin queens. The inhibitory power of different queens, as assessed by this bioassay, bears a positive relationship to the degree of ovary development (Willer and Fletcher 1986; Obin et al. 1988), suggesting that the quantity of pheromone produced by queens is related to fecundity.

Pheromonal inhibition of reproduction by virgin queens is remarkably effective, despite enormous colony size and complex nest structure. Mature monogyne colonies of *S. invicta* can contain 230,000 workers and tens of thousands of immature forms, all distributed over some 40 liters of nest volume (Markin et al. 1973; Tschinkel 1988). In addition, up to 5,000 virgin queens can be present at any one time (Vargo and Fletcher 1987). Yet the queen produces enough pheromone to prevent these virgin queens from producing males (Ross and Fletcher 1985).

As a first step toward locating the glandular source of the pheromone, Fletcher and Blum (1981a) tested parts of queen corpses for their inhibitory capability. Gasters alone were about as effective as whole corpses, whereas heads and thoraces together were only slightly inhibitory. These authors concluded that the pheromone is produced in the gaster, and that the head and thorax become contaminated with it.

Recently, I have been following up on this work by testing extracts of queens with the bioassay of Fletcher and Blum (1981a). I have found (unpublished data) that hexane extracts of whole queen bodies are active and that hexane extracts of queen abdomens exhibit the full activity of whole body extracts. Moreover, the activity appears to reside in the poison sac. In collaboration with Keith Slessor of Simon Fraser University in British Columbia, Canada, I have begun purification of the queen poison sac extract, which contains primarily alkaloids, especially *cis*-2-methyl-6-undecyl piperdine, and other minor components (Brand et al. 1973; Vander Meer and Morel 1995). We've begun by separating the poison sac extract into the major alkaloid-containing (basic) fraction and minor neutral fraction. Tests of these two fractions separately and in combination indicate that the basic fraction is not active alone, whereas the neutral fraction alone exhibits moderate activity, but the two combined yield full activity (unpublished data). Thus preliminary data would indicate that the major components are in the neutral fraction but that full activity requires the synergism of components present in both the neutral and basic fractions.

We also have been investigating the physiological mode of action of this pheromone. Based in part on earlier work by Barker (1978, 1979),

Fletcher and Blum (1983) suggested that the pheromone acts by suppressing JH titers. Following up on this work Vargo and Laurel (1994) obtained evidence favoring the JH suppression hypothesis. These authors found that topical application of virgin queens with the JH analog, methoprene, overrode the effects of the pheromone.

To obtain more precise information on the physiological processes underlying the inhibition of ovary development, Vargo and Laurel (1994) determined vitellogenin titers for virgin queens and functional (egg-laying) queens. Despite having undeveloped ovaries, virgin queens had vitellogenin titers that were as elevated as those of functional queens. This suggests that the effect of low JH titers resulting from the primer pheromone is on the uptake of vitellogenin by the oocytes rather than on vitellogenin synthesis.

Vargo and Laurel (1994) also investigated the possible mode of action of the primer pheromone. Virgin queens, whose antennae had been removed, dealated in the presence of the queen, suggesting that the pheromone acts by stimulating sensory cells in the antennae. A general model for the mode of action of the queen primer pheromone is presented in Figure 12.6.

The extent of pheromonal inhibition of reproduction by virgin queens in ants is not known, but Vargo and Porter (1993) have shown that it occurs in the closely related species *S. richteri* and the hybrid *S. invicta/richteri*, but apparently not in the more distantly related *S. geminata*.

Suppression of Reproduction by Functional Queens

In many social Hymenoptera, there is an inverse relationship between the number of reproducing individuals in the colony and the fecundity of individuals, a relationship termed the "reproductivity effect" by Michener (1964). Fletcher et al. (1980) and Greenberg et al. (1985) were the first to report that queens in polygyne colonies of *S. invicta* were less fecund than queens in monogyne colonies. The relationship between colony queen number and fecundity in field and standardized laboratory colonies was studied by Vargo and Fletcher (1989). These authors found that the number of queens in a colony had a strong negative effect on mean fecundity. I obtained evidence (Vargo 1992) for the involvement of pheromones in mutual inhibition among queens in polygyne colonies. Using small colony fragments, in which queens from polygyne colonies were isolated individually with workers and brood, the presence of queen corpses inhibited fecundity compared to controls receiving corpses of virgin queens. In addition, topical application of methoprene to individual queens increased fecundity

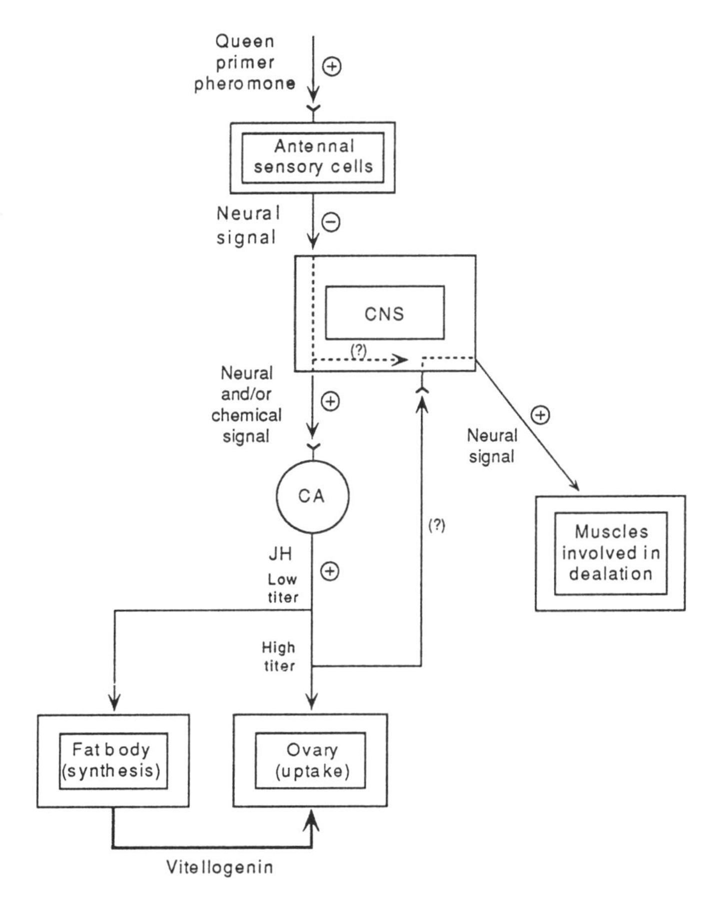

FIGURE 12.6. Proposed general model for the mode of action of the primer pheromone of queen fire ants that inhibits dealation and ovary development in virgin queens. In this model, the pheromone triggers antennal receptors which send inhibitory signals to the median neurosecretory cells in the brain. Largely inhibited, the median neurosecretory cells only weakly stimulate the corpora allata to synthesize JH, maintaining low titers of this hormone. At low levels, JH stimulates vitellogenin synthesis in the fat body. In the absence of the pheromone, the disinhibited neurosecretory cells send a stronger chemical and/or neural signal that triggers the corpora allata to produce larger quantities of JH. At higher titers, JH stimulates vitellogenin uptake by the ovaries and dealation. The latter process possibly involves an effect of JH on the nervous system. Dealation may result from a JH-independent pathway in the nervous system in lieu of, or in addition to, the JH-mediated pathway. These two possible pathways for control of dealation are flagged with question marks. (Redrawn with permission from Vargo and Laurel 1994).

suggesting that egg production in mature queens is regulated by endogenous levels of JH. Given the possible similarity in mode of actions, the primer pheromone responsible for mutual inhibition among functional queens and the primer pheromone that prevents reproductive development in virgin queens could well be one and the same.

It is not known how widespread mutual pheromonal inhibition in polygyne colonies could be in ants, but Bourke (1993) reported finding no evidence for it in colonies of the myrmicine, *Leptothorax acervorum*. However, as pointed out by the author, lack of evidence does not mean it doesn't occur, only that he did not detect it under his experimental conditions.

Concluding Remarks

The study of primer pheromones is important to our understanding of the structure and function of ant societies. Progress in this field has been slow due to the difficulty of developing effective bioassays. Nonetheless, advancement has been steady, and we now have a good idea of the various roles played by primer pheromones in the ant colony. I've combined these roles in the hypothetical ant society shown in Figure 12.7 to show the known ways that primer pheromones can affect reproduction and development of different colony members.

Recent studies of primer pheromones in the fire ant, *S. invicta*, have been particularly fruitful. I think it's fair to say that we now know more about the role of primer pheromones in this species than in any other social insect with the possible exception of the honey bee, *Apis mellifera*. With effective bioassays for *S. invicta* primer pheromones now at hand, we can look forward to continued progress on this species, in both the isolation and identification of active compounds and in the elucidation of their physiological mode of action.

Finally, because primer pheromones are basic to colony function in ants, their study has potentially important applications for control of pest species. With increasing emphasis on species-specific and ecologically sound means to control pest insect populations, there is renewed interest in novel approaches to pest management, including the use of pheromones. Ant primer pheromones hold promise for development as species-specific bioregulators to control pest ant populations. For example, it may be possible to introduce, to colonies of a pest ant, sufficient quantities of synthetic pheromone to inhibit completely egg production by queens. In addition, increased understanding of how primer pheromones act may reveal new avenues that could be exploited to control populations of selected pest ants.

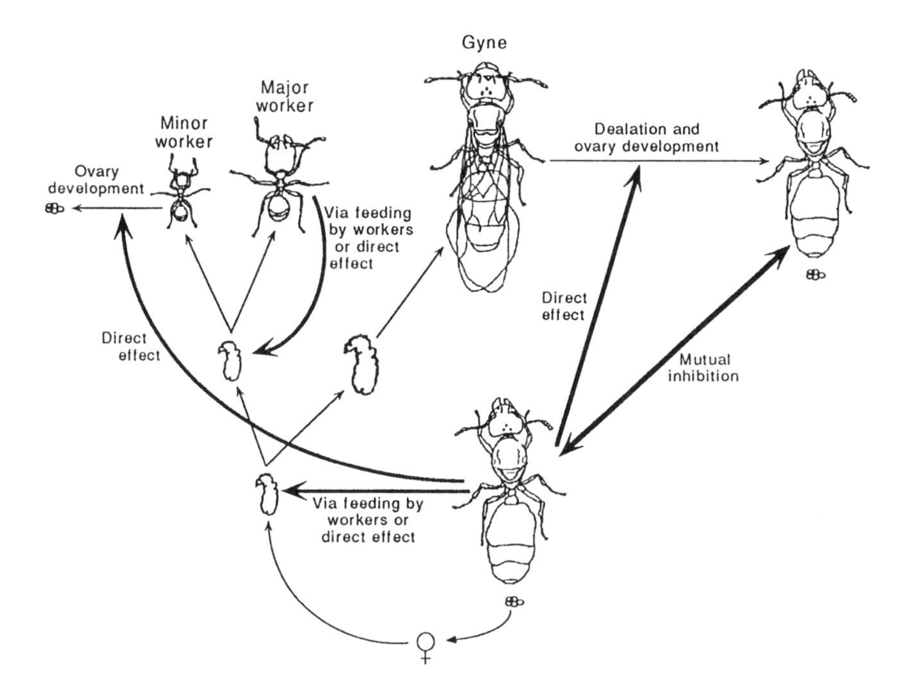

FIGURE 12.7. Role of primer pheromones in ant colonies. The illustrated colony is hypothetical, showing all the known effects of primer pheromones in ants; only a subset of these effects have been documented for any one species. The thin lines show developmental pathways and the thick lines indicate the effects of primer pheromones from the source individuals (queens or major workers) to the target individuals. Queen-produced primer pheromones can influence caste determination of bipotent female larvae, probably by affecting the feeding behavior of workers, although a direct effect on larval physiology cannot be ruled out. In species with a polymorphic worker caste, a pheromone produced by major workers may influence the developmental fate of worker-determined larvae, again probably through an indirect effect on worker feeding behavior, but a direct effect on larval development is possible. Queens may inhibit ovary development in workers or gynes through an effect on the endocrine system. Finally, when multiple egg-laying queens co-exist in a nest, they may exert mutual pheromonal inhibition, leading to lowered egg production.

Acknowledgements

This material is based in part upon work supported by the Texas Advanced Technology Program under Grant No. 003658134 and the USDA NRICGP (95-37302-1864).

Literature Cited

Banks, W.A., C.S. Lofgren, D.P. Jouvenaz, C.E. Stringer, P.M. Bishop, D.F. Williams, D.P. Wojcik and B.M. Glancey. 1981. Techniques for collecting, rearing and handling imported fire ants. United States Dept. Agric. Sci. Ed. Admin. AAT-S-21:1-9.

Barker, J.F. 1978. Neuroendocrine regulation of oocyte maturation in the imported fire ant *Solenopsis invicta*. *Gen. Comp. Endocrinol.* 35: 234-237.

Barker, J.F. 1979. Endocrine basis of wing casting and flight muscle histolysis in the fire ant *Solenopsis invicta*. *Experientia* 35: 552-554.

Benois, A. 1973. Incidences des facteurs écologiques sur le cycle annuel et l'activité saisonnière de la fourmi d'Argentine, *Iridomyrmex humilis* Mayr (Hymenoptera, Formicidae), dans la région d'Antibes. *Insectes Soc.* 20: 267-296.

Bourke, A.F.G. 1993. Lack of experimental evidence for pheromonal inhibition of reproduction among queens in the ant *Leptothorax acervorum*. *Anim. Behav.* 45: 501-509.

Brand, J.M., M.S. Blum and M.R. Barlin. 1973. Fire ant venoms: intraspecific and interspecific variation among castes and individuals. *Toxicon* 11: 325-331.

Brian, M.V. 1973a. Caste control through worker attack in the ant *Myrmica*. *Insectes Soc.* 20: 87-102.

Brian, M. V. 1973b. Feeding and growth in the ant *Myrmica*. *J. Anim. Ecol.* 42: 37-53.

Brian, M.V. 1975. Caste determination through a queen influence on diapause in larvae of the ant *Myrmica rubra*. *Ent. Exp. Appl.* 18: 429-442.

Brian, M.V. 1979. Caste differentiation and division of labor. In: *Social Insects* Vol. 1, H.R. Hermann (ed.), Academic Press, New York, pp. 121-222.

Brian, M. V. 1980. Social control over sex and caste in bees, wasps and ants. *Biol. Rev.* 55: 379-415.

Brian, M.V. and M.S. Blum. 1969. The influence of *Myrmica* queen head extracts on larval growth. *J. Insect Physiol.* 15: 2213-2223.

Brian, M.V. and J. Hibble. 1963a. Larval size and the influence of the queen on growth in *Myrmica*. *Insectes Soc.* 10: 71-82.

Brian, M.V. and J. Hibble. 1963b. 9-oxodec-*trans*-2-enoic acid and *Myrmica* queen extracts tested for influence on brood in *Myrmica*. *J. Insect Physiol.* 9: 25-34.

Brian, M.V. and J. Hibble. 1964. Studies of caste differentiation in *Myrmica rubra* L. 7. Caste bias, queen age and influence. *Insectes Soc.* 11: 223-238.

Carr, C.A.H. 1962. Further studies on the influence of the queen in ants of the genus *Myrmica*. *Insectes Soc.* 9: 197-211.

Choe, J.C. 1988. Worker reproduction and social evolution in ants (Hymenoptera: Formicidae). In: *Advances in Myrmecology*, J. C. Trager (ed.), E. J. Brill, New York, pp. 163-187.

Elmes, G.W. and J.C. Wardlaw. 1983. A comparison of the effect of a queen upon the development of large hibernated larvae of six species of the genus *Myrmica* (Hym. Formicidae). *Insectes Soc.* 30:134-148.

Fletcher, D.J.C. and M.S. Blum. 1981a. A bioassay technique for an inhibitory primer pheromone of the fire ant, *Solenopsis invicta* Buren. *J. Ga. Entomol. Soc.* 16: 352-356.

Fletcher, D.J.C. and M.S. Blum. 1981b. Pheromonal control of dealation and oogenesis in virgin queen fire ants. *Science* 212: 73-75.

Fletcher, D.J.C. and M.S. Blum. 1983. The inhibitory pheromone of queen fire ants: effects of disinhibition on dealation and oviposition by virgin queens. *J. Comp. Physiol. A* 153: 467-475.

Fletcher, D.J.C., M.S. Blum, T.V. Whitt and N. Tempel. 1980. Monogyny and polygyny in the fire ant *Solenopsis invicta* Buren. *Ann. Entomol. Soc. Am.* 73: 658-661.

Fletcher, D.J.C. and K. G. Ross. 1985. Regulation of reproduction in eusocial Hymenoptera. *Annu. Rev. Entomol.* 30: 319-343.

Greenberg, L., D.J.C. Fletcher and S. B. Vinson. 1985. Differences in worker size and mound distribution in monogynous and polygynous colonies of the fire ant *Solenopsis invicta* Buren. *J. Kansas Entomol. Soc.* 58: 9-18.

Gregg, R.E. 1942. The origin of castes in ants with special reference to *Pheidole morrisi* Forel. *Ecology* 23: 295-308.

Hölldobler, B. and E.O. Wilson. 1977. Weaver ants. *Sci. Amer.* 237: 146-154.

Hölldobler, B. and E.O. Wilson. 1983. Queen control in colonies of weaver ants (Hymenoptera: Formicidae). *Ann. Entomol. Soc. Am.* 76: 235-238.

Hölldobler, B. and E.O. Wilson. 1990. *The Ants.* Belknap Press of Harvard University Press, Cambridge, Mass.

Johnston, A.B. and E.O. Wilson. 1985. Correlates of variation in major/minor ratio in the ant, *Pheidole dentata* (Hymenoptera: Formicidae). *Ann. Entomol. Soc. Amer.* 78: 8-11.

Keller, L., L. Passera and J.-P. Suzzoni. 1989. Queen execution in the Argentine ant, *Iridomyrmex humilis* (Mayr). *Physiol. Entomol.* 14: 157-163.

Lofgren, C.S., W.A. Banks and B.M. Glancey. 1975. Biology and control of imported fire ants. *Annu. Rev. Entomol.* 20: 1-30.

Markin, G.P. 1970. The seasonal life cycle of the Argentine ant *Iridomyrmex humilis* (Hymenoptera, Formicidae), in southern California. *Ann. Entomol. Soc. Am.* 63: 1238-1242.

Markin, G.P., J.H. Dillier and H.L. Collins. 1973. Growth and development of colonies of the red imported fire ant, *Solenopsis invicta. Ann. Entomol. Soc. Am.* 66: 803-808.

Martinez, T. and D. Wheeler. 1991. Effect of the queen, brood and worker caste on haemolymph vitellogenin titre in *Camponotus festinatus* workers. *J. Insect Physiol.* 37: 347-352.

Michener, C.D. 1964. Reproductive efficiency in relation with colony size in hymenopterous societies. *Insectes Soc.* 11: 317-341.

Obin, M.S., B.M. Glancey, W.A. Banks and R.K. Vander Meer. 1988. Queen pheromone production and its physiological correlates in fire ant queens (Hymenoptera: Formicidae) treated with fenoxycarb. *Ann. Entomol. Soc. Am.* 81: 808-815.

Passera, L. 1969. Biologie de la reproduction chez *Plagiolepis pygmaea* Latreille et ses deux parasites sociaux *Plagiolepis grassei* Le Masne et Passera et *Plagioplepis xene* Stärcke (Hymenoptera, Formicidae). *Ann. Sci. Nat. Zool. Paris,* 12eme série 11: 327-482.

Passera, L. 1974. Différenciation des soldats chez la fourmi *Pheidole pallidula* (Nyl.) (Formicidae, Myrmicinae). *Insectes Soc.* 21: 71-86.

Passera, L. 1980. La fonction inhibitrice des reines de la fourmi *Plagiolepis pygmaea* Latr.: rôle des phéromones. *Insectes Soc.* 27: 212-215.

Passera, L. 1984 *L'Organisation Sociale des Fourmis.* Privat, Toulouse.

Passera, L., L. Keller and J.-p. Suzzoni. 1988. Queen replacement in dequeened colonies of the Argentine ant *Iridomyrmex humilis* (Mayr). Psyche 95: 59-65.

Porter, S.D., H.G. Fowler and W.P. Mackay. 1992. Fire ant mound densities in the United States and Brazil (Hymenoptera: Formicidae). *J. Econ. Entomol.* 85: 1154-1161.

Ross, K.G. and D.J.C. Fletcher. 1985. Comparative study of genetic and social structure in two forms of the fire ant *Solenopsis invicta* (Hymenoptera: Formicidae). *Behav. Ecol. Sociobiol.* 17: 349-356.

Tschinkel, W.R. 1988. Colony growth and the ontogeny of worker polymorphism in the fire ant, *Solenopsis invicta. Behav. Ecol. Sociobiol.* 22: 103-115.

Vander Meer, R.K. and L. Morel. 1995. Ant queens deposit pheromones and antimicrobial agents on eggs. *Naturwissenschaften* 82: 93-95.

Vargo, E.L. 1988. A bioassay for a primer pheromone of queen fire ants (*Solenopsis invicta*) which inhibits the production of sexuals. *Insectes Soc.* 35: 382-392.

Vargo, E.L. 1992. Mutual pheromonal inhibition among queens in polygyne colonies of the fire ant *Solenopsis invicta. Behav. Ecol. Sociobiol.* 31: 205-210.

Vargo, E.L. and D.J.C. Fletcher. 1986a. Evidence of pheromonal queen control over the production of sexuals in the fire ant, *Solenopsis invicta*. *J. Comp. Physiol.* A 159: 741-749.

Vargo, E.L. and D.J.C. Fletcher. 1986b. Queen number and the production of sexuals in the fire ant, *Solenopsis invicta* (Hymenoptera: Formicidae). *Behav. Ecol. Sociobiol.* 19: 41-47.

Vargo, E.L. and D.J.C. Fletcher. 1987. Effect of queen number on the production of sexuals in natural populations of the fire ant, *Solenopsis invicta*. *Physiol. Entomol.* 12: 109-116.

Vargo, E.L. and D.J.C. Fletcher. 1989. On the relationship between queen number and fecundity in polygyne colonies of the fire ant, *Solenopsis invicta*. *Physiol. Entomol.* 14: 223-232.

Vargo, E.L. and M. Laurel. 1994. Studies on the mode of action of a queen primer pheromone of the fire ant *Solenopsis invicta. J. Insect Physiol.* 40: 601-610.

Vargo, E. L. and L. Passera. 1991. Pheromonal and behavioral queen control over the production of gynes in the Argentine ant *Iridomyrmex humilis* (Mayr). *Behav. Ecol. Sociobiol.* 28: 161-169.

Vargo, E.L. and L. Passera. 1992. Gyne development in the Argentine ant *Iridomyrmex humilis*: role of overwintering and queen control. *Physiol. Entomol.* 17: 193-201.

Vargo, E.L. and S.D. Porter. 1993. Reproduction by virgin queen fire ants in queenless conditions: comparative study of three taxa (*Solenopsis richteri*, hybrid *S. invicta/richteri*, *S. geminata*) (Hymenoptera: Formicidae). *Insectes Soc.* 40: 283-293.

Vinson, S.B. And L. Greenberg. 1985. The biology, physiology, and ecology of imported fire ants. In: *Economic Impact and Control of Social Insects*, S. B. Vinson (ed.), Praeger, New York, pp. 193-226.

Wheeler, D.E. 1986. Developmental and physiological determinants of caste in social Hymenoptera: evolutionary implications. *Am. Nat.* 128: 13-34.

Wheeler, D. and T. Martinez. 1992. Oogenesis and its social regulation in worker ants. In: *Advances in Regulation of Insect Reproduction*, B. Bennettová, I. Gelbic and T. Soldán, (eds.), Institute of Entomology, Czech Academy of Science, pp. 311-316.

Wheeler, D.E. and H.F. Nijhout. 1984. Soldier determination in *Pheidole bicarinata*: inhibition by adult soldiers. *J. Insect Physiol.* 30: 127-135.

Willer, D.E. and D.J.C. Fletcher. 1986. Differences in inhibitory capability among queens of the ant *Solenopsis invicta*. *Physiol. Entomol.* 11: 475-482.

Wilson, E.O. 1971. *The Insect Societies.* Belknap Press of Harvard Univ. Press, Cambridge, Mass.

13

Primer Pheromones and Possible Soldier Caste Influence on the Evolution of Sociality in Lower Termites

Gregg Henderson

Introduction

The Order Isoptera, the most ancient society of animals, is unsurpassed among social insects in its regulation and maintenance of a highly organized and populous society. Termite colony populations exceed most other social insect colony populations (Wilson 1971). The termite soldier caste, the most highly specialized sterile caste of any insect society, is and always was common to every termite species (one species lacks soldiers, a phenomenon that evolved secondarily) (Noirot 1985). Morphological adaptations disallow soldiers from feeding without the aid of worker termites. However, despite this obvious incumbrance to the colony, the termite soldier caste is clearly essential to the termite society and is unequaled among other social insect societies in the high soldier to non-soldier ratio (Oster and Wilson 1978). The termite soldier caste may represent twenty percent or more of the colony population (Haverty 1977).

The regulation of caste structure in termite societies, composed of soldiers, workers (not a true caste in lower termites), and reproductives, has been of major interest to termitologists because it is key to our understanding of social insect evolution. This paper summarizes some of the key points in our current knowledge on caste development in lower termites and puts forth a new hypothesis on the role of the soldier caste in termite social evolution. Research on juvenile hormone, corpora allata activity, relatedness asymmetries, caste development, and seasonal fluctuations in caste proportions are brought together with a historical perspective on the termite soldier caste to bring a new awareness to a little discussed but fairly old idea. It will be argued that the evolution of termite sociality was, and continues to be, largely driven by primer pheromones. Moreover, the soldier caste, present at the origin of the Isoptera (Noirot 1985), may play an integral part in the termite society by lifting the primer pheromone's inhibition on worker maturation through juvenile hormone (JH) absorption. As a result, the only truly sterile caste of lower termites, the soldier, enhances its inclusive fitness and helps drive an evolutionarily stable social structure.

Definitions

Caste is defined as any of the various morphological types of individuals in an insect society in which one or more individuals of each sex are reproductively competent and are associated with all other members of the group (Castle 1934). Primer pheromones are chemicals dispersed by one or more individuals that cause a physiological change in nearby conspecifics by altering the endocrine and reproductive systems (Wilson 1971). A pheromone that causes caste differentiation is an example of a primer pheromone. In contrast, releaser pheromones cause immediate behavioral changes in the receiving conspecific that are readily observable.

Presently, the meaning of caste has been muddled in the literature as knowledge has accumulated about behavioral roles of individuals relative to age differences and interindividual variation. However, as argued by Peeters (1990), only morphologically distinct types should be considered true castes. In lower termites, the castes include the soldiers and the various reproductively competent kings and queens, which may be primary, secondary, or, in rare cases, tertiary. Primary reproductives are those individuals that develop into alates and fly from the nest to establish a new colony. Secondary reproductives develop from nymphs (forms having wing pads) but remain in the natal nest. Tertiary reproductives develop directly from workers. Soldiers are variously modified to function in defense of the colony. All soldiers possess sclerotized heads that are more heavily pigmented than the thorax or abdomen. Workers (properly called pseudergates in the lower termites) are not a true caste, since all are immatures (called larvae despite their hemimetabolous development) and may develop into a final form (soldier or reproductive), stay as workers through undifferentiated molts, or revert back to a worker through regressive molts from one or more of the nymphal instars. It appears that all workers eventually molt to a final form (Luykx 1993).

Termite Caste Differentiation

There have been several reviews on regulation of castes in the lower termites (Miller 1969, Stuart 1979, Noirot 1985). The pathways for caste differentiation generally fit the two schemes summarized by Castle (1934, Figures 13.1 and 13.2). Caste differentiation is probably a complex interaction of intrinsic and extrinsic factors that involve such variables as light, temperature, food, pheromones, endocrines, and genome (Greenberg and Tobe 1985). However, although far from being universally accepted, there is mounting evidence that juvenile hormone (JH), produced by the corpora allata glands (CA), is the primer pheromone that regulates caste expression in the lower termites (Lüscher and Springhetti 1960, Lüscher 1960, 1965, 1973, Lebrun 1969, Yin and Gillot 1975a, 1975b). Several studies using juvenile hormone analogues (JHAs) indicate that caste development is dependent on JH (Wanyonyi 1974, Yin and Gillot 1975b, Doki et al. 1984). The primer pheromone secreted by the royal pair appears to be the most important factor in inhibiting adult expression in workers and nymphs within a colony. Removal of the queen and king causes nymphs and larvae to develop into replacement queens and kings, whereas inhibition of adult expression is observed if they remain (Castle 1934, Lüscher 1961, Springhetti 1985). However, under natural conditions it is usually not until the colony becomes very large that workers can distance themselves from the royal pair's inhibitory pheromones and mature to reproductive status (McMahan 1966, Grassé 1982). Springhetti and Pinamonti (1977) showed the inhibitory role of the termite king and queen by feeding termites queen bee extracts. Drywood termite workers and nymphs confined with the royal pair and fed queen bee (*Apis mellifera* L.) extracts were more than twice as likely to molt to secondary reproductive than workers not given the extract (Figure 13.3, Springhetti and Pinamonti 1977). The extracts acted on the termite royal pair's fecundity (as queen substance does in honey bees), and as a function of a reduced synthesis of JH interfered with the king and queen's ability to inhibit worker-to-adult molts.

Laboratory studies on termite primer pheromones have largely focused on manipulating worker-to-soldier differentiation. Soldier competence is controlled by JH and JH mimics (Lüscher 1969, Hrdy and Krecek 1972, Wanyonyi and Lüscher 1973, Springhetti 1974, Wanyonyi 1974, Howard and Haverty 1978, Yin and Gillot 1975a). However, on

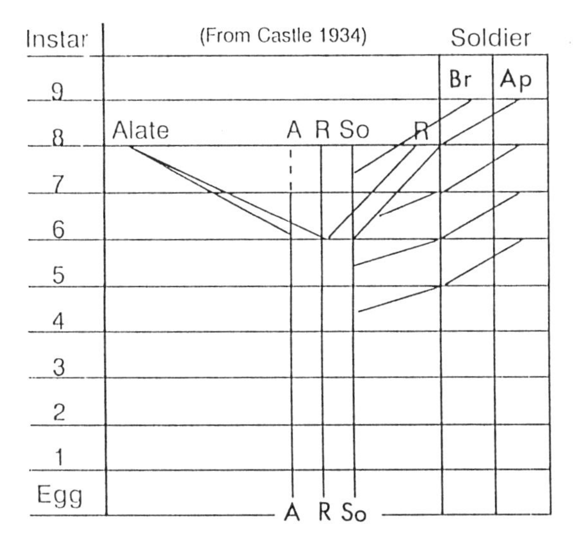

FIGURE 13.1 Potential caste directions of termite larvae in a colony with functional queen and king. A = alate, Ap = apterous, Br = brachypterous, R = reproductive, So = soldier. Castle (1934) did not believe that eggs had a particular destiny (potentiality) that might be suggested by this schematic. Although this question was debated at the time, Castle felt that termite eggs possessed the potentiality to go in any caste direction.

FIGURE 13.2 Potential caste directions of termite larvae in a colony without the functional queen and king. These potential directions are in addition to the schematic in Fig. 1. A = alate, B = brachypterous, Ap = apterous, FSo = fertile soldier, R = reproductive, So = soldier.

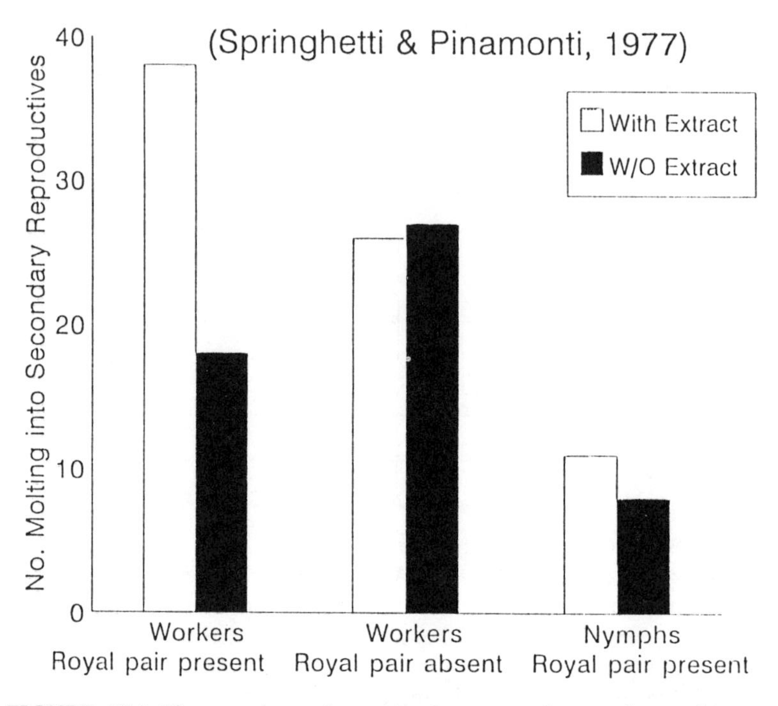

FIGURE 13.3 The number of termite larvae and nymphs molting to secondary reproductives when provided bee extracts. Interference of the royal pair's inhibitory effects is observed and worker-to-reproductive molt is significantly more likely to occur. Little difference is observed in nymphs with or without extracts, or workers when the royal pair is absent.

this point, a puzzling observation is made--soldiers develop in response to *increasing* amounts of JH. Adult (final form) expression in a hemimetabolous insect as a direct response to JH exposure is unique to termites. This phenomenon has led some researchers to hypothesize that a separate primer pheromone is responsible for soldier differentiation (Lüscher and Springhetti 1960, Lüscher 1972). In addition, observations that soldiers inhibit new soldier formation argued for an anti-JH being produced (Lüscher 1972, Lenz 1976, Bordereau 1985). However, as noted by Yin and Gillot (1975a, 1975b), there is no good evidence for the CA producing more than one hormone, and JH is almost certainly the sole hormone responsible in termite caste differentiation. Yin and Gillot (1975b) presented a series of graphs on JH hemolymph titers as they relate to caste differentiation likelihoods, providing a holistic map of our current understanding on the subject (see Figure 13.4). As noted by Lenz (1976) a specific JH titer is associated with each caste. In addition, JH esterase activities and

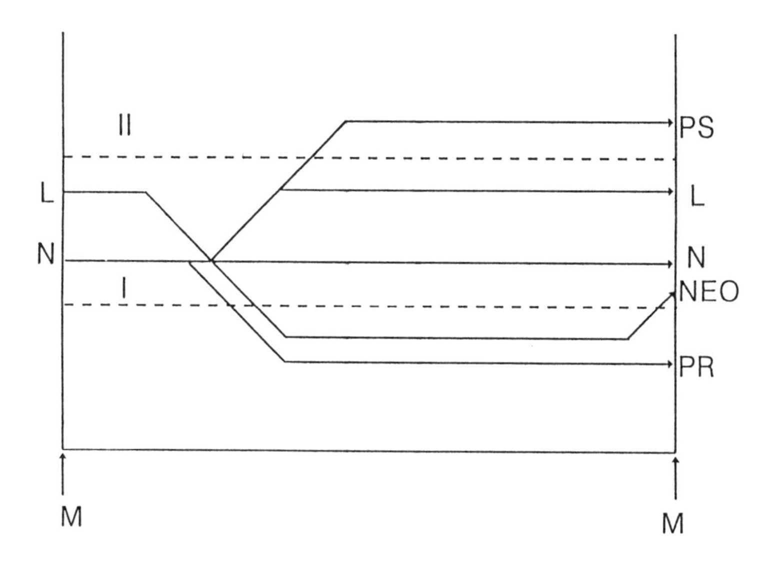

FIGURE 13.4 The likelihood of termite larvae (L) and nymphs (N) to molt into a presoldier (PS), primary reproductive (PR), or neotenic reproductive (NEO), when JH hemolymph titer reaches a high threshold (II) or low threshold (I). M = the molt cycle. Figure is slightly modified from Stuart (1979). Stuart modified the original series of graphs from Yin and Gillot (1975b).

JH hemolymph binding protein abundance show caste specificity (Wyss-Huber 1981).

Caste differentiation appears to be self-regulated by the various castes once a species-specific proportion is reached. Soldier termites regulate their own proportions by inhibiting the production of other soldiers (Castle 1934). Inhibition of the same caste is also reported for alates (Snyder 1915, Weesner 1953, Grassé and Noirot 1958, Miller 1942, Henderson and Delaplane 1994). Reproductives also inhibit alates and secondary reproductives of the nest (Springhetti 1985, Greenberg and Tobe 1985). Conversely, one caste can stimulate differentiation of another caste, as in the case of reproductives and nymphs stimulating the production of soldiers, and soldiers stimulating the production of supplementary reproductives (Springhetti 1985, Bordereau 1985). Self-regulation among alates may be a survival mechanism that increases the successful founding of new colonies by spreading out the number of competent swarmers available at any given time (Henderson and Delaplane 1994).

It would be a serious omission not to note that caste determination is also strongly influenced by nutrition (Jucci 1924, Esenther 1977, Lenz 1976, 1994). For example, Lenz (1976) examined the interaction of JHA on both soldier production and mortality in termite groups fed food sources of differing nutrition. He observed that soldier mortality and inhibition of new soldier formation was markedly reduced when fed a highly nutritious food source, such as fungus-decayed pine wood (Figure 13.5). However, the interaction of pheromones and nutrition are so intimately intertwined that the separation of either's role in caste determination is a nearly impossible task.

Form and Function of CA During Caste Determination

The CA play a critical role in caste differentiation in termites (Lüscher 1960, 1965, Yin and Gillot 1975a). These glands vary in size as a function of storage and secretion of JH (Yin and Gillot 1975a, 1975b). Gland size (volume) also varies with age and caste but not with sex

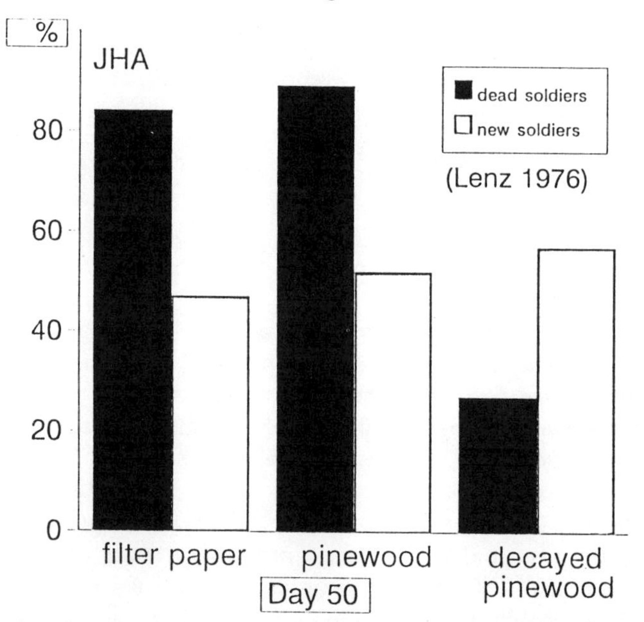

FIGURE 13.5 The percentage of termite larvae molting into soldiers and the percentage of "old" soldiers dying, when in the presence of a juvenile hormone analogue (JHA) and fed food sources of increasing nutrition. On day 50 (some points extrapolated from original graph of Lenz's), termites fed fungus-decayed wood showed a higher percentage of new soldiers and a low percentage of old soldier mortality.

(Yin and Gillot 1975a). Noirot (1969) noted that the enlarged CA of physogastric queens may even occupy space in the thorax. Lüscher (1965) found a seasonal variation in CA volume for workers, noting that the CA volume decreases in winter, a time when caste differentiation is largely suspended. Enlargement of the CA occurs during formation and maturation of soldiers, primary reproductives, and replacement reproductives from workers (Noirot 1969). A four-to-five fold increase in CA volume is observed during the formation of secondary reproductives (Lüscher 1965).

CA activity has been shown to be important in worker-to-soldier differentiation. Transplanting the CA of supplementary reproductives into workers invariably produces soldiers (Lüscher 1958, Lüscher and Springhetti 1960, Lebrun 1964). At the time of the first molt to a soldier (the nonfunctional white soldier), the volume of the CA may triple (Lüscher 1960, Lüscher 1965). White soldiers are completely helpless and act as larvae within the nest (Noirot 1985). The CA volume of functional soldiers is slightly larger than that of workers (Yin and Gillot 1975a).

Role of Soldier Caste in Termite Colony

The soldier termite's function is to protect the nest and peripheral galleries from attack by ants and specialized vertebrate predators (Wilson 1971). Soldier heads are thicker and stronger than heads of workers or alates, which gives them rigidity and helps power the highly musculated mandibles. This morphological adaptation allows a termite soldier to easily snap an equal-sized ant into two pieces but also causes it to be dependent on the worker caste for stomodeal and proctodeal liquid provisions.

The stereotypic description that defense is the only role of the termite soldier in colony function still largely pervades the literature today. However, little work has been devoted to caste function, or the long-term external exigencies that helped shape it (Oster and Wilson 1978). Among lower termites, subterranean termites tend to have a higher proportion of soldiers (mean = 9.2%) than do dampwood termites (mean = 4.6%) or drywood termites (4.4%) (means calculated from Haverty 1977). Of these behavioral categories of lower termites, as a rule, only subterranean termite colonies establish foraging trails to distant food sources and risk a high level of exposure to predation. Drywood and dampwood termite colonies, on the other hand, most often live in their food source and as a result might require fewer defenders. Thus, these field data support a proximate cause, the defense of the colony, in soldier caste evolution. Lenz (1976) and Oster and Wilson (1978) noted that caste proportions probably vary to optimize the most economical ratio in the face of environmental exigencies. Species-specific caste proportions are the result of variations in JH thresholds (Lenz 1976); thus, high soldier numbers may be the result of adaptable JH sensitivities. It is suggested here that species faced with high levels of predation may have an increased need for soldiers and therefore, over time, a reduced JH titer required for worker-to-soldier molts might evolve (Figure 13.6).

Another line of evidence in support of defense as a proximate cause of the soldier caste can be seen in the seasonal fluctuations of soldiers and alates. Reproductive (alates) and soldier castes fluctuate seasonally in abundance, but through physiological and behavioral mechanisms species-specific proportions are relatively stable from year to year (Haverty 1979). Fluctuations in soldier numbers are often positively correlated with alate numbers, both peaking just prior to their species-specific swarming period (Bodot 1969, Sen-Sarma and Mishra 1972, Howard and Haverty 1981, Waller and La Fage 1987). Soldiers typically stand guard around flight exit holes where predation can be high (Wilson 1971).

Possibly because soldier proportions are extremely high in some termite species, researchers have more recently searched for other roles that the soldier caste may play in colony function. Traniello (1981) found that soldiers in the higher termite, *Nasutitermes costalis* (Holmgren), communicate the location of new food sources to the workers. Likewise, Kaib (1990) noted that only minor soldiers initiate foraging to new locations in *Schedorhinotermes lamanianus*. These

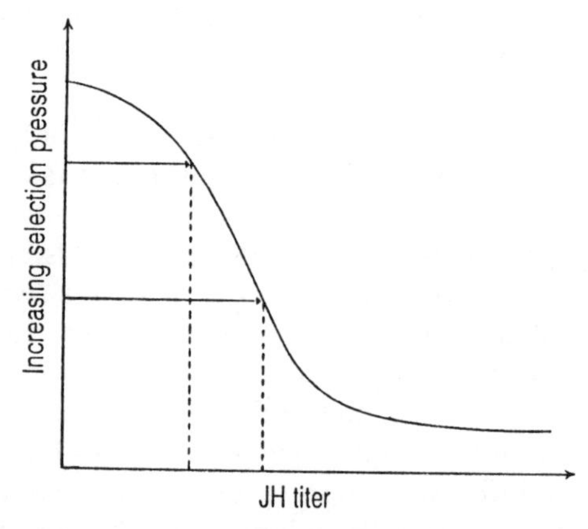

FIGURE 13.6 The role that juvenile hormone (JH) may have in fixing species-specific caste proportions. As selection pressure increases, such as predation risks, JH sensitivity for worker-to-soldier molt is reduced.

soldiers are twice as sensitive to trail pheromones as are the workers. Foraging by *S. lamanianus* workers only persists as long as soldiers reinforce the trail with their sternal gland secretions. Research on the trailing behavior of the lower termite, *Coptotermes formosanus* Shiraki, suggests a similar importance of soldiers for encouraging workers to move into new foraging areas (Wells and Henderson 1993). Finding suitable food resources, and keeping workers away from unsuitable food resources, may be important functions of the soldier caste (Lenz and Westcott 1985).

Another role of the soldier caste that has received little attention in the literature, is this caste's influence on siblings reaching reproductive competence. As noted earlier, soldiers inhibit workers from developing into soldiers. They also stimulate alate development and secondary reproductive development. It is suggested here that these facts are significant in the development and maintenance of sociality in lower termites (see below).

Soldier Caste Influence in Termite Sociality: A Testable Hypothesis

Theories on termite sociality have centered on diet constraints (summarized and added to in Nalepa 1994) and inbreeding effects (Bartz 1979, Myles and Nutting 1988). Until recently, a genetical theory of altruistic behavior in termites--in the absence of inbreeding--could not be addressed the way it has been in the Hymenoptera. With some rigorous mathematics, Hamilton (1964) argued that hymenopteran sociality could persist because of the sterile worker's high degree of relatedness among sisters. This led to a justification for the perpetuation of sterile workers, in that their altruistic act of being sterile actually increased their long-term genetic potential (inclusive fitness). However, in 1977, Syren and Luykx found that a higher degree of relatedness occurs also between termite siblings of the same-sex in many lower termites as a result of males being translocation heterozygotes with extensive multivalent formation occurring in meiosis. As pointed out by Luykx (1985, Figure 13.7), "...the analogy with the Hymenoptera is striking: individuals in the colony are more closely related to their same-sex siblings than they would be to their own offspring". Luykx (1985) speculated that it was possible male soldiers benefitted primarily male workers, with similar benefits occurring among female. Crozier and Luykx (1985), however, argued that such prejudicial treatment was an unlikely possibility, and referred to unpublished research of Luykx's that showed random mixing (distribution) of the sexes in a colony. However, a physiological

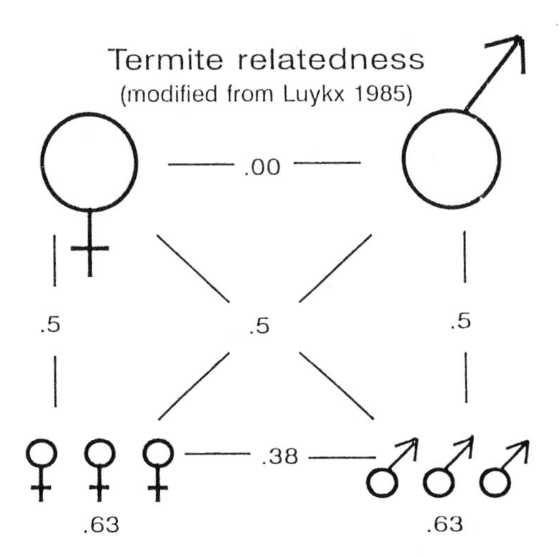

FIGURE 13.7 Relatedness asymmetries between same-sex offspring as a result of non-related (no inbreeding occurring) king a queen pairing of drywood termites. Values between female and male symbols represent degree of genetic relatedness.

prejudice that selectively influenced same-sex caste determination was not considered and remains an untested hypothesis. Though fragmentary in nature, some interesting data indicate that same-sex physiological prejudices occur in termite colonies. For example, female reproductives are more effective at inhibiting female worker differentiation, whereas male reproductives have little effect on male or female differentiation (Lüscher 1964). Miller (1969) noted that male and female reproductives give off sex-specific inhibiting pheromones, but the source of these data was not made clear. Ruppli (1969) found that the hostility in *Kalotermes flavicollis* (Fabricus) female reproductives is mostly directed at the males, which is probably a result of sex-specific recognition cues. Whether male soldiers preferentially advance male workers to reproductive competence, and female soldiers advance female workers reproductive competence, is an unknown. There is very little information on soldier sex-ratios in lower termites (Henderson and Rao 1993), let alone information on sex-linked treatment of siblings. However, recent finds in which only male alates were produced in a colony where the primary queen was killed (Lenz and Runko 1992) and in an old colony (Crosland et al. 1995), where the queen may have died, are quite suggestive and lead to some testable scenarios.

Regardless of proximate or ultimate causation considerations, research indicates that the soldier caste probably evolved in response to the high JH titers disseminated through the colony by the founding male and female at an early stage of isopteran history. The royal pair's chances of survival were clearly enhanced by maintaining a worker force, an effect of high levels of JH. However, for the society to persist in an evolutionary sense, reproductives needed to be produced (at least in numbers that the colony could afford to lose from the worker force). This occurs only when the workers are removed from the inhibitory influence of the queen's and king's JH. The removal of inhibitory effects can occur by workers distancing themselves from the royal pair (McMahan 1966), by the royal pair's death, or by the soldiers absorbing JH from the system (first suggested by Lüscher 1976). It is quite possible that worker-to-alate inhibition is lifted by white soldiers acting as JH sponges. The "sponge" that regulates JH hemolymph titer may involve CA absorption, proteins that bind-up JH, esterases that breakdown the hormone, or combinations of all three (de Kort et al. 1978). In *Zootermopsis nevadensis* Hagen (Termopsidae) castes, JH binding proteins are most abundant in workers, presoldiers (white soldiers) and soldiers (Okot-Kotber and Prestwich 1991). Thus, there may be more of a physiological dependency than realized to the fact that soldiers and alates generally peak at the same time of year--alates develop in response to low JH titers, soldiers develop in response to high JH titers. Such a society would represent a prime example of the interconnectedness of parental manipulation and kin selection (as opposed to being competing explanations) as modeled by Craig (1979), on the factors affecting the spread of alleles in a population.

At the white soldier instar the CA can triple in volume, and it may at that time be functioning as a storage unit, rather than a secretory unit as argued by Yin and Gillot (1975b). The high JH hemolymph titer in soldiers (Yin and Gillot 1975a, see Figure 4) could be the result of uptake of JH from nearby conspecifics during the white soldier stage. As noted earlier, the white soldier is a helpless individual that acts like a larva in the nest (Noirot 1985). Thus, interactions between workers and these individuals in the nest may be plentiful. It appears that further examination of termite worker JH titers and reproductive competence when in the presence of varying ratios and sexes of soldiers, would provide valuable information toward the hypothesis on the soldier's role in termite sociality. Although the JH connection to termite sociality is intriguing, it remains speculative and much research is needed. Prestwich et al. (1987) provide the framework for the necessary analytical work.

Soldiers are the only truly altruistic individuals in the lower termite colony, since workers always maintain reproductive competence. Thus, it is the soldier, and only the soldier, that must gain some degree of inclusive fitness within the colony for the social system to persist genetically. Inclusive fitness could be realized simply through the defensive aspects of competent soldiers, but by considering the idea of soldiers acting as JH sponges a puzzling question is also answered, i.e., why does a hemimetabolous insect molt to an adult form when exposed to JH? The social structure of termites is governed by, most probably, varying levels of JH. That termite sociality itself may have evolved through the hormones that regulate caste differentiation is an example of ergonomic optimization at its best.

Acknowledgment

Drs. Kenneth Grace (University of Hawaii), Barbara L. Thorne (University of Maryland) and Edward Vargo (University of Texas) provided important ideas and information that were incorporated in this paper. Drs. Seth Johnson, James Ottea, Gail Smith and Dorothy Prowell (LSU) made important editorial comments. Bob Vander Meer is thanked for provoking me to give some thought to primer pheromones. This paper was approved for publication by the Louisiana State University Agricultural Center and the Louisiana Agricultural Experiment Station as manuscript number 95-17-9274.

References Cited

Bartz, S.H. 1979. Evolution of eusociality in termites. *Proc. Natl. Acad. Sci. 76*: 5764-5768.

Bodot, P. 1969. Composition des colonies de termites: ses fluctuations au cours du temps. *Ins. Soc. 16*: 39-53.

Bordereau, C. 1985. The role of pheromones in termite caste differentiation In: Caste Differentiation in Social Insects (Watson, J.A.L., B.M. Okot-Kotber, and Ch. Noirot, Eds.), pp. 221-226. Pergamon Press, New York.

Castle, G.B. 1934. The damp-wood termites of western United States, genus *Zootermopsis* (formerly,*Termopsis*) In: Termites and Termite Control Kofoid, C. A., S.F. Light, A.C. Horner, M. Randell, W.B. Herms, and E.E. Bowe, Eds.), 2nd ed., revised, pp. 273-310. Univ. Calif. Press, Berkeley, California.

Craig, R. 1979. Parental manipulation, kin selection, and the evolution of altruism. *Evolution* 33: 319-334.

Crosland, M.W.J., G.X. Li, L.W. Huang, and J.R. Dai, 1995 (In press). Switch to single sex alate production in a colony of the termite, *Coptotermes formosanus*. *J. Entomol. Sci.* 29: 523-525.

Crozier, R.H. and P. Luykx, 1985. The evolution of termite eusociality is unlikely to have been based on a male-haploid analogy. *Amer. Nat. 126*: 867-869.

de Kort, C.A.D., S.J. Kramer, and M. Wieten, 1978. Regulation of juvenile hormone titres in the adult Colorado beetle: interaction with carboxylesterases and carrier proteins, pp. 507-510 *In*: Comparative Endocrinology (Gailord and Boer, eds), Elsevier Press, New York.

Doki, H., K. Tsunoda, and K. Nishimoto, 1984. Effect of juvenile hormone analogues on caste-differentiation of the termite, *Reticulitermes speratus* (Kolbe) (Isoptera: Rhinotermitidae). *Mater. und Org. 19*: 175-187.

Esenther, G.R. 1977. Fluctuation of survival, growth, and differentiation in biweekly samples of *Reticulitermes flavipes* workers. *Mater. und Org. 12*: 49-58.

Grassé, P.-P. 1982. Termitologia (Anatomie Physiologie Reproduction des Termites). Vol. 1, pp. 1-676. Masson, Paris, France.

Grassé, P.-P and Ch. Noirot, 1958. La société de *Calotermes flavicollis* (Insecte, Isoptère), de sa fondation au premier essaimage. *Compt. Rend. 246*: 1789-1795.

Greenberg, S. and S.S. Tobe, 1985. Adaptation of a radiochemical assay for juvenile hormone biosynthesis to study caste differentiation in a primitive termite. *J. Insect. Physiol. 31*: 347-352.

Hamilton, W.D. 1964. The genetical theory of social behaviour, I, II. *J. Theor. Biol. 7*: 1-52.

Haverty, M.I. 1977. The proportion of soldiers in termite colonies: a list and a bibliography (Isoptera). *Sociobiol. 2*: 199-216.

Haverty, M.I. 1979. Soldier production and maintenance of soldier proportions in laboratory experimental groups of *Coptotermes formosanus* Shiraki. *Ins. Soc. 26*: 69-84.

Henderson, G. and K.S. Delaplane, 1994. Formosan subterranean termite swarming behavior and alate sex-ratio (Isoptera: Rhinotermitidae). *Ins. Soc. 41*: 19-28.

Henderson, G. and K.S. Rao, 1993. Sexual dimorphism in soldiers of Formosan subterranean termites (Isoptera: Rhinotermitidae). *Sociobiol. 21*: 341-345.

Howard, R.W. and M.I. Haverty, 1978. Defaunation, mortality and soldier differentiation: concentration effects of methoprene in a termite. *Sociobiol. 3*: 73-78.

Howard, R.W. and M.I. Haverty, 1981. Seasonal variation in caste proportions of field colonies of *Reticulitermes flavipes* (Kollar). *Environ. Entomol. 10*: 546-549.

Hrdy, I. and J. Krecek. 1972. Development of superfluous soldiers induced by juvenile hormone in the termite, Reticulitermes lucifugus *santonensis*. *Ins. Soc. 19*: 105-109.

Jucci, C. 1924. Sulla differenziazione delle caste nella società dei termitidi. I. Neotenici (Reali veri e neotenici- l'escrezionse nei reali neotenici-las fisiologia e la biologia). *Atti Accad. Naz. Lincei. Mem., Classe Sci. Fis., Nat., Sez. III 14*: 269-500.

Kaib, M. 1990. Multiple functions of exocrine secretions in termite communication: exemplified by *Schedorhinotermes lamanianus* In: Social Insects and the Environment (Veeresh, G. K., B. Mallik, C. A. Viraktamath, Eds.) p. 37. *Proc. 11th Intr.* Congr.*IUSSI, India* , Oxford & IBH Publishing Co., New Delhi, India.

Lebrun, D. 1964. Le rôle des cops allates du termite à cou jaune, *Kalotermes* flavicollis.*Comptes Rendus de l'Académie des Sciences, Paris 259:* 4152-4155.

Lebrun, D. 1969. Glandes endocrines et biologie de *Calotermes flavicollis. Proc. VI Congr. IUSSI, Bern,* pp. 131-136.

Lenz, M. 1976. The dependence of hormone effects in termite caste determination on external factors. In: Phase and Caste Determination in Insects. Endocrine Aspects (M. Lüscher, Ed.). pp. 73-89. Pergamon Press, Oxford, England.

Lenz, M. 1994. Food resources, colony growth and caste development in wood-feeding termites. In: Nourishment & Evolution in Insect Societies (J. H. Hunt and C. A. Nalepa, Eds.), pp. 159-210. Westview Press, Boulder, Colorado.

Lenz, M. and S. Runko, 1992. Long-term impact of orphaning field colonies of *Coptotermes lacteus* (Froggatt) (Isoptera, Rhinotermitidae). *Proc. XIX Inter. Congr. Entomol., Bejing,* p. 244.

Lenz, M. and M. Westcott. 1985. Homeostatic mechanisms affecting caste composition in groups of *Nasutitermes nigriceps* (Isoptera: Termitidae) exposed to a juvenile hormone analogue, In: Caste Differentiation in Social Insects (Watson, J.A.L., B.M. Okot-Kotber, and Ch. Noirot, Eds.) pp. 251- 266. Pergamon Press, Oxford, England.

Lüscher, M. 1958. Experimentelle Erzeungung von Soldaten bie der Termite *Kalotermes flavicollis. Naturwis. 45*: 69-70.

Lüscher, M. 1960. Hormonal control of caste differentiation in termites. *Ann. New York Acad. Sci. 89*: 549-563.

Lüscher, M. 1961. Social control of polymorphism in termites. In: Insect Polymorphism (J. S. Kennedy, Ed.), pp. 57-67. Roy. Entomol. Soc., London.

Lüscher, M. 1964. Die spezifische Wirkung mannlicher und weiblicher Ersatzgeschlechtstiere auf die Entstehung von Ersatzgeschlechtstieren bei der Termite *Kalotermes flavicollis* (Fabr.). *Ins. Soc. 11*: 79-90.

Lüscher, M. 1965. Functions of the corpora allata in the development of termites. *Proc. 16th Intern. Congr. Zool., Washington, D. C., 1963,* Vol. 4, pp. 244-250. Nat. Hist. Press, Garden City, New York.

Lüscher, M. 1969. Die Bedeutung des Juvenilhormons für die Differenzierung der Soldaten bei der Termite *Kalotermes flavicollis. Proc. VI Congr. IUSSI, Bern,* pp. 165-170.

Lüscher, M. 1972. Environmental control of juvenile hormone (JH) secretion and caste differentiation in termites. *Gen and comp. Endocrinol. Suppl. 3*: 509-514.

Lüscher, M. 1973. The influence of the composition of experimental groups on caste development in *Zootermopsis* (Isoptera). *Proc. VII Congr. IUSSI, London,* pp. 253-256.

Lüscher, M. 1976. Evidence for an endocrine control of caste determination in higher termites, . In: Phase and Caste Determination in Insects. Endocrine Aspects (M. Lüscher, Ed.). pp. 91-103. Pergamon Press,Oxford, England.

Lüscher, M. and A. Springhetti, 1960. Untersuchungen über die Bedeutung der Corpora allata für die Differenzierung der Kasten bei der Termite *Kalotermes flavicollis* F. *J. Insect Physiol. 5*: 190-212.

Luykx, P. 1985. Genetic relations among castes in lower termites In: Caste Differentiation in Social Insects (Watson, J. A. L., B. M. Okot-Kotber, Ch. Noirot, eds.), pp. 17-26. Pergamon Press, New York.

Luykx, P. 1993. Turnover in termite colonies: a genetic study of colonies of *Incisitermes schwarzi* headed by replacement reproductives. *Ins. Soc. 40*: 191-205.

McMahan, E.A. 1966. Food transmission with the *Cryptotermes brevis* colony (Isoptera: Kalotermitidae). *Ann. Entomol. Soc. Amer. 59*: 1131-1137.

Miller, E.M. 1942. The problem of castes and caste differentiation in *Prorhinotermes simplex* (Hagen). *Miami Univ. Bull. 15*: 3-27.

Miller, E.M. 1969. Caste differentiation in the lower termites. In: Biology of Termites (Krishna, K. and F. M. Weesner, Eds.). Vol. 1, pp. 283-310. Academic Press, New York.

Myles, T.G. and W.L. Nutting, 1988. Termite eusocial evolution: a re- examination of Bartz's hypothesis and assumptions.*Quart. Rev. Biol. 63*: 1-23.

Nalepa, C.A. 1994. Nourishment and the origin of termite eusociality In: Nourishment & Evolution in Insect Societies (J. H. Hunt and C. A. Nalepa, Eds.), pp. 57-104. Westview Press, Boulder, Colorado.

Noirot, Ch. 1969. Glands and secretions. In: Biology of Termites (K. Krishna and F. M. Weesner, Eds.). Vol. 1. pp. 89-119. Academic Press, New York.

Noirot, Ch. 1985. Pathways of caste development in the lower termites In: Caste Differentiation in Social Insects (Watson, J. A. L., B. M. Okot-Kotber, Ch. Noirot, eds.), pp. 59-74. Pergamon Press, New York.

Okot-Kotber, B.M. and G.D. Prestwich, 1991. Juvenile hormone binding proteins of termites detected by photoaffinity labeling: comparison of *Zootermopsis nevadensis* with two rhinotermitids. *Arch. Biochem. Physiol. 17*: 119-128.

Oster, G.F. and E.O. Wilson, 1978. Caste and Ecology in the Social Insects. Princeton University Press, Princeton, New Jersey.

Peeters, C. 1990. The double meaning of the term caste. In: Social Insects and the Environment (G.K. Veeresh, B. Mallik, C.A. Viraktamath, eds.), p. 192. *Proc. 11th Intr. Congr. IUSSI, India* , Oxford & IBH Publishing Co., New Delhi, India.

Prestwich, G.D., S. Robles, and M. Mohamed, 1987. A biochemical basis for caste differentiation in termites, pp. 314-315 *In*: Chemistry and Biology of Social Insects, (Eder, J. and H. Rembold, eds.). Verlag J. Peperny, Munich.

Ruppli, E. 1969. Die Eli mination überzähliger Ersatzgeschlechtstiere bie der Termite *Kalotermes flavicollis* (Fabr.). *Insectes Sociaux* 16: 235-248.

Sen-Sarma, P.K. and S.C. Mishra, 1972. Seasonal fluctuations of colony composition and population in *Neotermes bosei* Snyder (Insecta: Isoptera: Kalotermitidae). *J. Ind. Acad. Wood Sci. 3*:43-48.

Springhetti, A. 1974. The influence of farnesoic acid ethyl ester on differentiation of *Kalotermes flavicollis* Fabr. (Isoptera) soldiers. *Experientia* 30: 1197-1198.

Springhetti, A. 1985. The function of the royal pair in the society of *Kalotermes lavicollis* (Fabr.) (Isoptera: Kalotermitidae). *In:* Caste Differentiation in Social Insects (Watson, J.A.L., B.M. Okot-Kotber, Ch. Noirot, eds.). pp. 165-176. Pergamon Press, New York.

Springhetti, A. and S. Pinamonti. 1977. Some effects of queen bee extracts (*Apis mellifera* L.) on *Kalotermes flavicollis* Fabr. (Isoptera). *Ins. Soc.* 24: 61-70.

Snyder, T.E. 1915. Insects injurious to forests and forest products. Biology of the termites of the eastern United States with preventive and remedial measures. *U. S. Dept. Agric., Bull. Bur. Entomol. 94*: 13-85.

Stuart, A.M. 1979. The determination and regulation of the neotenic reproductive caste in lower termites (Isoptera): with special reference to the genus *Zootermopsis* (Hagen). *Sociobiol. 4*: 223-237.

Syren, R.M. and P. Luykx, 1977. Permanent segmental interchange complex in the termite *Incisitermes schwarzi. Nature 266*: 167-168.

Traniello, J.F.A. 1981. Enemy deterrence in the recruitment strategy of a termite: soldier-organized foraging in *Nasutitermes costalis. Proc. Nat. Acad. Sci. USA 78*: 1976-1979.

Waller, D.A. and J.P. La Fage, 1987. Seasonal patterns in foraging groups of *Coptotermes formosanus* (Rhinotermitidae). *Sociobiol. 13*: 173-181.

Wanyonyi, K. 1974. The influence of juvenile hormone analogue ZR 512 (Zoecon) on caste development in *Zootermopsis nevadensis* (Hagen) (Isoptera), *Ins. Soc 21:* 35-44.

Wanyonyi, K. and M. Lüscher, 1973. The action of juvenile hormone analogues on caste development in *Zootermopsis* (Isoptera). *Proc. VII Congr. IUSSI.* pp. 392-395.

Wells, J.D. and G. Henderson, 1993. Fire ant predation on native and introduced subterranean termites in the laboratory: effect of high soldier number in *Coptotermes formosanus . Ecol. Entomol.* 18: 270-274.

Weesner, F.M. 1953. Biology of *Tenuironstritermes tenuirostris* (Desneux) with emphasis on caste development. *Univ. Calif. (Berkeley) Publ. Zool. 57*: 251-302.

Wilson, E.O. 1971. The Insect Societies. Belknap Press/ Harvard Univ. Press. Cambridge, Massachusetts.

Wyss-Huber, M. 1981. Caste differences in hemolymph proteins in two species of termites. *Insectes Sociaux 28*: 71-86.

Yin, C.-M. and C. Gillot, 1975a. Endocrine activity during caste differentiation in *Zootermopsis angusticollis* Hagen (Isoptera): a morphometric and autoradiographic study. *Can J. Zool. 53*: 1690-1700.

Yin, C.-M. and C. Gillot, 1975b. Endocrine control of caste differentiation in*Zootermopsis angusticollis* Hagen (Isoptera). *Can. J. Zool. 53*: 1701-1708.

14

Royal Flavors: Honey Bee Queen Pheromones

Keith N. Slessor, Leonard J. Foster, and Mark L. Winston

Introduction

Pheromones produced by social insect queens perform both primer and releaser functions in controlling various aspects of colony behavior. Releaser pheromones elicit a stimulus-response mediated by the nervous system whereas primer pheromones function by physiologically altering the endocrine and/or reproductive systems (Pankiw, 1995). Much of the past work with queen pheromones has focused on the effects of live mated queens on worker or soldier behavior, using queenless versus queenright colonies, because almost no primer pheromones have been chemically identified for any social insect.

The single exception is the honey bee *Apis mellifera* L., in which the queen mandibular pheromone's identity is known (Slessor *et al.* 1988). The first component of this pheromone, 9-keto-2(E)-decenoic acid (9ODA) was identified independently by Callow and Johnston (1960) and Barbier and Lederer (1960), and was one of the first pheromones identified for any insect. 9ODA alone, however, does not nearly duplicate the attractive pheromonal activity present in a queen's mandibular gland. We recently demonstrated that the queen's mandibular pheromone (QMP) is a five-component blend consisting of 9ODA and four other components, 9(R)- and 9(S)-hydroxy-2-(E)-decenoic acid (HDA), methyl *p*-hydroxybenzoate (HOB) (Callow *et al.* 1964), and a new substance, 4-hydroxy-3-methoxyphenylethanol (HVA) (Slessor *et al.* 1988, 1990). This pheromone is both a releaser and a primer, acting as a releaser to attract workers to their queen and signal her presence. As a primer pheromone, QMP inhibits queen rearing through an as-yet unknown mechanism and suppresses the ontogeny of worker foraging by diminishing the release of Juvenile Hormone in worker bees (Winston and Slessor, 1992; Pankiw 1995). Nevertheless, although synthetic QMP duplicates the pheromonal activity of the queen's mandibular gland for all the functions we have examined (Slessor *et al.* 1988; Winston and Slessor, 1992), quantitative bioassays of this known pheromone do not duplicate the activity of a live queen, suggesting the presence of a second important queen pheromone. In this paper we describe investigations that indicate its presence and source in the queen, and show that it is active in at least the releaser function of attracting worker bees.

Honey Bee Queen Mandibular Pheromone

Our early work demonstrated that five components of the queen mandibular gland attract workers to form a retinue around a queen honey bee (Slessor *et al.* 1988, 1990), or a glass lure baited with synthetic components (Kaminski *et al.* 1990). The five components act synergistically, the mixture being many times more active in initiating retinue formation than any one of the components. The composition and quantity of QMP varies widely between queens, but 200 ug of ODA, 80 ug of HDA, 20 ug of HOB and 2 ug of HVA is an average value and is referred to as a queen equivalent (Qeq) (Pankiw *et al.* 1997). The two chiral isomers of HDA are present in varying amounts in individual queens. Young queens typically contain the 2 isomers in nearly equal proportions, whereas in mature, laying queens the (R)-(-)enantiomer predominates (Slessor et al. 1990). This bias may arise in the biosynthesis of the precursor of HDA, 17-hydroxyoctadecanoic acid, by asymmetric hydroxylation of octadecanoic acid during the functionalization step, or in the further conversion of HDA by preferential oxidation of the (S)-(+) enantiomer during the formation of ODA (Plettner *et al.* 1996).

Queens secrete approximately 1 Qeq each 24 h period (Naumann *et al.* 1991; Winston and Slessor, 1992), and much of this material is spread over the exterior of the queen's body where it is removed by retinue bees. At any time a typical queen has approximately 10^{-3} Qeq of QMP on the surface of her body (Slessor *et al.* 1990), but this amount varies widely between individuals. Distribution of the pheromone is mainly through contact with licking and antennating workers in the retinue, although some QMP is transferred to the wax comb (Naumann *et al.* 1991). Using radioactively labelled components Naumann *et al.* (1992) demonstrated that QMP moves as a unit.

Investigations into the function of this queen pheromone reveal that not only does QMP mediate retinue formation around the queen, but also influences several colony activities, including direct reproductive activities such as the inhibition of queen production (Winston *et al.* 1989, 1990) and swarming (Pettis *et al.* 1995; Winston *et al.* 1991). The queen also inhibits worker ovarian development in concert with a brood signal (Jay, 1968). Surprisingly, regular applications of QMP to queenless colonies of workers did not affect ovarian development, suggesting that QMP may not be the primer queen pheromone active in this context (Willis *et al.* 1990). The releaser effect of QMP is primarily the attraction of workers, either naturally to the queen in the colony or a swarm during swarming (Schmidt *et al.* 1993), or artificially to a surface, as observed in pollination stimulated by QMP sprayed on blooming crops (Higo *et al.* 1995; Winston and Slessor, 1993). The decenoic acid components of QMP also attract drones during queen mating flights (Loper *et al.* 1995).

A Second Queen Attractant Pheromone

Comparisons of worker honey bee responses to synthetic QMP and to queen mandibular extract show no significant differences between colony and laboratory situations. However, when responses to live queens and synthetic QMP were compared, the queen was invariably superior. Studies of the attraction of swarms, retinue formation, emergency queen cell formation, and inhibition of worker ovarian development all showed the queen to be more effective in promoting/inhibiting these queen-initiated behaviors (Willis *et al.* 1990; Winston *et al.* 1989, 1990, 1991). Thus, the queen possesses another source of attraction and/or behaviors that promote the appropriate responses.

Renner and Baumann (1964), Velthuis (1967, 1970), and Vierling and Renner (1977) described a potential source of a second attractive queen pheromone in the subepidermal complexes of glandular cells on the dorsal surface of the abdomen. These glands, the tergite glands, were reported to be an important source of a worker-attractive queen pheromone. In addition, a drone attractant emanating from the tergite gland of young queens and active over short distances was also reported by Renner and Vierling (1977). Workers discriminate the volatile emissions of queen tergite glands but not mandibular glands when analyzed in a metabolic bioassay, indicating that this might be a possible source of kin recognition labels (Moritz and Crewe, 1988, 1991). Virgin and young queen tergite glands contain considerable quantities of decanoate esters, especially decyl decanoate (Espelie *et al.* 1990), a substance that workers can discriminate from some of the other esters and from hexadecane (Breed *et al.* 1992).

From 1988-93 we investigated the composition and activity of extracts of the tergite gland by gas chromatography and monitored activity with the laboratory retinue bioassay (Kaminski *et al.* 1990). Petri dishes containing 15 workers from a colony were presented with a glass lure coated with extract and the workers' attendance to the lure monitored for 5 min. Every 30 s the number of workers contacting the lure was recorded and the sum of contacts used as a measure of the extract's or fraction's attraction. Evaluation of the attractiveness of tergal extracts by the retinue bioassay was complicated by the natural presence of QMP externally on the queen's abdomen, since the tergite glands are very near the abdominal surface. Bioassay responses often were only marginally more active than the corresponding QMP response. Methanol extracts provided the strongest activity, although still surprisingly weak, giving maximal response at only the 10^{-1} to 10^{-2} Qeq level (QMP is active at the 10^{-5} - 10^{-6} Qeq level). Responses to methanol extracts were variable and the extracts often lost activity after storage in methanol at -20°C for a few months. Tergite gland extracts prepared by careful removal and gentle swabbing with methanol on the outer surface in an attempt to remove contaminating QMP gave even poorer responses in the retinue bioassay. Such carefully prepared extracts would have been expected to provide a rich source of the new pheromone if its origin was the queen's tergite glands.

Low (QMP) Responding Worker Bees

To avoid the problem of QMP contamination in the tergite extracts, we took advantage of a result from our early work with QMP (Kaminski *et al.* 1990) that showed some colonies of worker bees responded poorly to QMP in the laboratory retinue bioassay. Subsequent breeding and cross-fostering experiments demonstrated a strong genetic component to this low QMP response (Pankiw *et al.* 1994). This low response was maintained throughout the year and was neither age nor dose dependent (Figure 14.1), and was not correlated with the levels of QMP found in their queen. However, bees in these colonies, which we referred to as *low (QMP) responders* showed no diminished attention to their own queen (Pankiw *et al.* 1995). Pankiw suggested that low-responding colonies might be responding preferentially to the secondary queen attractant that we were attempting to identify, and that low-responding colonies could be used in the retinue bioassay to minimize the effect of contaminating QMP.

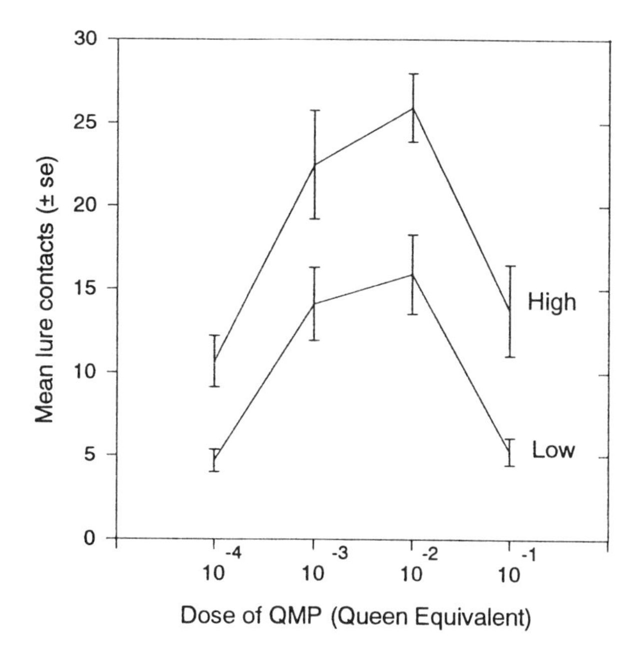

FIGURE 14.1 Dose response of high (N = 5)- and low (N = 6)-responding colonies to various doses of QMP. High-responding colonies contacted the lure significantly more than low-responding colonies ($P < 0.05$). (Reprinted with permission from *J. Insect Behav.* **7**: 1-15 (1994)).

During the 1990-3 seasons, we measured the contaminating QMP levels of these extracts, and lures containing this matching level of QMP were used as standards in each experiment. Colonies of workers that had been selected for their low response to QMP were utilized for the retinue bioassays and responded to fresh queen tergal extracts significantly better than to a lure matched with the amount of QMP contained in that extract. We considered this enhanced retinue activity a response to the putative secondary attractant queen pheromone, although its presence in extracts of the tergal glands could be due to translocation. Gas chromatographic analysis of the extracts, crude and fractionated, both derivatized with bis(trimethylsilyl)trifluoroacetamide (Gehrke and Leimer, 1971) and underivatized, were recorded. No correlation was apparent between the pseudoqueen contact activity to these extracts and over 270 peaks that were present at levels >100pg/queen. Thus, six years of work had demonstrated the possible presence of a new queen-borne attractant, but indicated that the material was ephemeral, unstable and, if detectable by gas chromatography, not obvious by its presence.

Reevaluation of the Source of the Second Queen Attractant

In 1994, we questioned our basic hypothesis that honey bee queens have a second source of attractive pheromone, and that this source is the tergite gland. A new extract, prepared by washing several intact queen bodies, was compared to a matching concentration of QMP and found to be significantly more attractive to low QMP responding workers in the retinue bioassay (Figure 14.2). A worker body wash with the matching quantity of QMP found in the queen wash failed to provide the attraction of the full queen extract. Queens apparently do have a secondary source of attraction and, as expected, this is not found on workers.

We then conducted a similar experiment using washes from various parts of queen's bodies, to determine the location of the secondary attractant's source. Extracts prepared from each body part and solutions containing the same amount of QMP found in each extract were prepared and matched to their body part. When bioassayed, it was clear that the abdominal wash was not the source of the secondary attractant, as it was no more attractive than the corresponding amount

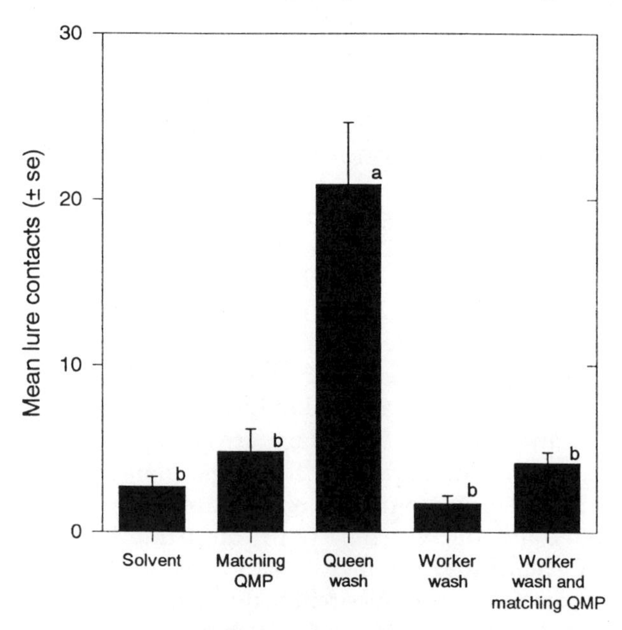

FIGURE 14.2 Mean response of workers from a low-responding colony to washes of intact queen and worker bodies compared to the amount of QMP found in the queen wash (N = 10). Bars superscripted by the same letter are not significantly different (Tukey's test, $P < 0.05$).

of QMP it contained (Figure 14.3). Comparisons cannot be made between the responses to different queen parts as they contain different quantities of QMP and the putative new pheromone. Both the head and the thorax/legs washes out-performed their matched QMP's, suggestive of a pheromone source anterior to the abdomen. The location of the pheromone source cannot be deduced from these results as translocation of exuded pheromone can spread the material over the cuticular surface and contaminate cuticular washes, as has been demonstrated with QMP. However, the results clearly indicate that the tergite gland found on the abdomen was not the source of the secondary queen attractant produced by mated queens.

Contamination of a sample with pheromone moving from a glandular source over the insect's cuticular surface can be minimized by preparing a body part extract, rather than a wash. If the pheromone gland is present in that portion of the body extracted, pheromone components should be found in the extract at a much higher concentration than the contaminant. To establish the source of the second queen pheromone, extracts were made of both head (minus the mandibular glands) and thorax plus legs and tested at two concentrations, 10^{-2} or 10^{-4} queen equivalents with matching controls of QMP (Figure 14.4). At both concentrations, the head extract showed

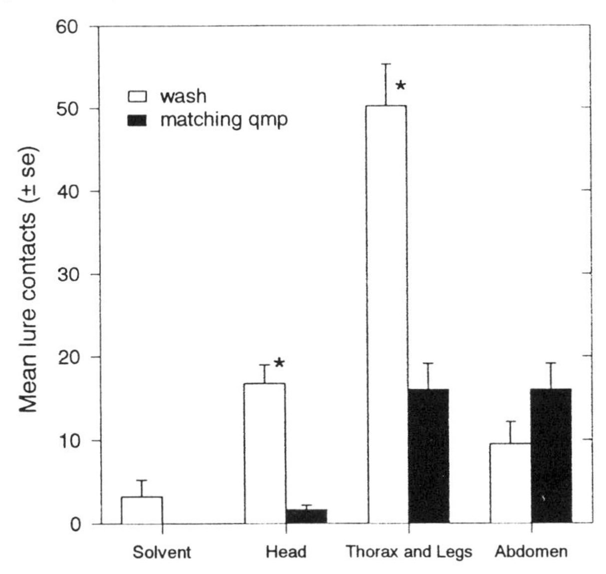

FIGURE 14.3 Mean response of workers from a low-responding colony to washes of queen body parts compared to a matching QMP concentration in each wash (N = 10). Bars superscripted with an asterisk are significantly different than their matching QMP, (Tukey's test, $P < 0.05$).

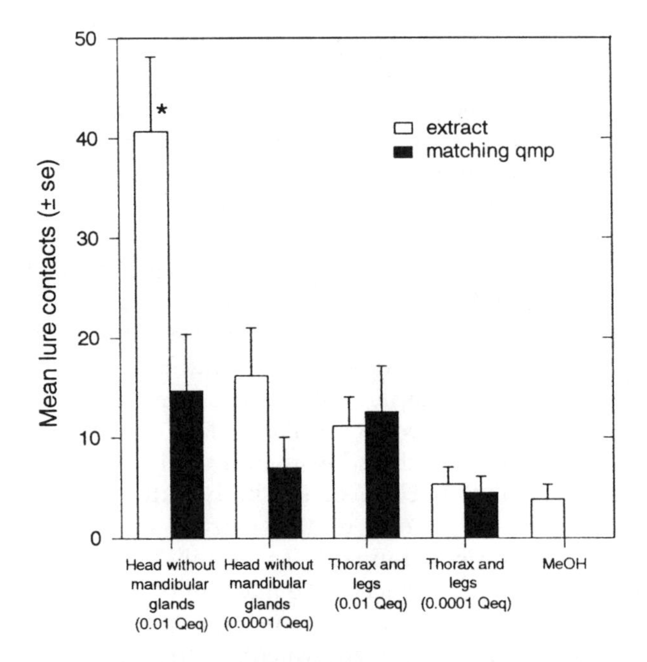

FIGURE 14.4 Mean response of workers from a low-responding colony to an extract of queen head (with mandibular glands removed) and an extract of thorax and legs at 2 dilutions compared to matching QMP concentrations in each extract (N = 10). Bars superscripted with an asterisk are significantly different than their matching QMP, (Tukey's test, $P < 0.05$).

more activity than a lure containing the matching amount of QMP, with the 10^{-2} extract significantly more active. The thorax plus legs extract only showed activity attributable to the level of QMP present in the extract.

QMP Revisited

We then retested whether an unrecognized component of the mandibular gland was responsible for this added attraction. Mandibular glands were removed and extracts prepared from the glands and the remaining portion of the head. These extracts were bioassayed with low-responding workers at 10^{-2} and 10^{-4} queen equivalents and again matched against the QMP they contained (Figure 14.5). At both concentrations, the response to the mandibular gland extract was not significantly different than its matched QMP (confirming results in Slessor *et al.* 1988), whereas the complete head extract was significantly more active than the matched QMP concentration.

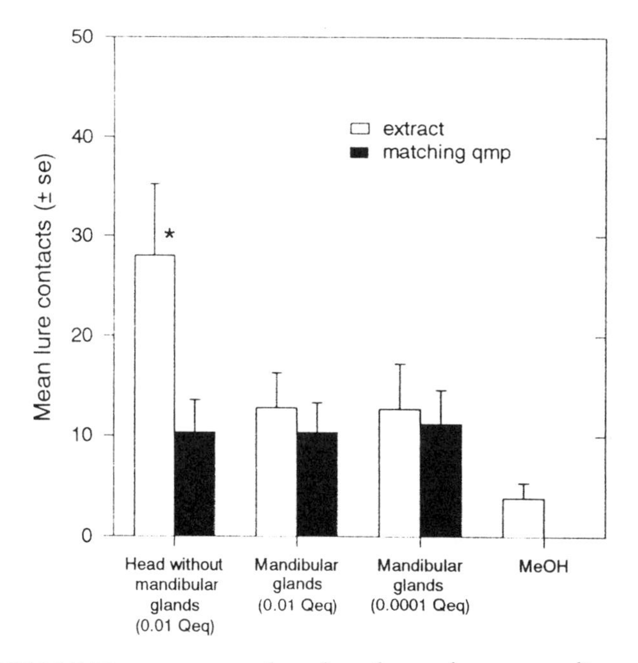

FIGURE 14.5 Mean response of workers from a low-responding colony to an extract made from queen heads (with mandibular glands removed) and an extract of the mandibular glands at 2 dilutions compared to matching QMP concentrations in each extract (N = 10). Bars superscripted with an asterisk are significantly different than their matching QMP, (Tukey's test, $P < 0.05$).

The glandular source of the new pheromone in the queen honey bee head remains unknown. Blum (1993) notes three glands present in the head of a honey bee queen, in addition to the mandibular gland; the postgenal, cephalic labial, and hypopharyngeal glands would seem to be potential sources of this new pheromone, although the latter gland is reported to be poorly developed in queens. Once we have established the chemical identity of the pheromone, its glandular source will be determined.

Stability of the Putative Pheromone

Our experience with crude extracts of the second queen attractant pheromone indicates a component lability, as activity is lost on storage and on fractionation under a wide variety of conditions. The reason for the loss of activity remains unknown at the present time. We do not know whether this lability precludes observation of the pheromone's components by gas chromatography, whether derivatized or not.

Preliminary chemical evidence indicates that the pheromone may be comprised of at least an acidic component as some activity is maintained in material extracted into base and re-extracted on acidification.

Conclusions

The results presented here substantiate the hypothesis that the queen honey bee has a second semiochemical attraction system and that the source of this pheromone is located in the head of the queen honey bee, but not in the mandibular gland. We currently are attempting to isolate the active material(s) using low responders in the retinue bioassay with concentration-matched synthetic QMP as a control. When identification is complete, we hope to locate the glandular source of this elusive semiochemical, synthesize it and determine its role in the sociochemistry of the honey bee colony.

Our research to date has found the queen honey bee primer and releaser pheromones to be complex in both composition and function. For composition, there are a minimum of two major sources in the queen's head, the five-component QMP as well as the as-yet unidentified material discussed here. Both of these pheromones act in a number of releaser functions, and at least QMP acts as a primer pheromone that is of vital significance to proper colony functioning. Further, this basic research on honey bee pheromones has led to numerous commercial applications that have enhanced our ability to manipulate honey bees for honey production and crop pollination (Winston and Slessor, 1993).

We believe that primer pheromones of other social insects can provide a similarly rich source of basic biological and chemical information, as well as economically important tools to manage both beneficial and harmful social species. This type of research is highly challenging because of its technical difficulty and the need for close collaboration between biologists and chemists. We hope the example of the honey bee we have provided here will stimulate and encourage research into other social insect primer pheromones, to further elucidate the fascinating and economically significant world of social insect semiochemistry.

Acknowledgments

We thank Erika Plettner and Tanya Pankiw for valuable comments on this manuscript. Phero Tech Inc., formulated and supplied QMP. We are indebted to Tanya Pankiw, Greg Sutherland, Peter Chua, Heather Higo, Steve Mitchell, and Phil Laflamme for their assistance. These studies have been supported by the Natural Sciences and Engineering Research Council of Canada Strategic and Operating Grants and Simon Fraser University.

References Cited

Barbier, J., and E. Lederer, 1960. Structure chimique de la substance royale de la reine d'abeille (*Apis mellifica* L.). *Comptes Rendus des Seances de L'Academie de Science Paris,* 251: 1131-35.

Blum, M.S., 1993. Honey Bee Pheromones. In *The Hive and the Honey Bee*, (J.M. Graham, Ed.), Dadant and Sons, Hamilton, Illinois, pp. 376-378.

Breed, M.D., T.M. Stiller, M. Blum, and R.E. Page Jr., 1992. Honeybee nestmate recognition: Effects of queen feces pheromones. *J. Chem. Ecol.* 18:1633-1640.

Callow, R.K., and N.C. Johnston, 1960. The chemical constitution and synthesis of queen substances of the honeybees (*Apis mellifera*). *Bee World*, 41: 152-53.

Callow, R.K., J.R. Chapman, and P.N. Paton, 1964. Pheromones of the honeybee: chemical studies of the mandibular gland secretion of the queen. *J. Apic. Res.* 3: 77-89.

Espelie, K.E., V.M. Butz, and A. Dietz, 1990. Decyl decanoate: A major component of the tergite glands of honeybee queens, (*Apis mellifera* L.). *J. Apic. Res.* 29: 15-19.

Free, J.B., 1987. *Pheromones of Social Bees.* Cornell Univ. Press, Ithaca, NY.

Gehrke, C.W., and K. Leimer, 1971. Trimethylsilylation of amino acids. Derivatization and chromatography. *J. Chromatogr.* 57: 219-238.

Higo, H.A. M.L. Winston, and K.N. Slessor, 1995. Mechanisms by which honey bee (Hymenoptera: Apidae) queen pheromone sprays enhance pollination. *Ann. Entomol. Soc. Amer.* 88: 366-373.

Jay, S.C., 1968. Factors influencing ovary development of worker honeybees under natural conditions. *Can. J. of Zoology*, 46: 345-347.

Kaminski, L.-A., K.N. Slessor, M.L. Winston, N.W. Hay, and J.H. Borden, 1990. Honey bee response to queen mandibular pheromone in laboratory bioassays. *J. Chem. Ecol.* 16: 841-850.

Loper, G. M., O. R. Taylor, Jr., M. L. Winston, L. J. Foster, and J. Kochansky, 1995. Relative attractiveness of queen mandibular pheromone components to honey bees (*Apis mellifera* L.) drones. *J. Apic. Res.* in press.

Moritz, R.F.A., and R.M. Crewe, 1988. Chemical signals of queens in kin recognition of honeybees, *Apis mellifera* L. *J. Comp. Physiol. A.* 164: 83-89.

Moritz, R.F.A., and R.M. Crewe 1991. The volatile emissions of honeybee queens (*Apis mellifera* L.). *Apidologie* 22: 205-212.

Naumann, K., M.L. Winston, K.N. Slessor, G.D. Prestwich, and F.X. Webster, 1991. The production and transmission of honey bee queen (*Apis mellifera* L.) mandibular gland pheromone. *Behavioural Ecology and Sociobiology* 29: 321-332.

Naumann, K., M.L. Winston, K.N. Slessor, G.D. Prestwich, and B. Latli, 1992. Intra-nest transmission of aromatic honey bee queen mandibular gland pheromone components: movement as a unit. *Can Entomol.* 124: 917-934.

Pankiw, T., 1995. Worker responses to, and queen production of, honey bee (*Apis mellifera*) queen mandibular pheromone. Ph.D. dissertation, Simon Fraser University, Burnaby, B.C. Canada.

Pankiw, T., M.L. Winston, and K.N. Slessor, 1994. Variation in worker response to honey bee (*Apis mellifera* L.) queen mandibular pheromone (Hymenoptera: Apidae). *J. Insect Behav.* 7: 1-15.

Pankiw, T., M.L. Winston, E. Plettner, K.N. Slessor, J.S. Pettis, and O.R. Taylor, 1997. Mandibular gland components of European and Africanized honey bee queens (*Apis mellifera* L.). *J. Chem. Ecol.* In press.

Pankiw, T., M.L. Winston, and K.N. Slessor, 1995. Queen attendance behavior of worker honey bees (*Apis mellifera* L.) that are high and low responding to queen mandibular pheromone. *Ins. Soc.* 41: 1-8.

Pettis, J.S., M.L. Winston, and A.M. Collins, 1995. Suppression of queen rearing in European and Africanized honey bees *Apis mellifera* L. by synthetic queen mandibular gland pheromone. *Ins. Soc.* 42: 1-9.

Plettner, E., K.N. Slessor, M.L. Winston, and J.E. Oliver, (1996). Caste-Selective Pheromone Biosynthesis in Honeybees. *Science* 271: 1851-1853.

Renner, M., and M. Baumann, 1964. Uber Komplexe von subepidermalen Drusenzellen (Druftdrusen) der Bienenkonigin. *Naturwissenschaften*, 51: 68-69.

Renner, M., and G. Vierling, 1977. Die Rolle des Taschendrusenpheromons beim Hochzeitsflug der Bienenkonigin. *Behavioural Ecology and Sociobiology* 2: 329-338.

Schmidt, J.O., K.N. Slessor, and M.L. Winston, 1993. Roles of Nasonov and queen pheromones in attraction of honeybee swarms. *Naturwissenschaften* 80: 573-575.

Slessor, K.N., L.-A. Kaminski, G.G.S. King, J.H. Borden, and M.L. Winston, 1988. The semiochemical basis of the retinue response to queen honey bees. *Nature* 332: 354-356.

Slessor, K.N., L.-A. Kaminski, G.G.S. King, and M.L. Winston, 1990. Semiochemicals of the honey bee mandibular glands. *J. Chem. Ecol.* 16: 851-860.

Velthuis, H.H.W., 1967. On abdominal pheromones; the queen honeybee. *Proceedings of the 21st International Beekeeping Congress, College Park, Maryland, USA*, p. 472.

Velthuis, H.H.W., 1970. Queen substance from the abdomen of the honeybee queen. *Zeitschrift fur vergleichende Physiologie* 70: 210-222.

Vierling, G., and M. Renner, 1977. Die Bedeutung des Sekretes der Tergittaschendrusen fur die Attraktivitat der Bienenkonigin gegenuber jungen Arbeiterinnen. *Behavioural Ecology and Sociobiology* 2: 185-200.

Willis, L.G., M.L. Winston, and K.N. Slessor, 1990. Queen honey bee mandibular pheromone does not affect worker ovary development. *Can. Ent.* 122: 1093-1099.

Winston, M.L., 1987. *The Biology of the Honey Bee.* Harvard University Press, Cambridge, MA.

Winston, M.L., and K.N. Slessor, 1992. An essence of royalty: Honey bee queen pheromone. *American Scientist,* July - Aug. 374-385.

Winston, M.L., and K.N. Slessor, 1993. Applications of queen honey bee mandibular pheromone for beekeeping and crop pollination. *Bee World* 74: 111-128.

Winston, M.L., K.N. Slessor, L.G. Willis, K. Naumann, H.A. Higo, M.H. Wyborn, and L.-A. Kaminski, 1989. The influence of queen mandibular pheromones on worker attraction to swarm clusters and inhibition of queen rearing in the honey bee (*Apis mellifera* L.). *Ins. Soc.* 36:15-27.

Winston, M.L., H.A. Higo, and K.N. Slessor, 1990. Effect of various dosages of queen mandibular gland pheromone on the inhibition of queen rearing in the honey bee (Hymenoptera: Apidae). *Ann. Entomol. Soc. Amer.* 83: 234-238.

Winston, M.L., H.A. Higo, S.J. Colley, T. Pankiw, and K.N. Slessor, 1991. The role of queen mandibular pheromone and colony congestion in honey bee (*Apis mellifera* L.) reproductive swarming. *J. Insect Behavior* 4: 649-659.

Contributors

Dr. Leeanne E. Alonso
Department of Zoology
University of Oklahoma
730 Van Vleet Ova 1, Rm 314
Norman, OK 73019-0235

Dr. Anne-Geniève Bagnères
Centre National de la Recherche Scientifique
Laboratoire de Neurobiologie
UPR 9024
31 Chemin J. Aiguier
13402 Marseille Cedex 20 FRANCE

Professor Johan Billen
Zoological Institute
University of Leuven
Naamsestraat 59
B-3000 Leuven BELGIUM

Professor Gary J. Blomquist
Department of Biochemistry
University of Nevada
Reno, NV 89557

Professor Christian Bordereau
Department de Zoologie
Universite de Bourgogne
UA CNRS 674
6 Bd Gabriel
21000 Dijon FRANCE

Professor Michael D. Breed
EPO Biology
Campus Box 334
University of Colorado
Boulder, CO 80309

Dr. Jean-Luc Clément
Centre National de la Recherche Scientifique
Laboratoire de Neurobiologie
UPR 9024
31 Chemin J. Aiguier
13402 Marseille Cedex 20 FRANCE

Professor Karl E. Espelie
Department of Entomology
University of Georgia
Athens, GA 30602

Professor Leonard J. Foster
Department of Chemistry
Simon Fraser University
Burnaby BC V5A 1S6 CANADA

Professor George J. Gamboa
Department of Biology
Oakland University
Rochester, NY 48309

Professor Abraham Hefetz
G.S Wise Faculty of Life Sciences
Department of Zoology
Tel Aviv University
69978 Tel-Aviv ISRAEL

Professor Robert L. Jeanne
Department of Entomology
University of Wisconsin
Madison, WI 53706

Professor Gregg Henderson
Department of Entomology
Louisiana State University
Baton Rouge, LA 70803

Professor Bert Hölldobler
Zoologie II
University of Würzburg
Biozentrum
Am Hubland
D-97074 Würzburg GERMANY

Dr. Peter J. Landolt
Research Entomologist
USDA-ARS
5230 Konnowac Pass Rd
Wapato, WA 98951

Dr. Laurence Morel
Department of Pathology and Laboratory Medicine
University of Florida,
JHMHC Box 275
Gainesville, Fl 32610

Professor E. David Morgan
Department of Chemistry
Keele University
Staffordshire, ST5 5BG U.K.

Professor Shuping Mpuru
Department of Biochemistry
University of Nevada
Reno, NV 89557

Professor Jacques Pasteels
Laboratoire de Biologie Animale et Cellulaire
Faculte des Sciences
Université Libre de Bruxelles
Avenue F.D. Roosevelt, 50
B-1050 Bruxelles BELGIUM

Professor Hal Reed
Department of Biology
Oral Roberts University
Tulsa, Oklahoma 74171

Dr. Justin O. Schmidt
Carl Hayden Bee Research Center
2000 E. Allen Road
Tucson, AZ 85719

Professor Steven J. Seybold
Department of Biochemistry
University of Nevada
Reno, NV 89557

Dr. Theresa L. Singer
Coop. Exten. Service
University of Arkansas
P.O. Box 357
Lonoke, ARK 72086

Professor Keith N. Slessor
Department of Chemistry
Simon Fraser University
Burnaby BC V5A 1S6 CANADA

Professor Julie A. Tillman
Department of Biochemistry
University of Nevada
Reno, NV 89557

Dr. Robert K. Vander Meer
Research Chemist
USDA/ARS
P.O. Box 14565
Gainesville, FL 32604

Dr. Edward L. Vargo
Department of Zoology
University of Texas
Austin, TX 78712-1064

Professor Mark L. Winston
Department of Biological Sciences
Simon Fraser University
Burnaby BC V5A 1S6 CANADA

Author Index

Taxonomic Index

Subject Index

Geranylcitronellol 240, 250